Narrative as Virtual Rea

PARALLAX: RE-VISIONS OF CULTURE AND SOCIETY
Stephen G. Nichols, Gerald Prince, and Wendy Steiner
SERIES EDITORS

Narrative as Virtual Reality

Immersion and Interactivity in Literature
and Electronic Media

Marie-Laure Ryan

THE JOHNS HOPKINS UNIVERSITY PRESS | BALTIMORE AND LONDON

The Johns Hopkins University Press
2715 North Charles Street
Baltimore, Maryland 21218-4363
www.press.jhu.edu

Library of Congress Cataloging-in-Publication Data will be
found at the end of this book.
A catalog record for this book is available from the British Library.

ISBN 0-8018-6487-9

ɣ

To the *genius loci* that dwells between the
waterfalls of Cascadilla and Fall Creeks,
and to the one that haunts the Devil's Gulch

CONTENTS

FIGURES AND TABLES

FIGURES

TABLES

ACKNOWLEDGMENTS

If by *village* we understand a community of minds, and by *community* all the minds that are linked to the author's without necessarily knowing of each other's existence, it indeed takes a village, perhaps even several, to write a book. There is the global village of the minds that the author has encountered in books; there is the local village of friends, family, and personal acquaintances; and there is the virtual village of people met in cyberspace or through regular mail. The list of Works Cited that appears at the end of this book identifies some of the most influential members of the global village. Here I would like to express my gratitude for the support offered by both the virtual and the local village.

The longtime interest in my work of Thomas Pavel, Gerald Prince, Uri Margolin, Lubomir Doležel, Harold Mosher, Monika Fludernik, and Katharine Young is one of the reasons I have persevered in the "career" of independent scholar. To the support of N. Katherine Hayles I owe the opportunity to have written half of this in Ithaca, a place to which not only Ulysses yearns to return. Maura Burnett, my editor at the Johns Hopkins University Press, was a dedicated and efficient advocate of the project. Thanks to the sharp eye of Mary Yates, my copy editor, "the signifier" doesn't get confused with "the signified," and all improperly raised negatives are lowered down to their rightful logical position. Mary's editing skills will benefit not only this book but my future writing as well. The collective intelligence of the electronic discussion list of the Society for the Study of Narrative Literature served me as prosthetic memory when I could not remember references. Through the list, Janet Galligani Casey, Deborah Martinsen, Antje Schaum Anderson, Caroline Webb, Edina Szalay, Charlotte Berkowitz, Laura Beard, and Ted Mason shared with me their experiences of fearing fiction and generously let me use their input in this book.

Many of the people who started out as members of the virtual village crossed over into the sphere of personal friends. This includes two eminent narratologists, Emma Kafalenos and David Herman, who kept me constantly stimulated by sending me materials and references relevant to my interests, by commenting on this project in all its phases of development, and by providing inspiring e-mail exchanges. Among those who were never virtual friends, I am grateful to my hiking buddies in the local village for showing me all the sides of the mountains, especially to Jon Thiem and Jacques Rieux, who provided not only useful materials but also challenging views on literature and cyberculture—views all the more challenging because Jon pretends to despise computers, and Jacques (faithful reader of the technological pages of the *New York Times*) categorically refuses to own one.

And finally I would like to express my gratitude to my immediate family for putting up with a body whose mind was usually in virtual worlds. Phil and Caitlin lured this body to places in Colorado where the world displayed itself from such a lofty perspective that the topics discussed in this book no longer seemed to matter, while Duncan, my teen-aged system manager, made sure that our computers, Amadeus and Hieronymus, were always equipped with the latest software, whether or not I needed it. As Duncan's expertise deepened, the disasters became rarer, and the benefits of his constant reconfigurations of the system grew more substantial. Besides forcing me into a state of permanent learning, Duncan also generously dispensed his wisdom on computer games and, when all the electronic devices were turned off, serenaded me with his viola, in his view the most perfect instrument ever engineered.

The writing of this book benefited from a grant in 1998 from the National Endowment for the Humanities (grant FB-34667-98) and from a postdoctoral fellowship in 1999 from the Cornell Society for the Humanities. The efforts of Tim Murray, acting director of the society, in organizing an exhibit of digital art, "Contact Zones: The Art of CD-ROM," had a significant impact on the views presented in the book. I am also grateful to Richard Gilligan of the Cornell Theory Center for introducing me to the VR CAVE.

Though this book was conceived as a global project, not put together out of a collection of articles, many of the chapters include material from previously published essays. The seed for the entire book is "Immersion versus Interactivity: Virtual Reality and Literary Theory," published in electronic form in *Postmodern Culture*, September 1994, reprinted by permission of the Johns Hopkins University Press. An expanded print form appeared in *Sub/Stance 89*, 28, no. 2 (1999): 110–37, reprinted by permission of the University of Wisconsin Press.

Chapter 1 contains passages from "Cyberspace, Virtuality, and the Text," published in *Cyberspace Textuality: Computer Technology and Literary Theory*, edited by Marie-Laure Ryan (Bloomington: Indiana University Press, 1999), 78–107. Reprinted by permission of Indiana University Press.

The interlude to chapter 5 is excerpted from "Allegories of Immersion: Virtual Narration in Postmodern Fiction," published in *Style* 29, no. 2 (1995): 262–86. Reprinted by permission of the journal.

Chapter 6 is based on "The Text as World versus the Text as Game: Possible Worlds, Semantics and Postmodern Theory," published in *Journal of Literary Semantics* 27, no. 3 (1998): 137–63. Reprinted by permission of Julius Groos Verlag.

Chapters 8 and 10 expand "Interactive Drama: Narrativity in a Highly Interactive Environment," published in *Modern Fiction Studies* 43, no. 3 (1997): 677–707, Purdue Research Foundation. Reprinted by permission of the Johns Hopkins University Press.

I also owe a debt of gratitude to Michael Joyce for allowing me to quote and reproduce art from his hypertext fiction *Twelve Blue: Story in Eight Bars*, published by Postmodern Culture and Eastgate Systems, 1996 and 1997. Joyce's book *Of Two Minds* was the inspiration for the architecture of the present one.

Narrative as Virtual Reality

Introduction

Few of us have actually donned an HMD (head-mounted display) and DGs (data gloves), and none has entered the digital wonderland dangled before our eyes by the early developers of virtual reality: a computer-generated three-dimensional landscape in which we would experience an expansion of our physical and sensory powers; leave our bodies and see ourselves from the outside; adopt new identities; apprehend immaterial objects through many senses, including touch; become able to modify the environment through either verbal commands or physical gestures; and see creative thoughts instantly realized without going through the process of having them physically materialized.

Yet even though virtual reality as described above is still largely science fiction, still largely what it is called—a virtual reality—there is hardly anybody who does not have a passionate opinion about the technology: VR will someday replace reality; VR will never replace reality; VR challenges the concept of reality; VR will enable us to rediscover and explore reality; VR is a safe substitute for drugs and sex; VR is pleasure without risk and therefore immoral; VR will enhance the mind, leading us to new powers; VR is addictive and will enslave us; VR is a radically new experience; VR is as old as Paleolithic art; VR is basically a computer technology; all forms of representation create a VR experience; VR undermines the distinction between fiction and reality; VR is the triumph of fiction over reality; VR is the art of the twenty-first century, as cinema was for the twentieth; VR is pure hype and ten years from now will be no more than a footnote in the history of culture and technology.

We may have to wait until the new century reaches adulthood to see whether these promises and threats will materialize. But since the *idea* of VR is very much a part of our cultural landscape, we don't have to wait that long to explore the perspectives it opens on representation. Approaching VR as a semiotic phenomenon, I propose in this book to rethink textuality, mimesis, narrativity, literary theory, and the cognitive processing of texts in the light of the new modes of artistic world construction that have been made possible by recent developments in electronic technology.

VR has been defined as an "interactive, immersive experience generated by a computer" (Pimentel and Teixeira, *Virtual Reality*, 11). As a literary theorist I am primarily interested in the two dimensions of the VR experience as a novel way to describe the types of reader response that may be elicited by a literary text of either the print or the electronic variety. I propose therefore to transfer the two concepts of immersion and interactivity from the technological to the literary domain and to develop them into the cornerstones of a phenomenology of reading, or, more broadly, of art experiencing. In the course of this investigation we will visit both traditional literary texts and the new genres made possible by the digital revolution of the past two decades, such as hypertext, art CD ROMs, synchronic role-playing games (MOOs), the largely virtual genre of interactive drama, and its embryonic implementations in electronic installation art. My purpose will be twofold: to revisit print literature, more specifically the narrative kind, in terms of the concepts popularized by digital culture, and, conversely, to explore the fate of traditional narrative patterns in digital culture.

The history of Western art has seen the rise and fall of immersive ideals, and their displacement, in the twentieth century, by an aesthetics of play and self-reflexivity that eventually produced the ideal of an active participation of the appreciator—reader, spectator, user—in the production of the text. This scenario affects both visual and literary art, though the immersive wave peaked earlier in painting than in literature.

In pre-Renaissance times painting was more a symbolic representation of the spiritual essence of things than an attempt to convey the illusion of their presence. Its semiotic mode was signification rather than simulation. More attentive to what Margaret Wertheim (*Pearly Gates*, 87) calls "the inner eye of the soul" than to the "physical eye of the body," medieval artists painted objects as they believed them to be, not as they appeared to easily deceived senses. (The same can be said of children's drawings that represent the sky as a thin line at the top of the page rather than as a background behind figures.) Pictorial space was a strictly two-dimensional surface from which the body of the spectator was excluded, since bodies are three-dimensional objects.

All this changed when the discovery of the laws of perspective allowed the projection of a three-dimensional space onto a two-dimensional surface. This projection opens up a depth that assigns spatial coordinates—the center of projection, or physical point of view—to the body of the spectator. Perspective painting immerses a virtual body in an environment that stretches in imagination far beyond the confines of the canvas. From its spatial point of view the embodied gaze of the spectator experiences the depicted objects as virtually present, though the flat surface of the painting erects an invisible wall that prevents physical interaction. This strictly visual immersion reached its high point in the incredible trompe l'oeil effects of the Baroque age. The frescoes of Baroque churches blur the distinction between physical and pictorial space by turning the latter into a continuation of the former.

The illusion of a penetrable space received a first challenge when impressionism disoriented the eye with visible brushstrokes that directed attention to the surface of the canvas, and with shimmering light effects that blurred the contours of objects. Though impressionistic space is still three-dimensional, it opens itself to virtual bodies only after the mind completes a complex process of interpretation and construction of sensory data. For the spectator who has assimilated the lesson of impressionism, visual space can no longer be taken for granted.

In the early twentieth century, pictorial space either folded down into a play of abstract shapes and colors on a canvas that openly displayed its two-dimensionality, or exploded into the multiple perspectives of cubist experiments. Whereas the return to flat representation expelled the body from pictorial space, the cubist approach shattered the physical integrity of both space and the body by forcing the spectator to occupy several points of view at the same time. If abstract and cubist paintings lure the spectator into a game of the imagination, this game is no longer the projection of a virtual body in a virtual space but the purely mental activity of grouping shapes and colors into meaningful configurations. As art became more and more conceptual, the eye of the mind triumphed once again over the eye of the body.

But the appeal of a pictorial space imaginatively open to the body is hard to kill off, and in the second third of the twentieth century, immersive ideals made a notorious comeback with the sharply deline-ated dreamscapes of surrealism. The art scene is now split between conceptual schools that engage the mind, hyperrealistic images that insist on the presence of objects to the embodied eye, and three-dimensional installation art in which the actual body is placed in an intellectually challenging environment. By letting the user walk around the display, and occasionally take physical action to activate data, installation art offers a prefiguration of the combination of immersion and interactivity that forms the ideal of VR technology.

In the literary domain, no less than in the visual arts, the rise and fall of immersive ideals are tied to the fortunes of an aesthetics of illusion, which implies transparency of the medium. The narrative style of the eighteenth century maintained an ambiguous stance to-ward immersion: on one hand, it cultivated illusionist effects by sim-ulating nonfictional narrative modes (memoirs, letters, autobiogra-phies); on the other, it held immersion in check through a playful, intrusive narrative style that directed attention back and forth from the story told to the storytelling act. The visibility of language acted as a barrier that prevented readers from losing themselves in the story-world.

The aesthetics of the nineteenth-century novel tipped this balance in favor of the story-world. Through techniques that are examined in greater detail in chapters 4 and 5 of this book, high realism effaced the narrator and the narrative act, penetrated the mind of characters, transported the reader into a virtual body located on the scene of the action, and turned her into the direct witness of events, both mental and physical, that seemed to be telling themselves. Readers not only developed strong emotional ties to the characters, they were held in constant suspense by the development of the plot. The immersive quality of nineteenth-century narrative techniques appealed to such a wide segment of the public that there was no sharp distinction be-tween "popular" and "high" literature: wide strata of society wept for Little Nell or waited anxiously for the next installment of Dickens's serial novels.

The rest of the story has been told many times: how literature, cross-fertilized with the New Criticism, structuralism, and deconstruction, took a "linguistic turn" in the mid-twentieth century, privileged form over content, emphasized spatial relations between words, puns, intertextual allusion, parody, and self-referentiality; how the novel subverted plot and character, experimented with open structures and permutations, turned into increasingly cerebral wordplay, or became indistinguishable from lyrical prose. This evolution split literature into an intellectual avant-garde committed to the new aesthetics and a popular branch that remained faithful to the immersive ideals and narrative techniques of the nineteenth century. (Ironically, the high branch turned out to be heavily dependent on the resources of the low branch for its game of parody.) As happened in the visual arts, immersion was brought down by a playful attitude toward the medium, which meant in this case the exploitation of such features as the phonic substance of words, their graphic appearance, and the clusters of related or unrelated senses that make up their semantic value field. In this carnivalesque conception of language, meaning is no longer the stable image of a world in which the reader projects a virtual alter ego, nor even the dynamic simulation of a world in time, but the sparks generated by associative chains that connect the particles of a textual and intertextual field of energies into ever-changing configurations. Meaning came to be described as unstable, decentered, multiple, fluid, emergent—all concepts that have become hallmarks of postmodern thought.

Though this game of signification needs nothing more than the encounter between the words on the page and the reader's imagination to be activated, it is easy to see how the feature of interactivity conferred upon the text by electronic technology came to be regarded as the fulfillment of the postmodern conception of meaning. Interactivity transposes the ideal of an endlessly self-renewable text from the level of the signified to the level of the signifier. In hypertext, the prototypical form of interactive textuality (though by no means the most interactive), the reader determines the unfolding of the text by clicking on certain areas, the so-called hyperlinks, that bring to the screen other segments of text. Since every segment contains several such hyperlinks, every reading produces a different text, if by *text* one understands a particular set and sequence of signs scanned by the reader's eye. Whereas the reader of a standard print text constructs personalized interpretations out of an invariant semiotic base, the reader of an interactive text thus participates in the construction of the text as a visible display of signs. Although this process is restricted to a choice among a limited number of well-charted alternatives—namely, the branching possibilities designed by the author—this relative freedom has been hailed as an allegory of the vastly more creative and less constrained activity of reading as meaning formation.

These analogies between postmodern aesthetics and the idea of interactivity have been systematically developed by the early theorists of hypertext, such as George Landow, Jay David Bolter, Michael Joyce, and Stuart Moulthrop. These authors were not only literary scholars, they had also contributed to the development of hypertext through the production of either software, instructional databases, or literary works,[1] and they had a stake in the promotion of the new mode of writing. They chose to sell hypertext to the academic community—an audience generally hostile to technology but also generally open to postmodern theory—by hyping their brainchild as the fulfillment of the ideas of the most influential French theorists of the day, such as Barthes, Derrida, Foucault, Kristeva, Deleuze, Guattari, and Bakhtin—the latter an adopted ancestor. Many of those who came to electronic textuality from literary theory happily joined in the chorus. To cite a few particularly telling examples of this rhetoric, Bolter calls hypertext a "vindication of postmodern theory," as if postmodern ideas were the sort of propositions that can be proved true or false ("Literature in the Electronic Space," 24); Richard Lanham speaks of an "extraordinary convergence" of postmodern thought and electronic textuality (*Electronic Word*, chap. 4);[2] and Ilana Snyder argues that hypertext teaches "deconstructive skills" that readers supposedly do not acquire from standard texts (*Hypertext*, 119).[3] Though all these comments describe hypertext, not interactivity per se, it was the interactive nature of the genre that inspired these pronouncements.

The list of the features of hypertext that supports the postmodernist approach is an impressive one. It is headed by Roland Barthes and Julia Kristeva's notion of *intertextuality*, the practice of integrating a variety of foreign discourses within a text through such mechanisms as quotation, commentary, parody, allusion, imitation, ironic transformation, rewrites, and decontextualizing/recontextualizing operations. Whether intertextuality is regarded as a specific aesthetic program or as the basic condition of literary signification, it is hard to deny that the electronic linking that constitutes the basic mechanism of hypertext is an ideal device for the implementation of intertextual relations. Any two texts can be linked, and by clicking on a link the reader is instantly transported into an intertext. By facilitating the creation of polyvocal structures that integrate different perspectives without forcing the reader to choose between them, hypertext is uniquely suited to express the aesthetic and political ideals of an intellectual community that has elevated the preservation of diversity into one of its fundamental values.

The device also favors a typically postmodern approach to writing closely related to what has been described by Lévi-Strauss as *bricolage* (tinkering, in Sherry Turkle's translation). In this mode of composition, as Turkle describes it (*Life on the Screen*, 50–73), the writer does not adopt a "top-down" method, starting with a given idea and breaking it down into constituents, but proceeds "bottom-up" by fitting together reasonably autonomous fragments, the verbal equivalent of *objets trouvés*, into an artifact whose shape and meaning(s) emerge through the linking process. The result is a patchwork, a collage of disparate elements, what Gilles Deleuze and Félix Guattari have called a "machinic assemblage" (*A Thousand Plateaus*, 332–35). As Silvio Gaggi has shown, this broken-up structure, as well as the dynamic reconfiguration of the text with every new reading, proposes a metaphor for the postmodern conception of the subject as a site of multiple, conflicting, and unstable identities.

While hypertext can bring together the heterogeneous, it can also break apart elements traditionally thought to belong together. The dismantling effect of hypertext is one more way to pursue the typically postmodern challenge of the epistemologically suspect coherence, rationality, and closure of narrative structures, one more way to deny the reader the satisfaction of a totalizing interpretation. Hypertext thus becomes the metaphor for a Lyotardian "postmodern condition" in which grand narratives have been replaced by "little stories," or perhaps by no stories at all—just by a discourse reveling in the Derridean performance of an endless deferral of signification. Through its growth in all directions, hypertext implements one of the favorite notions of postmodernism, the conceptual structure that Deleuze and Guattari call a "rhizome." In a rhizomatic organization, in opposition to the hierarchical tree structures of rhetorical argumentation, the imagination is not constrained by the need to prove a point or to progress toward a goal, and the writer never needs to sacrifice those bursts of inspiration that cannot be integrated into a linear argument.

Building interactivity into the object of a theoretical mystique, the "founding fathers" of hypertext theory promoted the new genre as an instrument of liberation from some of the most notorious bêtes noires of postmodern thought: linear logic, logocentrism, arborescent hierarchical structures, and repressive forms of power. George Landow writes, for instance, that hypertext embodies the ideal of a nonhierarchical, decentered, fundamentally democratic political system that promotes "a dialogic mode of collective endeavor" (*Hypertext 2.0*, 283): "As long as any reader has the power to enter the system and leave his or her mark, neither the tyranny of the center nor that of the majority can impose itself" (281). Over twenty years ago Roland Barthes identified the figure of the author as one of these oppressive forms of authority from which readers must be liberated: "We know to give writing its future, it is necessary to overthrow the myth [of the author]: the birth of the reader must be at the cost of the death of the Author" ("Death of the Author," 78). The purpose of new forms of writing—such as what Barthes called "the scriptible"—is "to make the reader no longer a consumer but a producer of text" (*S/Z*, 4).

For the critics mentioned above, interactivity is just what the structuralist doctor (would have) ordered: "There is no longer one author but two, as reader joins author in the making of the text," writes Bolter ("Literature in the Electronic Space," 37). For Michael Joyce, hypertexts are "read when they are written and written as they are read" (*Of Two Minds*, 192). Or to quote again Landow: "Electronic linking reconfigures our experience of both author and authorial property, and this reconception of these ideas promises to affect our conceptions of both the authors (and authority) of texts we study and *of ourselves as authors*" (*Hypertext 2.0*, 25; my italics). In *Grammatron*, a hypertextual novel-cum-theory that challenges traditional generic distinctions, Mark Amerika takes the cult of interactivity to new extremes, by hailing what he calls "hypertextual consciousness" as the advent of a new stage, perhaps the final one, in the political, spiritual, and artistic growth of mankind:

> The teleportation of Hypertextual Consciousness (HTC) through the smooth space of discourse networks creates an environment where conceptions of authorship, self, originality, narrative and commentary take on different meanings. One can now picture a cyborg-narrator creating a discourse network that serves as a distribution point for various lines of flight to pass through and manipulate data linked together by the collective-self. Directing a site (giving birth to a node) will be one way to reconfigure our notion of authorship but in reconfiguring this notion aren't we in effect radically-altering (killing) the author-as-self and opening up a more fluid vista of potential-becomings? (Fragment "Teleport")

To the skeptical observer, the accession of the reader to the role of writer—or "wreader," as some agnostics facetiously call the new role—is a self-serving metaphor that presents hypertext as a magic elixir: "Read me, and you will receive the gift of literary creativity." If taken literally—but who really does so?—the idea would reduce writing to summoning words to the screen through an activity as easy as one, two, three, click. Under these conditions no writer would ever suffer from the agony of the blank page. Call this writing if you want; but if working one's way through the maze of an interactive text is suddenly called writing, we will need a new word for retrieving words from one's mind to encode meanings, and the difference with reading will remain. One wonders what conclusions would have been drawn about the political significance of hypertext and the concept of reader-author if the above-mentioned critics had focused on the idea of *following* links, or on the limitation of the reader's movements to the paths designed by the author. Perhaps they would have been more inclined to admit that aesthetic pleasure, like political harmony, is a matter not of unbridled license but of controlled freedom.

While interactivity has been hyped as a panacea for evils ranging from social disempowerment to writer's block, the concept of immersion has suffered a vastly different fate. At best it has been ignored by theorists; at worst, regarded as a menace to critical thinking. (A notable exception is Janet Murray, who devotes a chapter of her book *Hamlet on the Holodeck* to immersion as part of a more general discussion of the aesthetics of the electronic medium.) If we believe some of the most celebrated parables of world literature, losing oneself in a book, or in any kind of virtual reality, is a hazard for the health of the mind. Immersion began to work its ravages as early as the first great novel of European literature. "In short," writes Cervantes in *Don Quixote*, "he so immersed himself in those romances that he spent whole days and nights over his books; and thus with little sleeping and much reading, his brains dried up to such a degree that he lost the use of his reason" (58). The situation does not seem to be better in the virtual realities of the electronic kind: we hear tales of people suffering from AWS (Alternate World Syndrome), a loss of balance, feeling of sickness, and general "body amnesia" (Heim, *Virtual Realism*, 52), when they leave VR systems; of MOO addicts who cannot adapt to ROL (Sherry Turkle's acronym for "the rest of life"); or of children who experience emotional trauma when they inadvertently let their virtual pets die.

The major objection against immersion is the alleged incompatibility of the experience with the exercise of critical faculties. The semiotic blindness caused by immersion is illustrated by an anecdote involving the eighteenth-century French philosopher Diderot. As Wallace Martin reports, "He tells us how he began reading *Clarissa* several times in order to learn something about Richardson's techniques, but never succeeded in doing so because he became personally involved in the work, thus losing his critical consciousness" (*Recent Theories*, 58). According to Jay Bolter, the impairment of critical consciousness is the trademark of both literary and VR immersion: "But is it obvious that virtual reality cannot in itself sustain intellectual or cultural development. . . . The problem is that virtual reality, at least as it is now envisioned, is a medium of percepts rather than signs. It is virtual television" (*Writing Space*, 230). "What is not appropriate is the absence of semiosis" (231).

The cause of immersion has not been helped by its resistance to theorization. Contemporary culture values those ideas that produce brilliant critical performances, that allow the critic to deconstruct the text and put it back together again in the most surprising configurations, but what can be said about immersion in a textual world except that it takes place? The self-explanatory character of the concept is easily interpreted as evidence that immersion promotes a passive attitude in the reader, similar to the entrapment of tourists in the self-enclosed virtual realities of theme parks or vacation resorts. This accusation is reinforced by the association of the experience with popular culture. "Losing oneself in a fictional world," writes Bolter, "is the goal of the naive reader or one who reads as entertainment. It is particularly a feature of genre fiction, such as romance or science fiction" (*Writing Space,* 155). Through its reliance on stereotypes, popular literature indeed turns the reading experience into something like taking a dip in a Jacuzzi: it is easy to get in, but you cannot stay in very long, and you feel tired once you get out.

But this does not mean that immersive pleasure is in essence a lowbrow, escapist gratification, as Bolter seems to imply. At its best, immersion can be an adventurous and invigorating experience comparable to taking a swim in a cool ocean with powerful surf. The environment appears at first hostile, you enter it reluctantly, but once you get wet and entrust your body to the waves, you never want to leave. And when you finally do, you feel refreshed and full of energy. As for the allegedly passive character of the experience, we need only be reminded of the complex mental activity that goes into the production of a vivid mental picture of a textual world. Since language does not offer input to the senses,[4] all sensory data must be simulated by the imagination. In "The Circular Ruins" Jorge Luis Borges writes of the protagonist, who is trying to create a human being by the sheer power of his imagination, "He wanted to dream a man: he wanted to dream him with minute integrity and insert him into reality" (*Ficciones,* 114). Similarly, we must dream up textual worlds with "minute integrity" to conjure up the intense experience of presence that inserts them into imaginative reality. Is this the trademark of a passive reader?

To counter these two trends it will be necessary to take a more critical look at interactivity, and a more sympathetic one at immersion. This attitude is admittedly no less biased than the approaches I want to avoid, but it offers an alternative to both the rapturous celebrations of digital literature and the Luddite laments for the book that have greeted the recent explosion of information technologies. If I appear harsher on interactive than on immersive texts, it is not because I view the intrusion of the computer into literary territory as a threat to humanistic values, as does Sven Birkerts, the most eloquent champion of immersion, but because interactivity is still in an experimental stage while literature has already perfected the art of immersive world construction. It is precisely its experimental nature that makes interactivity fascinating. I am interested in the device not as a ready-made message-in-the-medium, as its postmodern advocates read it, but as a language and a *design* problem whose solutions will always be in the making. In my discussion of interactivity I therefore avoid allegorical readings and concentrate instead on the expressive properties of the feature, its potential and limitations, its control of the reader, and its problematic relation to immersion.

The organization of this book grew out of the very definition that inspired the whole project: "virtual reality is an immersive, interactive experience generated by a computer." We will begin by visiting the virtual as philosophical concept, move on to VR as technology, explore its two components, immersion and interactivity, and conclude the itinerary by considering what is for me the ultimate goal of art: the synthesis of immersion and interactivity. This book, then, is as much about virtual literature—literature that could be—as about the actual brand. But since we cannot even begin to envision the virtual without an eye on the real, my presentation interleaves theoretical chapters on the problematics of immersion and interactivity with short case studies of actual texts, labeled interludes, that anticipate, allegorize, or concretely implement one or both of the dimensions of the archetypal VR experience.

Judging by their current popularity in both theory and advertising language, the terms *virtual* and *virtuality* exert a powerful magnetism on the contemporary imagination, but as is always the case when a word catches the fancy of the general public, their meaning tends to dissolve in proportion to the frequency of their use. In its everyday usage the word *virtual* is ambiguous between (1) "imaginary" and (2) "depending on computers." (A third, more philosophical sense, does not seem as influential on the popular usage.) When we speak of "virtual pets" we mean the computer image of corporeally nonexistent animal companions, but when we speak of "virtual technologies" we certainly do not mean something that does not exist, or we would not spend hundreds of dollars for computer software. Virtual technologies fabricate objects that are virtual in sense 1 but they are themselves virtual in sense 2. When N. Katherine Hayles characterizes the condition of contemporary mankind as "virtual," and further defines this condition as "the cultural perception that material objects are interpenetrated by information patterns" ("Condition of Virtuality," 69), she makes a culturally well accepted, but philosophically less evident, association: Why should information be regarded as virtual, or at least as meaningfully connected with virtuality? Is it because information enables us to build "virtual realities"—digital images that offer simulacra of physically habitable environments? Is it because informational patterns contain *in potentia* new forms of life (as in biological engineering), new forms of art, and, for the dreamers of the coupling of man and machine, new forms of humanity? Is it because information lives principally these days in the silicon memory of computers, invisible and seemingly inexistent until the user summons it to the screen?

I have suggested here three distinct senses of *virtual:* an optical one (the virtual as illusion), a scholastic one (the virtual as potentiality), and an informal technological one (the virtual as the computer-mediated). All three are involved in VR: the technological because VR is made of digital data generated by a computer; the optical because the immersive dimension of the VR experience depends on the reading of the virtual world as autonomous reality, a reading facilitated by the illusionist quality of the display; and the scholastic because as interactive system, VR offers to the user a matrix of actualizable possibilities. In the first chapter of this book I explore the optical and the scholastic interpretation of the virtual by relating them to the work of two prominent French theorists: Jean Baudrillard for the virtual as illusion and Pierre Lévy for the virtual as potentiality. I dwell on these two versions of the virtual not only for the sake of their involvement with VR technology but also because each of them presents important implications for literary theory and the phenomenology of reading.

In the second chapter I turn to VR proper. Though the current state of the technology falls way short of the expectations raised at the time of its first introduction to the general public, the "myth" matters as much as the technological reality for a project that uses VR as metaphor, and I therefore move back and forth between the exalted vision of the early prophets and the more sober descriptions of the technical literature. Immersion in a virtual world is discussed from both a technological and a phenomenological point of view. Whereas the technological approach asks what features of digital systems produce an immersive experience, the phenomenological issue analyzes the sense of "presence" through which the user feels corporeally connected to the virtual world. I look for answers to this second question in the writings of a philosopher acutely aware of the embodied nature of perception, Maurice Merleau-Ponty. If these concerns seem to showcase immersion to the detriment of interactivity, it is not because VR subordinates one to the other—it may or it may not, depending on its ultimate purpose—but because immersion is by far the more problematic concept. We all know instinctively what interactivity consists of in a computer program—submitting input and receiving output—but it is much harder to tell what it means to feel immersed in a virtual world, and how digital technology and interface design can promote this experience.

The phenomenological idea of consciousness as a sense of being-in-the-world—or in this case, in a simulated world—is at the core of the theory and poetics of immersion presented in the second part of the book. The term *immersion* has become so popular in contemporary culture that people tend to use it to describe any kind of intensely pleasurable artistic experience or any absorbing activity. In this usage, we can be immersed in a crossword puzzle as well as in a novel, in the writing of a computer program as well as in playing the violin. Here, however, I would like to single out and describe a specific type of immersion, one that presupposes an imaginative relationship to a *textual world*—an intuitive concept to be refined in chapter 3. In the phenomenology of reading, immersion is the experience through which a fictional world acquires the presence of an autonomous, language-independent reality populated with live human beings.

For a text to be immersive, then, it must create a space to which the reader, spectator, or user can relate, and it must populate this space with individuated objects. It must, in other words, construct the setting for a potential narrative action, even though it may lack the temporal extension to develop this action into a plot. This fundamentally *mimetic* concept of immersion remains faithful to the VR experience, since the purpose of VR technology is to connect the user to a simulated reality. It applies to novels, movies, drama, representational paintings, and those computer games that cast the user in the role of a character in a story, but not to philosophical works, music, and purely abstract games such as bridge, chess, and Tetris, no matter how absorbing these experiences can be.

Immersion may not have been particularly popular with the "textual" brands of literary theory—those schools that describe the text as a system of signs held together by horizontal relations between signifiers—but this does not mean that the experience has been totally ignored since these theories became mainstream. Chapter 3 discusses the work of some scholars working on the outskirts of literary studies—cognitive psychology, empirical approaches to literature, or analytic philosophy—who have addressed the issue that I call immersion, though they have done so under a variety of other names: Victor Nell's analysis of the psychological state of being "lost in a book"; Richard Gerrig's concept of transportation; the possible-worlds approach to the semantics of fictionality and its description of the phenomenology of reading fiction as an imaginative "recentering" of the universe of possibilities around a new actual world; Kendall Walton's theory of fiction as game of make-believe and his concept of "mental simulation"; and in an interlude, the spiritual exercise recommended by St. Ignatius of Loyola of a reading discipline involving all the senses in the mental representation of the textual world. These theories show that, far from promoting passivity, as its opponents have argued, immersion requires an active engagement with the text and a demanding act of imagining.

Whether textual worlds function as imaginary counterparts or as models of the real world, they are mentally constructed by the reader as environments that stretch in space, exist in time, and serve as habitat for a population of animate agents. These three dimensions correspond to what have long been recognized as the three basic components of narrative grammar: setting, plot, and characters. The "poetics" proposed in chapters 4 and 5 associates these narrative elements with three distinct types of immersion—spatial, temporal, and emotional—and analyzes the narrative devices that favor each of them. In my discussion of temporal and emotional immersion I seek explanations for two closely related immersive paradoxes that have generated lively debate among philosophers and cognitive psychologists for a number of years: how readers can experience suspense the second or third time they read a text, even though they know how it ends; and how the fate of fictional characters can generate emotional reactions with physical symptoms, such as crying, even though readers know fully well that these characters never existed.

Chapter 6 examines the change of metaphor that marked the transition from immersion to interactivity as artistic ideals. Whereas the aesthetics of immersion implicitly associates the text with a "world" that serves as environment for a virtual body, the aesthetics of interactivity presents the text as a game, language as a plaything, and the reader as the player. The idea of verbal art as a game with language is admittedly not a recent invention; ancient literatures and folklore are full of intricate word games, and the novel of the eighteenth century engaged in very self-conscious games of narration. But it is only in the middle of the twentieth century, after the concept of game rose to prominence as a philosophical and sociological issue and began infiltrating many other disciplines, that literary authors developed the metaphor into an aesthetic program. The concept of "game" covers, however, a wide variety of activities, and it is too often used in a generic sense by literary critics. Chapter 6 narrows down the metaphor by exploring what kind of games and what specific features pertaining to these games provide meaningful analogies with the literary domain.

No less intuitively meaningful than immersion, the concept of interactivity can be interpreted figuratively as well as literally. In a figural sense, interactivity describes the collaboration between the reader and the text in the production of meaning. Even with traditional types of narrative and expository writing—texts that strive toward global coherence and a smooth sequential development—reading is never a passive experience. As the phenomenologist Roman Ingarden and his disciple Wolfgang Iser have shown, the construction of a textual world or message is an active process through which the reader provides as much material as he derives from the text. But the inherently interactive nature of the reading experience has been obscured by the reader's proficiency in performing the necessary world-building operations. We are so used to reading classic narrative texts—those with a well-formed plot, a setting we can visualize, and characters who act out of a familiar logic—that we do not notice the mental processes that enable us to convert the temporal flow of language into a global image that exists all at once in the mind. Postmodern narrative deepens the reader's involvement with the text by proposing new reading strategies, or by drawing attention to the construction of meaning. Through their experimental and self-referential character, these texts stand as the illustration of a strong figural version of interactivity.

But the type of interactivity that receives the greatest attention in these pages is the one that largely owes its existence to electronic technology: the textual mechanisms that enable the reader to affect the "text" of the text as a visible display of signs, and to control the dynamics of its unfolding. Here again we encounter a contrast between a weak and a strong form. In the weak literal sense, discussed in chapters 7 and 8, interactivity is a choice between predefined alternatives. In chapters 9 and 10 I consider a stronger form in which the reader—more aptly called the interactor—performs a role through verbal or physical actions, thus actually participating in the physical production of the text. (By *text* I do not necessarily mean something that is permanently inscribed.)

Symmetry would demand that I split my coverage of interactivity into a theory and a poetics chapter, as I do for immersion, but in the case of interactivity the two concepts are much more entangled, and the scope and purpose of theory much more problematic. As a type of reading experience, immersion is a relatively speculative idea that needs to be defined. Its theorization depends on a particular conception of the literary text, while its poetics is a typology of its various manifestations. Interactivity, by contrast, is an empirical feature of certain types of text, and its plain existence is no more in need of demonstration in texts than in VR. We can debate endlessly what it means to be immersed, but if we stick to what I call a literal conception of interactivity, the mechanism is easily defined. What distinguishes the pure theory from the poetics of interactivity, in the current literature, is mainly a matter of ideological slant: we may call "theory" the postmodern/deconstructionist readings of interactivity discussed above, while a "poetics" would be a more descriptive and empirical approach that keeps its mind open as to what the uses and effects of interactivity might be. Most work on the subject of electronic textuality is a blend of the two approaches, but I would place the work of Landow, Bolter, Joyce, and Moulthrop on the theory end, though these scholars did make important contributions to both areas, while the more recent books of Espen Aarseth and Janet Murray clearly occupy the poetics end of the spectrum.

Bypassing theory, then, I present in chapter 7 a list of lists that examine a variety of concrete rhetorical problems associated with interactivity: the forms and functions of the device; the relations between interactivity, electronic support, and ergodic design (a concept proposed by Aarseth); the properties of the electronic medium and their exploitation in the creation of new modes of interface between the text and the reader; and the metaphors through which hypertext readers conceptualize interactivity.

Chapter 8 narrows down the inquiry to the possibility of creating genuinely narrative structures in an interactive environment. If narrativity is a reasonably universal semantic structure, a cognitive framework in which we arrange information to make sense of it as the representation of events and actions, it consists of a certain repertory of basic elements arranged into specific logical and temporal configurations. Several scholars have raised the question of narrativity in conjunction with hypertext, but the paradox of maintaining a reasonably solid semantic structure in a fluid environment has been generally avoided in favor of more discourse-oriented issues. (I am alluding here to the classic narratological distinction between discourse, the "expression plane of narrative" [Prince, *Dictionary*, 21], and story, the "content plane," the "what," the "narrated.") Aarseth, for instance, proposes a narratological reading of hypertext and computer games that remains entirely focused on the relevance of the parameters of Gérard Genette's model of the fictional narrative act: author, reader, narrator, and narratee. Landow discusses hypertext as a "reconfiguration of narrative" (*Hypertext 2.0*, chap. 6), but the interactive presentation that he has in mind is either a novel discourse phenomenon that leaves the narrative deep structure intact, or a fundamentally antinarrative device that results in the breaking apart of this deep structure. Literature can admittedly achieve significance by challenging narrative coherence and traditional plot structures, as postmodernism has amply demonstrated, but in giving up well-formed narrative content it also renounces the most time-tested formula for creating immersion.

The realization of the ideal of immersive interactivity is therefore crucially dependent on the development of what Janet Murray (*Hamlet*, chap. 7) has called "multiform plot" or "storytelling system": a collection of textual fragments and combinatory rules that generate narrative meaning for every run of the program, much in the way a Chomsky-type grammar produces a vast number of well-formed sentences by combining words according to syntactic rules. In such a "kaleidoscopic system," as Murray also calls it, the user's actions would create unforeseen combinations of elements, but the pieces would always interlock into a narratively meaningful picture. Murray illustrates the idea of the storytelling system with the example of the bards of oral culture who built ever-new narrative performances out of a fixed repertory of phrases, epithets, similes, and episodes, but the example cannot be directly transferred to the domain of electronic text design because oral epics are not interactive on the level of plot. Though live oral performance reacts to subtle clues from the audience—facial expressions, laughter, and the particular quality of the atmosphere—the bard does not normally consult the audience on how to continue the tale; and even if he did, the audience, knowing the plot, would probably ask for an episode that would readily fit into the global structure. In chapter 8 I look into designs that provide feasible solutions to the problem of interactive narrativity. This leads to an examination of the options between which the interactive text will have to choose in order to survive as an art form when the interest due to its novelty recedes.

Even when narrative coherence is maintained, though, immersion remains an elusive experience in interactive texts. In the last two chapters I argue that the marriage of immersion and interactivity requires the imagined or physical presence of the appreciator's body in the virtual world—a condition easily satisfied in a VR system but problematic in hypertext because every time the reader is asked to make a choice she assumes an external perspective on the worlds of the textual universe. In VR we act within a world and experience it from the inside, but in interactive texts of the selective variety we choose a world, more or less blindly, out of many alternatives, and we are not imaginatively committed to any one of them, because the interest of branching texts lies in the multiplicity of paths, not in any particular development.

As chapter 9 shows, VR is not the only environment that offers an experience both immersive and interactive: children's and adults' games of make-believe, fairs and amusement parks, ritual, Baroque art and architecture, and certain types of stage design in the theater propose an active participation of either an actual or virtual body in a reality created by the imagination. The study of these experiences should therefore provide valuable guidelines for the design of electronic texts. Chapter 10 expands the search for immersive interactivity to digital projects, such as computer games, MOOs, automated dialogue systems, installation art, and even a virtual form of VR—a blueprint for future projects—called interactive drama. It is symptomatic of the utopian nature of this quest for the ultimate artistic experience that the most perfect synthesis of immersion and interactivity should be found not in a real work but in a fictional one: the multimedia "smart" book described in Neal Stephenson's science-fiction novel *The Diamond Age.*

By proposing to read VR as a metaphor for total art, I do not mean to suggest that the types of art or entertainment discussed in these last two chapters are superior to the mostly immersive forms of part II or the mostly interactive ones of part III. If aesthetic value could be judged by numerical coefficients, as in certain "artistic" sports such as equestrian dressage or figure skating, a text that scored 10 on immersion and 1 on interactivity—a good realistic novel—would place higher than a text that scored 3 for each criterion. Whether or not future VR installations will be able to offer more than mediocrity on both counts, however, we can still use the *idea* of VR as a metaphor for the fullest artistic experience, since in the Platonic realm of ideas VR scores a double 10.

But why should the synthesis of immersion and interactivity matter so much for aesthetic philosophy? In its literal sense, immersion is a corporeal experience, and as I have hinted, it takes the projection of a virtual body, or even better, the participation of the actual one, to feel integrated in an art-world. On the other hand, if interactivity is conceived as the appreciator's engagement in a play of signification that takes place on the level of signs rather than things and of words rather than worlds, it is a purely cerebral involvement with the text that downplays emotions, curiosity about what will happen next, and the resonance of the text with personal memories of places and people. On the shiny surface of signs—the signifier—there is no room for bodies of either the actual or the virtual variety. But the recipient of total art, if we dare to dream such a thing, should be no less than the subject as Ignatius of Loyola defined it: an "indivisible compound" of mind and body.[5] What is at stake in the synthesis of immersion and interactivity is therefore nothing less than the participation of the whole of the individual in the artistic experience.

PART I   Virtuality

The Two (and Thousand) Faces of the Virtual

I dwell in Possibility
A fairer House than Prose,
More numerous of Windows,
Superior for Doors.

Of Chambers, as the Cedars—
Impregnable of eye;
And for an everlasting Roof
The Gambrels of the Sky.

Of Visitors—the fairest—
For Occupation—This—
The spreading wide my narrow Hands
To gather Paradise—
—EMILY DICKINSON

In the popular imagination of the last decade of the twentieth century, the word *virtual* triggers almost automatically the thought of computers and digital technology. This association was built in several steps, though the early ones have largely fallen into oblivion. Nowadays we label virtual everything we experience or meet in "cyberspace," the imaginary place where computers take us when we log on to the Internet: virtual friends, virtual sex, virtual universities, virtual tours of virtual cities. Before the Internet forced itself, almost overnight, into our daily lives, the virtuality of digital technology was associated with the concept of VR, introduced to the public in the late 1980s. Computers were credited with the power to create artificial worlds, and though the Internet is a far cry from the three-dimensional, multisensory, immersive, and interactive environments envisioned by the promoters of VR, we projected onto cyberspace the dreams that the VR industry had awakened but largely failed to deliver.

Earlier in the history of the semantic liaison, *virtual* was a technical term of computer architecture that expressed the discrepancy between the physical machine and the machine with which users and high-level programmers think they are communicating. Computer programs are written in a quasi-human language made up of a large number of powerful modules and commands, but the actual processor can understand only a small number of instructions coded in zeros and ones. It takes a translator, known as a compiler or an interpreter, to turn the instructions typed by the user into executable code. In the same vein, the term *virtual* was applied to memory to refer to a type of storage, such as a floppy disk, that is not physically part of the computer's active memory but whose contents can easily be transferred back and forth to the brain of the machine, so that from the point of view of the user this storage behaves as if it were an integral and permanent part of computer memory.

Yet another virtual feature of computers resides in their versatility. As a machine, a computer has no intrinsic function. Through its software, however, it can simulate a number of existing devices and human activities, thus becoming a virtual calculator, typewriter, record player, storyteller, babysitter, teacher, bookkeeper, or adviser on various matters. Or even, as VR suggests, a virtual world and living space. The software industry exploited these technical uses—of which there are many others—by metonymically promoting its products as "virtual technologies." For the general public, the narrow technical meaning meant nothing; but the label *virtual* became a powerful metaphor for the accelerating flight of technology into the unknown. The term gave an almost science-fictional aura to the products of a culture that had to be hatching something fundamentally new, since it was approaching the mythical landmark of the turn of the millennium.

Let us backtrack even further, in this hopeless but tempting search for pure and original meaning, by asking what is virtual about artificial worlds and pseudo-memory and versatile machines. Etymology tells us that *virtual* comes from the Latin *virtus* (strength, manliness, virtue), which gave to scholastic Latin the philosophical concept of *virtus* as force or power. (This sense survives today in the expression "by virtue of.") In scholastic Latin *virtualis* designates the potential, "what is in the power *[virtus]* of the force." The classic example of virtuality, derived from Aristotle's distinction between potential and actual existence *(in potentia* vs. *in actu)*, is the presence of the oak in the acorn. In scholastic philosophy "actual" and "virtual" exist in a dialectical relation rather than in one of radical opposition: the virtual is not that which is deprived of existence but that which possesses the potential, or force, of developing into actual existence. Later uses of the term, beginning in the eighteenth and nineteenth centuries, turn this dialectical relation to actual into a binary opposition to real: the virtual becomes the fictive and the nonexistent. This sense is activated in the optical use of the term. According to Webster's dictionary, a virtual image, such as a reflection in a mirror, is one made of virtual foci, that is, of points "from which divergent rays of light seem to emanate but do not actually do so." Exploiting the idea of fake and illusion inherent to the mirror image, modern usage associates the virtual with that which *passes as* something other than what it is. This passing involves an element of illegitimacy, dishonesty, or deficiency with respect to the real. A virtual dictator may be "as good"— or in this case as bad—as a real dictator, but he remains inferior to a "legitimate" one, to use an oxymoron, because he is not officially recognized as such. (He could, in principle, be indicted for abuse of power.) Yet the deficiency of the virtual with respect to the real may be so small that "for all practical purposes" the virtual becomes the real.

As we see from these lexical definitions, the meaning of *virtual* stretches along an axis delimited by two poles. At one end is the optical sense, which carries the negative connotations of double and illusion (two ideas combined in the theme of the treacherous image); at the other is the scholastic sense, which suggests productivity, openness, and diversity. Somewhere in the middle are the late-twentieth-century associations of the virtual with computer technologies. For convenience's sake I will call one pole the virtual as fake and the other the virtual as potential. (See table 1 for a list of the connotations I collected in the course of my various readings on virtuality.) Both of these interpretations have found influential and eloquent spokesmen in recent French theory: Jean Baudrillard for the virtual as fake, Pierre Lévy for the virtual as potential.

BAUDRILLARD AND THE VIRTUAL AS FAKE

The philosophy of Baudrillard presents itself as a meditation on the status of the image in a society addicted to "the duplication of the real

TABLE 1 | The Meaning of Virtual

Actual	Virtual
enacted	potential
factual	counterfactual
accomplished	possible
closed	open
material	mental
concrete	abstract
particular	general
complete	incomplete
determinate	indeterminate
corporeal	spectral
bound body or object	aura
kernel	irradiation
temporal	atemporal
inscribed in space	deterritorialized
singular	plural
manufactured object	blueprint, code
present	past and future
here	there
us	them, the other (as imagined)
solid, tangible	evanescent, nontangible
figure	ground
visible	latent
presence	absence
presence	telepresence
face to face	mediated
mechanical, printed	electronic
matter	information
space	cyberspace
that which counts	that which does not count
being	presenting
identity	passing as, role-playing
serious behavior	make-believe
lived experience	fantasy and dreams
fact	fiction
essence	appearance
authenticity	fake, simulation
truth	illusion, falsity
original	copy, double
represented, referent	image
virtual	real (theme of the virtual and the real exchanging places)
.....	real (theme of the disappearance of the real)

NOTE: At the top, the scholastic opposition of virtual to actual. At the bottom, the optical opposition of virtual to real. In the middle, the popular association of virtual with computer technology.

by means of technology" (Poster, "Theorizing," 42). Once, the power to automatically capture and duplicate the world was the sole privilege of the mirror; now this power has been emulated by technological media—photography, movies, audio recordings, television, and computers—and the world is being filled by representations that share the virtuality of the specular image. The general tone and content of

Baudrillard's meditation on this state of affairs are given by the epigraph to his most famous essay, "The Precession of Simulacra," a quotation attributed to Ecclesiastes nowhere to be found in the Bible (hence, evidently, the lack of reference to a verse number). True to its message and subject matter, the essay thus opens with a simulacrum: "The simulacrum is never what hides the truth—it is truth that hides the fact that there is none. The simulacrum is true" (1). A simulacrum, for Baudrillard, is not the dynamic image of an active process, as are computer simulations, but a mechanically produced, and therefore passively obtained, duplication whose only function is to *pass as that which it is not:* "To simulate is to feign to have what one doesn't have" (3). Baudrillard envisions contemporary culture as a fatal attraction toward simulacra. This "will to virtuality," to borrow Arthur Kroker and Michael Weinstein's evocative term (*Data Trash*, chap. 3), precludes any dialectical relation and back-and-forth movement between the real and its image. Once we break the second commandment, "Thou shalt not make images," we are caught in the gravitational pull of the fake, and the substance of the real is sucked out by the virtual, for as Baudrillard writes in *The Perfect Crime*, "There is no place for both the world and its doubles" (34). In the absence of any Other, the virtual takes the place of the real and becomes the hyperreal. In Baudrillard's grandiose evolutionary scheme, we have reached stage 4 in the evolution of the image:

1. "It is the reflection of a profound reality."
2. "It masks and denatures a profound reality."
3. "It masks the absence of a profound reality."
4. "It has no relation to any reality whatsoever: it is its own pure simulacrum." ("Precession," 6; numbering mine)

Does the seemingly inevitable historical evolution from stage 1 to stage 4 represent a fall into inauthenticity, an abdication of representational responsibility, and a cynical betrayal of the Real, or, on the contrary, a gradual discovery of the True Nature of the image? Has the culture of illusion committed a "perfect crime" that killed reality without leaving any traces, as Baudrillard suggests in the later book by that title, or has it definitively slain the illusion of the real and reached the ultimate semiotic wisdom? Oscillating between the roles

of modern-day Ecclesiastes and solemn theorist of semiotic nihil-
ism—and obviously enjoying himself in both roles—Baudrillard is
careful to maintain an ambiguous stance.

The word *virtual* itself is absent from "The Precession of Simulacra," an essay written in the late 1970s, when the principal channel
of "the image" and the main threat to the real was television. But
when computer technology began to impose the notion of virtuality,
in the late 1980s, Baudrillard suddenly discovered a new culprit for
modern society's "crime against reality." It is as if technology had
caught up with the theory and turned it into prophecy by delivering
its missing referent. As Mark Poster writes, "Baudrillard's writing
begins to be sprinkled with the terms 'virtual' and 'virtual reality' as
early as 1991. But he uses these terms interchangeably with 'simulation,' and without designating anything different from the earlier
usage" ("Theorizing," 45). In *The Perfect Crime* (1996) virtual reality is
treated not as just another way to produce simulacra but as the ultimate triumph of the simulacrum:

> With the Virtual, we enter not only upon the era of the liquida-
> tion of the Real and Referential, but that of the extermination of
> the Other.
>
> It is the equivalent of an ethnic cleansing which would not
> just affect particular populations but unrelentingly pursue all
> forms of otherness.
>
> The otherness . . .
>
> Of the world—dispelled by Virtual Reality. (*Perfect Crime,*
> 109)

According to Baudrillard, we don't live in a world where there is
something called VR technology, we are immersed in this technology,
we live and breathe virtual reality. All the concepts and buzzwords
associated with VR provide easy fuel for Baudrillard's insatiable theoretical machine. Consider the following passages from "Aesthetic Illusion and Virtual Reality."[1]

*On the transparency of the medium, one of the acknowledged goals of
VR developers:*

> And if the level of reality decreases from day to day, it's because
> the medium itself has passed into life, and become a common

ritual of transparency. It is the same for the virtual: all this digital, numerical and electronic equipment is only the epiphenomenon of the virtualization of human beings in their core. (*Art and Artefact*, 20)

By the same logic that denies a place for both the world and its doubles, there is no place in the mind for both life and the lifelikeness of transparent media. Our fascination with the latter turns us into "virtual beings" through a reasoning that skips several intermediary steps in one powerful leap: (1) VR technology (and modern media in general) aims toward transparency; (2) transparency allows immersion; (3) by a metonymic transfer, immersion in a virtual world leads to a virtualization of the experiencer. One must assume that this virtualization involves a loss of humanity, as we offer ourselves as data and as servants to the machine.

On the project of creating three-dimensional environments with which the user can interact:

> For example some museums, following a sort of Disneyland processing, try to put people not so much in front of the painting—which is not interactive enough and even suspect as pure spectacular consumption—but into the painting. Insinuated audiovisually into the virtual reality of the *Déjeuner sur l'herbe* [by Renoir], people will enjoy it in real time, feeling and tasting the whole Impressionist context, and eventually interacting with the picture. The masses usually prefer passive roles and avoid representation. This must change, and they must be made interactive partners. It is not a question of free speaking or free acting—just break their resistance and destroy their immunities. (22)

In this passage Baudrillard's a priori commitment to the idea that we are prisoners of our own technologies of representation allows only one interpretation of interactivity: it is a simulacrum of activity that conceals the fundamental passivity of the user, just as the world outside prisons is for Baudrillard a simulacrum of freedom that conceals the fundamentally carceral nature of society ("Precession," 12).

On the digital coding of information:

> Now what exactly is at stake in this hegemonic trend towards virtuality? What is the idea of the virtual? It would seem to be

the radical actualization, the unconditional realization, of the
world, the transformation of all our acts, of all historical events,
of all material substance and energy into pure information. The
ideal would be the resolution of the world by the actualization
of all facts and data. (23)

If reality has become an edifice of digital information, any bug or
virus can bring the end of the world. We have seen the effects of the
literalization of this belief in the millenarian hysteria of Y2K cultism.
On telepresence:

Artificial intelligence, tele-sensoriality, virtual reality and so
on—all this is the end of illusion. The illusion of the world—not
its analytical countdown—the wild illusion of passion, of thinking, the aesthetic illusion of the scene, the psychic and moral
illusion of the other, of good and evil (of evil especially, perhaps), of true and false, the wild illusion of death, or of living at
any price—all this is volatilized in psychosensorial telereality, in
all these sophisticated technologies which transfer us to the virtual, to the contrary of illusion: to radical disillusion. (27)

Why is virtual reality the end of illusion? Because it is the deliberate
and cynical choice of the virtual as fake over the world, as if we faced
an absolute binary choice: live in the real, or live in the virtual, and as
if we were seduced by the virtual into making the wrong choice. In
this black-or-white vision, once we enter the virtual worlds of modern media they close down upon us, and there is no way back to the
real. Further on, however, Baudrillard seems to switch sides, gleefully
warning us that "fortunately all this is impossible" (27), as if we had
invested our hopes in this dystopian vision. Because of technological
limitations, VR will never deliver on its promise to provide a perfect
duplicate of reality. So what is there to fear? For the numerous admir-
ers of Baudrillard, the value of his thought is less as a description of
the real—or of the place of the virtual in the real—than as a theory of
the *what if:* What if VR were perfectly realized? Would we spend our
entire lives inside a Disneyland of digital data? Would images became
our world? How would we tell the difference between simulation and
reality? If we could not do so, would this mean that simulations *had
become* reality—or alternatively, that reality *was* a simulation?

By asking us to entertain hypothetical situations and dystopic possibilities Baudrillard theorizes the triumph of the virtual as fake as something contained in the virtual as potential, but his language creates a fake all of its own, by hyperbolically couching the potential in the language of actuality: the real does not *threaten* to disappear in Baudrillard's text, it has already been killed *as a matter of fact*. Readers have no problems undoing the hyperbole and linking the theory to real-world tendencies rather than to terminal states of affairs: the invasion of culture by visual representations; the control of the mind by the media; the voracious appetite of modern society for images, an appetite that sometimes tempts us to kill the real in order to produce simulacra; and last but not least a cultural fascination with the hyper-real, a copy more real than the real that destroys the desire for the original. For many cultural critics who draw inspiration from Baudrillard, the real has not disappeared; it has merely exchanged places with the virtual, in the admittedly watered-down sense that time spent in the virtual seems to grow at an alarming rate in lives that ought to be, or used to be, rooted in a solid reality. Our gods are virtual, like the Golden Calf—the image that broke the second commandment at the very moment it was given—when they should be real.

But Baudrillard himself has another idea of the ontological status of his discourse. If we have reached stage 4 in the evolution of the image, this means that his theory falls under the scope of its own pronouncement and becomes one of these simulacra that engender their own reality. In "Radical Thought," an essay from *The Perfect Crime*, Baudrillard distinguishes two kinds of thought and leaves no doubt as to where he situates his own:

> A certain form of thought is bound to the real. It starts out from the hypothesis that ideas have referents and that there is a possible ideation of reality. A comforting polarity, which is that of tailor-made dialectical and philosophical solutions. The other form of thought is eccentric to the real, a stranger to dialectics, a stranger even to critical thought. It is not even a disavowal of the concept of reality. It is illusion, power of illusion, or, in other words, a playing with reality, as seduction is a playing with desire, as metaphor is a playing with truth. (96)

The ultimate is for an idea to disappear as idea, to become a thing among things. (100)

So, for example, you put forward the idea of the simulacrum, without really believing in it, even hoping that the real will refute it (the guarantee of scientificity for Popper).

Alas, only the fanatical supporters of reality react: reality, for its part, does not seem to wish to prove you wrong. Quite the contrary, every kind of simulacrum parades around it. And reality, filching the idea, henceforth adorns itself with all the rhetoric of simulation. It is the simulacrum which ensures the continuity of the real today, the simulacrum which conceals not the truth, but the fact that there isn't any [stage 3]—this is to say, the continuity of the nothing. . . .

It's terrifying to see the idea coincide with the reality. (101)

Radical thought encounters no resistance from the real, because in contrast to the "regular" brand—the kind that does not understand itself and that lives in the illusion of referentiality—it conceives its mode of operation as declarative rather than as descriptive. Like fictional discourse, it inhabits not the true-or-false but the true by say-so. In *The Perfect Crime,* as Poster observes, "Baudrillard has become virtual and knows himself to be such: he argues that his critical theory of simulation has become the principle of reality" ("Theorizing," 46). As the representation, or virtual, that becomes reality, Baudrillard's theory embodies, literally, the paradoxical idea of *virtual* reality.

For those who are prevented by an enduring sense of the presence and alterity of the real from accepting the idea that it derives from Baudrillard's discourse—should we call this sense simply "common"?—there remains fortunately the alternative of a nonradical interpretation. We live in simulacra because we live in our own mental models of reality. What I call "the world" is my perception and image of it. Therefore, what is real for me is the product of my copy-making, virtual-producing, meaning-making capability. The copies that make up my world cannot be perfect duplications, but this does not make them necessarily false, deceptive, or deprived of referent. In this interpretation, the absolutely real has not disappeared; it is, rather, as Slavoj Žižek defines it, "a surplus, a hard kernel which resists any process of modeling, simulation, or metaphorization" (*Tarrying*, 44). We know that this "other" real exists, and often we butt into it, but we do not live in it, except perhaps in some moments of thoroughly private and nearly mystical experience, because the human mind is an indefatigable fabricator of meaning, and meaning is a rational simulacrum of things. Disarming the other of its otherness by representing it and building "realities" as worlds to inhabit are one and the same thing. It is simply thinking.

LÉVY AND THE VIRTUAL AS POTENTIAL

Becoming Virtual, the English title of Pierre Lévy's *Qu'est-ce que le virtuel*, may seem at first sight to confirm Baudrillard's most pessimistic prediction for the future of humanity. But the impression is dispelled as early as the second page of the introduction to Lévy's treatise:

> The virtual, strictly defined, has little relationship to that which is false, illusory, or imaginary. The virtual is by no means the opposite of the real. On the contrary, it is a fecund and powerful mode of being that expands the process of creation, opens up the future, injects a core of meaning beneath the platitude of immediate physical presence. (16)

Lévy outlines his concept of virtuality—inspired in part by Gilles Deleuze's ideas on the topic—by opposing two conceptual pairs: one static, involving the possible and the real, and the other dynamic, linking the actual to the virtual.[2] The possible is fully formed, but it resides in limbo. Making it real is largely a matter of throwing the dice of fate. In the terminology of modal logic, this throw of the dice may be conceived as changing the modal operator that affects a proposition, without affecting the proposition itself. All it takes to turn the possibility into the actuality of a snowstorm is to delete the symbol ◇ (possibility operator) in front of the proposition "It is snowing today." The operation is fully reversible, so that the proposition p can pass from mere possibility to reality back to possibility. In contrast to the predictable realization of the possible, the mediation between the virtual and the actual is not a deterministic process but a form-giving force. The pair virtual/actual is characterized by the following features:

1. The relation of the virtual to the actual is one-to-many. There is no limit on the number of possible actualizations of a virtual entity.
2. The passage from the virtual to the actual involves transformation and is therefore irreversible. As Lévy writes, "Actualization is an *event*, in the strongest sense of the term" (171).
3. The virtual is not anchored in space and time. Actualization is the passage from a state of timelessness and deterritorialization to an existence rooted in a here and now. It is an event of contextualization.
4. The virtual is an inexhaustible resource. Using it does not lead to its depletion.

These properties underscore the essential role of the virtual in the creative process. For Lévy, the passage from the virtual to the actual is not a predetermined, automatic development but the solution to a problem that is not already contained in its formulation:

> [Actualization] is the creation, the invention of a form on the basis of a dynamic configuration of forces and finalities. Actualization involves more than simply assigning reality to a possible or selecting from among a predetermined range of choices. It implies the production of new qualities, a transformation of idea, a true becoming that feeds the virtual in turn. (25)

As this idea of feedback suggests, the importance of Lévy's treatment of virtuality resides not merely in its insistence on the dynamic nature of actualization but in its conception of creativity as a two-way process involving both a phase of actualization and a phase of virtualization. The complementarity of the two processes is symbolized in Lévy's text by the recurrent image of the Moebius strip, an image that stands in stark contrast to Baudrillard's vision of a fatal attraction toward the virtual.

While actualization is the invention of a concrete solution to answer a need, virtualization is a return from the solution to the original problem. This movement can take two forms. Given a certain solution, the mind can reexamine the problem it was meant to resolve, in order to produce a better solution; cars, for instance, are a more efficient way to solve the problem of transportation than horse-drawn carriages. Virtualization can also be the process of reopening the field of problems that led to a certain solution, and finding related problems to which the solution may be applied. A prime example of this process is the evolution of the computer from a number-crunching automaton to a world-projecting and word-processing machine.

The concept of virtualization is an extremely powerful one. It involves any mental operation that leads from the here and now, the singular, the usable once-and-for-all, and the solidly embodied to the timeless, abstract, general, multiple, versatile, repeatable, ubiquitous, immaterial, and morphologically fluid. Skeptics may object that Lévy's concept of virtualization simply renames well-known mental operations such as abstraction and generalization; but partisans will counter that the notion is much richer because it explains the *mechanisms* of these operations. If thought is the production of models of the world—that is, of the virtual as double—it is through the consideration of the virtual as potential that the mind puts together representations that can act upon the world. While a thought confined to the actual would be reduced to a powerless recording of facts, a thought that places the actual in the infinitely richer context of the virtual as potential gains control over the process of becoming through which the world plays out its destiny.

The power of Lévy's concept of virtualization resides precisely in its dual nature of timeless operation responsible for all of human culture, and of trademark of the contemporary *Zeitgeist*. In our dealing with the virtual, we are doing what mankind has always done, only more powerfully, consciously, and systematically. The stamp of postmodern culture is its tendency to virtualize the nonvirtual and to virtualize the virtual itself. If we live a "virtual condition," as N. Katherine Hayles has suggested (*How We Became Posthuman*, 18), it is not because we are condemned to the fake but because we have learned to live, work, and play with the fluid, the open, the potential. In contrast to Baudrillard, Lévy does not seem alarmed by this exponentiation of the virtual because he sees it as a productive acceleration of the feedback loop between the virtual and the actual rather than as a loss of territory for the real.

Lévy's examples of virtualization include both elementary cultural activities and contemporary developments. Among the former are toolmaking and the creation of language. Toolmaking involves the virtual in a variety of ways. The concrete, manufactured object extends our physical faculties, thus creating a virtual body. It is reusable, thus transcending the here and now of actual existence. Other virtual dimensions of tools are inherent to the design itself: it exists outside space and time; it produces many physically different yet functionally similar objects; it is born of an understanding of the recurrence of a problem (if I need to drive *this* nail here and now I will need to drive nails in other places and at other times); and it is not worn out by the process of its actualization.

Language originates in a similar need to transcend the particular. The creation of a system of reusable linguistic types (or *langue*) out of an individual or communal experience of the world is a virtualizing process of generalization and conceptualization. In contrast to a proper name, a noun like *cat* can designate not only the same object in different contexts but also different objects in different contexts with different properties: my cat, your cat, the bobcats in the mountains, and the large cats of Africa. It is this recyclable character of linguistic symbols that enables speakers to embrace, if not the whole, at least vast expanses of experience with a finite vocabulary.[3] Whereas the creation of language is the result of the process of virtualization, its use in an act of *parole* is an actualization that turns the types into concrete tokens of slightly variable phonic or graphic substance and binds utterances to particular referents. Even in its manifestation as *parole*, however, language exercises a virtualizing power. Life is lived in real time, as a succession of presents, but through its ability to refer to physically absent objects, language puts consciousness in touch with the past and the future, metamorphoses time into a continuous spread that can be traveled in all directions, and transports the imagination to distant locations.

As examples of more specifically contemporary forms of virtualization, Lévy mentions the transformations currently undergone by the economy and by the human body. In the so-called information age, the most desirable good is no longer solid manufactured objects but knowledge itself, an eminently virtual resource since it is not depleted by use and since its value resides in its potential for creating wealth.

On the negative side, the virtualization of the economy has encouraged the pyramid schemes that currently plague the industries of sales and investments. As for the body, it is virtualized by any practice and technology that aims at expanding its sensorium, altering its appearance, or pushing back its biological limits. In a fake-theory of the virtual, the virtualization of the human body is represented by the replacement of body parts with prostheses; it finds its purest manifestation in the implant of artificial organs and cosmetic surgery. In a potential-theory, the virtualization of the body is epitomized by performance- and perception-enhancing devices, such as the running sneaker and the telescope. The inspiration for these practices is the fundamentally virtualizing question "To what new problems can I apply this available resource, the body I was born with?" as well as the actualizing one "How should I refashion this body to make it serve these new functions?"

The development of simulation technologies such as VR illustrates yet another tendency of contemporary culture: the virtualization to a second degree of the already virtual. Consider computers. They are virtual objects by virtue of being an idea and a design out of which particular machines can be manufactured. These machines are virtual, as we have seen, in the sense that they can run different software programs that enable them to emulate (and improve on) a number of different other machines. Among their applications are simulative programs whose purpose is to test formal models of objects or processes by exploring the range of situations that can develop out of a given state of affairs. The knowledge gained by trying out the potential enables the user to manage the possible and to control the development of the real. If all tools are virtual entities, computer simulations are doubly or perhaps even triply virtual, since they run on virtual machines, and since they incorporate the virtual into their mode of action.

THE TEXT AS DOUBLE AND AS FAKE

As they are implicated in thought, the two faces of the virtual are also implicated in texts, the inscription and communicable manifestation of the thinking process. Descriptions of the text, especially of the artistic text, as image functioning as a double of the real go back at least as far as Aristotle:

> Imitation comes naturally to human beings from childhood (and in this they differ from other animals, i.e. in having a strong propensity for imitation and in learning their earliest lessons through imitation); so does the universal pleasure in imitation. What happens in practice is evidence of this: we take delight in viewing the most accurate possible images of objects which themselves cause distress when we see them (e.g. the shapes of the lower species of animals, and corpses). The reason for this is that understanding is extremely pleasant, not just for philosophers but for others too in the same way, despite their limited capacity for it. This is the reason why people take delight in seeing images; what happens is that as they view them they come to understand and work out what each thing is. (*Poetics* 3.1, 6)

We can read this passage as the expression of a classic view of representation, Baudrillard's stage 1 in the evolution of the image. It is because the work of art provides "understanding" of objects in the world, as it replaces raw sensory experiences with intelligible models of things, that we derive pleasure from the process of artistic duplication. But in stressing the innate propensity of human beings for imitation and the "delight" caused by images from early childhood on, the *Poetics* fragment suggests to the modern reader a much less didactic type of gratification: we enjoy images precisely because they are not "the real thing," we enjoy them for the skill with which they are crafted. This pleasure presupposes that the readers or spectators of artistic texts do not fall victim to a mimetic illusion; it is because they know in the back of their minds that the text is a mere double that they appreciate the illusionist effect of the image, the fakeness of the fake.

Baudrillard and Umberto Eco describe this attitude as a typically postmodern attraction for the hyperreal. In his *Travels in Hyperreality*, for instance, Eco suggests that visitors to Disneyland experience far greater fascination with automata that reproduce pirates or jungle animals than they would with live crocodiles or flesh and blood actors. Both Baudrillard and Eco lament this attraction to the image as a loss of desire for the original. But the Disneyland tourist, beloved scapegoat of cultural critics, deserves credit for the ability to appreciate the art that goes into the production of the fake. Rather than ridiculing the tourist's attitude for its lack of intellectual sophistication, I would suggest that we regard this attitude as the admittedly embryonic manifestation of a fundamental and timeless dimension of the aesthetic experience.

In the literary domain, the "fake" interpretation of the virtual entertains obvious affinities with the concept of fictionality. The feature of inauthenticity describes not only the irreal character of the reference worlds created by fiction but also, as John Searle has suggested, the logical status of fictional discourse itself. Some literary theorists, most notably Barbara Herrnstein Smith and Mary Louise Pratt, propose to regard fiction as the imitation of a nonfictional genre, such as chronicle, memoir, letter, biography, or autobiography. Without going this far—for many fictional texts do not seem to reproduce any identifiable type of reality-based discourse—we can profitably describe fiction as a virtual account of fact, or, with Searle, as a pretended speech act of assertion, since even though the fictional text evokes imaginary characters and events, or attributes imaginary properties to counterparts of real-world individuals, it does so in a language that logically presupposes the actual existence of its reference world.

This idea of the text, and in fact of the work of art in general, as a *virtual something else* has been systematically explored by Susanne K. Langer in *Feeling and Form*, a work published in 1953. Langer's interpretation of the virtual foregrounds the optical illusion: "The most striking virtual objects in the natural world are optical—perfectly definite visible 'things' that prove to be intangible, such as rainbows and mirages" (48). And also: "An image is, indeed, a purely virtual 'object.' Its importance lies in the fact that we do not use it to guide us to something tangible and practical, but treat it as a complete entity with only *visual* attributes and relations. It has no others; its visible character is its entire being" (ibid.). To extend the optical concept of virtuality to nonvisual forms of art, without resorting to worn-out and medium-insensitive metaphors such as "painting" with words or with sound, Langer detaches the notion of image from any individuated content. Though the work of art is an essentially mimetic text, this mimeticism resides more in the production of an equivalent of one of the fundamental, almost Kantian, a priori categories of human experience than in the reproduction of concrete aspects of life or singular objects. The virtual images of art are not primarily images of bodies, flowers, animals, characters, and events, or the abstract expression of feelings, but what we might call today dynamic simulations of abstract objects of thought, such as space, time, memory, and action.

One of the virtual aspects of the artistic image—and perhaps of the image in general—resides in its detachment from any particular spatio-temporal context. As a real object inscribed in space and time, the work of art is *in* the world, but as a virtual object that creates its own space and time, it is not *of* the world. This discontinuity between the artistic image and the surrounding world is particularly prominent in Langer's account of the visual arts as virtual space:

> The space in which we live and act is not what is treated in art at all. The harmoniously organized space in a picture is not experiential space, known by sight and touch, by free motion and restraint, far and near sounds, voices lost and re-echoed. It is an entirely visual affair; for touch and hearing and muscular action it does not exist. For them it is a flat canvas. . . . This purely visual space is an illusion, for our sensory experiences do not agree on it in their report. . . . Pictorial space is not only organized by means of color . . . it is created; without the organizing shapes it is simply not there. Like the space "behind" the surface of the mirror, it is what the physicists call "virtual space"—an intangible image. (72)

As one might expect, there is a form of art that parallels in the time dimension the virtualization of space that takes place in the visual arts. This art form is music. The effect of music, according to Langer, is to create a "virtual time" that differs from what may be called "clock-time" or "objective time" in that it gives form to the succession of moments and turns its own passing—transfigured as *durée*—into sensory perception:

> The direct experience of passage, as it occurs in each individual life is, of course, something actual, just as actual as the progress of the clock or the speedometer; and like all actuality it is only in part perceived. . . . Yet it is the model for the virtual time created in music. There we have its image, completely articulated and pure. . . . The primary illusion of music is the sonorous image of passage, abstracted from actuality to become free and plastic and entirely perceptible. (113)

The remaining equivalencies in Langer's systematic description of the arts as "virtual-something-else" describe dance as virtual gesture; poetry as virtual life; narrative as virtual memory; drama as virtual history; and film as virtual dream.[4] Though all of these equivalencies offer provocative insights on the genre under consideration, one cannot avoid the impression that Langer is forced into some categorizations by the tyranny of the pattern and the desire to avoid duplicate labels. In several cases the characterization could describe several art forms. The "virtual gesture" of dance, for instance, is defined in such a way that the label applies equally well to the other types of performance art, such as mime, drama, and even film. The respective characterizations of drama and narrative as "virtual history" and "virtual memory" would be better expressed in terms of "mimetic" versus "diegetic" modes of presentation, for what Langer has in mind is the fact that dramatic action takes place in the present while narrative typically (but not necessarily) encodes the result of the narrator's retrospective act of memory. The weakest equivalence of all, in my view, is the description of poetry as "virtual life" on the ground that a poem creates "a world of its own" (228). Why should poetry be more of a simulation of life than drama and narrative, two genres generally credited with far greater world-creating power than lyric art? The entire discussion of poetry seems symptomatic of the belief, widespread in the era of New Criticism—when Langer's book was written—that poetry embodies the essence of language art. As the most sublime of literary genres, it had to virtualize the most "vital" principle, the spark of life itself.

THE TEXT AS POTENTIALITY

As an analytical concept, the virtual as potential is no less fecund for literary and textual theory than the virtual as fake. Here again we must begin with Aristotle: "The function of the poet is not to say what *has* happened, but to say the kind of things that *would* happen, i.e. what is possible in accordance with probability and necessity" (*Poetics* 5.5, 16). This pronouncement may seem to restrict literature unduly to the representation of events and objects that could occur in the real world, given its physical, logical, and perhaps psychological and economic laws. A narrow interpretation of possibility would leave out not only fairy tales, science fiction, the fantastic, and magical realism but also the absurd, the symbolic, the allegorical, and the dreamlike. All these literary landscapes can be reclaimed by broadening the horizon of "probability and necessity" to the territories covered by a purely imaginative brand of possibility. The task of the poet is not necessarily to explore the alternative worlds that can be put together by playing with the laws of the real but to construct imaginary worlds governed by their own rules. These rules—which may overlap to various degrees with the laws of the real—must be sufficiently consistent to afford the reader a sense of what is and isn't possible in the textual world as well as an appreciation of the imaginative, narrative, and artistic "necessity" of what ends up being actualized.

The virtual as potential also lies at the core of the conception of the text developed by the two leading figures of reader-response criticism, the Polish phenomenologist Roman Ingarden and his German disciple Wolfgang Iser. Ingarden conceives the literary work of art, in its written form, as an incomplete object that must be actualized by the reader into an aesthetic object. This actualization requires of the reader a filling in of gaps and places of indeterminacy that can take a highly personal form, since every reader completes the text on the basis of a different life experience and internalized knowledge. Rather than associating the written or oral signs that make up the text with a specific possible world, it is therefore more appropriate to speak, with David Lewis, of a plurality of textual worlds. In this power to unfold into many worlds resides for Iser the virtuality of the work of art and the condition for the aesthetic experience: "It is the virtuality of the work that gives rise to its dynamic nature, and this in turn is the precondition for the effects that the work calls forth" ("Reading Process," 50).

For Pierre Lévy, the virtual as potential represents not only the mode of being of the literary text but the ontological status of all forms of textuality. "Since its Mesopotamian origin," writes Lévy, "the text has been a virtual object, abstract, independent of any particular substrate" (*Becoming Virtual*, 47). Paradoxically, this virtual object originates in an actualization of thought. The act of writing taps into, and enriches in return, a reservoir of ideas, memories, metaphors, and linguistic material that contains potentially an infinite number of texts. These resources are textualized through selection, association, and linearization. But if the text is the product of an actualization, it reverts to a virtual mode of existence as soon as the writing is over. From the point of view of the reader, as reader-response theorists have shown, the text is like a musical score waiting to be performed. This potentiality is not just a matter of being open to various interpretations or of forming the object of infinitely many acts of perception; otherwise texts would be no more and no less virtual than works of visual art or things in the world such as rocks and tables. The virtuality of texts and musical scores stems from the complexity of the mediation between what is there, physically, and what is made out of it. Color and form are inherent to pictures and objects, but sound is not inherent to musical scores, nor are thoughts, ideas, and mental representations inherent to the graphic or phonic marks of texts. They must therefore be constructed through an activity far more transformative than interpreting sensory data. In the case of texts, the process of actualization involves not only the process of "filling in the blanks" described by Iser but also simulating in imagination the depicted scenes, characters, and events, and spatializing the text by following the threads of various thematic webs, often against the directionality of the linear sequence.

As a generator of potential worlds, interpretations, uses, and experiences, the text is thus always already a virtual object. But the marriage of postmodernism and electronic technology, by producing the freely navigable networks of hypertext, has elevated this built-in virtuality to a higher power. "Thought is actualized in a text and a text in the act of reading (interpretation). Ascending the slope of actualization, the transition to hypertext is a form of virtualization" (Lévy, *Becoming Virtual*, 56). This virtualization of the text matters cognitively only because it involves a virtualization of the act of reading. "Hypertextualization is the opposite of reading in the sense that it produces, from an initial text, a textual reserve and instrument of composition with which the navigator can project a multitude of other texts" (54). In hypertext, a double one-to-many relation creates an additional level of mediation between the text as produced by the author—*engineered* might be a better term—and the text as experienced by the reader. This additional level is the text as displayed on the screen. In a traditional text, we have two levels:

1. The text as collection of signs written by the author
2. The text as constructed (mentally) by the reader

The object of level 1 contains potentially many objects of level 2. In a virtualized text, the levels are three:

1. The text as written or "engineered" by the author
2. The text as presented, displayed, to the reader
3. The text as constructed (mentally) by the reader

In this second scheme, which is also valid for the print implementations of what Eco calls "the open work," the textual machinery becomes "a matrix of potential texts, only some of which will be realized through interaction with a user" (Lévy, *Becoming Virtual*, 52). As a virtualization of the already virtual, hypertext is truly a hyper-text, a self-referential reflection of the virtual nature of textuality.

When Lévy speaks of the virtualization of the text, the type of hypertext he has in mind is not so much a "work" constructed by a single mind as the implementation of Vannevar Bush's idea of the Memex: a gigantic and collectively authored database made up of the interconnection and cross-reference of (ideally) all existing texts.[5] It is, properly speaking, the World Wide Web itself. In this database the function of the links is much more clearly navigational than in the standard forms of literary hypertext. The highlighted, link-activating key words capture the topic of the text to be retrieved and enable readers to customize the output to their own needs. In Lévy's words, the screen becomes a new "typereader [machine à lire], the place where a reserve of possible information is selectively realized, here and now, for a particular reader. Every act of reading on a computer is a form of publishing, a unique montage" (54). As the user of the electronic reading machine retrieves, cuts, pastes, links, and saves, she regards text as a resource that can be scooped up by the screenful. Electronic technology has not invented the concept of text as resource, or database, but it has certainly contributed to the current extension of this approach to reading. The attitude promoted by the electronic reading machine is no longer "What should I do with texts?" but "What *can* I do with them?" In a formula that loses a lot in translation, Lévy writes, "Il y a maintenant du texte, comme on dit de l'eau et du sable" (Now there is only text, as one might say of water and sand [62]). If text is a mass substance rather than a discrete object, there is no need to read it in its totality. The reader produced by the electronic reading machine will therefore be more inclined to graze at the surface of texts than to immerse himself in a textual world or to probe the mind of an author. Speaking on behalf of this reader Lévy writes, "I am no longer interested in what an unknown author thought, but ask that the text make me think, here and now. The virtuality of the text nourishes my actual intelligence" (63). The non-holistic mode encouraged by the electronic reading machine tends to polarize the attitude of the reader in two directions: reading becomes much more utilitarian, or much more serendipitous, depending on whether the user treats the textual database as what Gilles Deleuze and Félix Guattari (A Thousand Plateaus) call a striated space, to be traversed to get somewhere, or as a smooth space, to be explored for the pleasure of the journey and for the discoveries to be made along the way.

TWO

Virtual Reality as Dream and as Technology

Viewing 3-D graphics on a screen is like looking into the ocean from a glass-bottom boat. We see through a flat window into an animated environment; we experience being on a boat.

Looking into a virtual world using a stereographic screen is like snorkeling. We are at the boundary of a three-dimensional environment, seeing into the depth of the ocean from its edge; we experience being between at the surface of the sea.

Using a stereoscopic HMD [head-mounted display] is like wearing scuba gear and diving into the ocean. Immersing ourselves in the environment, moving among the reefs, listening to the whale song, picking up shells to examine, and conversing with other divers, we invoke our fullest comprehension of the scope of the undersea world. We're There.

—MEREDITH BRICKEN

When virtual reality technology burst into public view in the early 1990s, it was less through a revolutionary computer system than through a grand flourish of rhetoric. The idea of VR sprang fully formed from the brain of its prophets, and it was presented by the media to the general public as being in a state of perfect implementation. The popular perception of VR was primarily shaped by the declarations of a charismatic developer, as well as musician and visual artist, Jaron Lanier, who coined the term *virtual reality*,[1] and of an imaginative journalist, Howard Rheingold, whose 1991 book *Virtual Reality* took readers on a tour of the underground operations wherein brave new worlds of digital data were rumored to be secretly hatching. Little did many of us realize in the early 1990s that their enthusiastic and quite precise descriptions of VR applications were largely castles in the air, and how much separated "real" VR from the virtual brand. Rheingold's suggestion (345–52) that VR technology might be put in the service of teledildonics, sex in a bodysuit with a computer-simulated partner, did more to put VR on the cultural map than any demonstration of HMD, data glove, or three-dimensional visual display. But it was an interview with Lanier, published in 1988 in *Whole Earth Review* (reprinted in Zhai, *Get Real*), that provided the most vivid input for the image of the role and potential of VR technology that prevails in the public at large.

Lanier's VR was not a space that we would visit for short periods of time, as is the case with current installations, but a technology that would play a major role in our daily lives and deeply transform the conditions of our material existence. In this sense it would be real rather than virtual. A computer would be installed in our house, the so-called Home Reality Engine; we would turn it on, don a minimal VR outfit—nothing more encumbering than glasses to see and gloves to manipulate—and presto, we would be surrounded by a virtual world in which the material objects that furnish the house would take on whatever appearance we specified. (Projecting virtual appearances onto real objects, rather than creating virtual objects *ex nihilo*, solves the annoying problem of bumping into things.) In this world of our creation we would take on any identity we wished, but our virtual body would be controlled by the movements of the real body, and we would interact with the virtual world through physical gestures. The computer would keep track of all our past actions and creations, and since time spent in the system would be a significant part of our lives, these digital archives would become a substitute for memory. We would be able to relive earlier experiences by simply rerunning the software.

The boldness of this vision was quickly lapped up by VR theorists. Even scientists adopted Lanier's vision as a goal to shoot at (wasn't he, after all, a computer wizard?)—a situation that often makes it difficult to separate the science-fictional from the scientific and the futurological from the technological in the literature devoted to the idea. The discourse of contemporary culture intertwines at least three points of view on VR, those of dreamers, developers, and philosophers, and in this chapter I attempt to give each of them its due.

DREAMS OF VR, AND SOME REALITIES

At the first Conference on Cyberspace, held in Austin, Texas, in 1990, imaginations were turned loose and metaphors flowed freely. The term *cyberspace* is now mainly associated with the Internet, but for the participants in the conference it covered a wide range of applications of digital technology that included computer-generated environments—virtual worlds proper—as well as networking. For Marcos Novak, an architect, "cyberspace is poetry inhabited, and to navigate through it is to become a leaf on the wind of a dream" ("Liquid Architecture," 229). Nicole Stenger, an artist and poet using computer technology, declared, "Without exaggeration, cyberspace can be seen as the new bomb, a pacific blaze that will project the imprint of our disembodied selves on the walls of eternity" ("Mind," 51). Michael Heim, well-known VR philosopher, spoke eloquently of the "erotic ontology of cyberspace," widening *erotic* to a Platonic sense ("Erotic," 59). Michael Benedikt, the organizer of the conference and also an architect, described the lure of VR as a timeless attraction to other worlds, insisting on the spiritual and artistic implications of this fascination:

> Cyberspace's inherent immateriality and malleability of content provides the most tempting stage for the acting out of mythical realities, realities once "confined" to drug-enhanced ritual, to theater, painting, books, and such media that are always, in themselves, somewhat less than what they reach, mere gateways. Cyberspace can be seen as an extension, some might say an inevitable extension, of our age-old capacity and need to dwell in fiction. ("Introduction," 6)

This idea of dwelling in fiction evokes a popular theme of recent film and literature: walking into a story and becoming a character. We have seen this theme in *Alice in Wonderland* (Alice falls through the hole into narrative scripts that have been running for quite some time), in Woody Allen's story "The Kugelmass Episode" (a professor at an American college steps into the world of *Madame Bovary*), in the philosophical dialogues of Douglas Hofstadter's *Gödel, Escher, Bach* (Achilles and the Tortoise are pushed into the paintings of Escher), and in the 1998 movie *Pleasantville* (two teenagers from the 1990s are transported into their favorite TV show from the 1950s). Of all the versions of the scenario—or is it a postmodern myth?—none is more familiar to the general public than the Holodeck of the TV series *Star Trek: The Next Generation*. Several authors, including Michael Heim and Janet Murray, have exploited the Holodeck association to give a concrete face to VR. An imagination easily aroused is usually quick to feel let down, and as Lanier observed in late 1998, the Holodeck analogy turned out in the long run to be a double-edged sword for the PR of VR: "As for the waning of virtual reality from public attention, I bear some of the blame for it. I always talked about virtual reality in its ultimate implementation and when that didn't happen, interest declined. Because everyone wanted the Holodeck from *Star Trek*, virtual reality couldn't fulfill its promise so quickly" (quoted in Ditlea, "False Starts"). Precisely because of its utopian character, however, the Holodeck scenario provides a convenient approach to the dreams that were invested in the VR project. In the words of Michael Heim, the Holodeck is

> a virtual room that transforms spoken commands into realistic landscapes populated with walking, talking humanoids and detailed artifacts appearing so life-like that they are indistinguishable from reality. The Holodeck is used by the crew of the starship Enterprise to visit faraway times and places such as medieval England and 1920s America. Generally, the Holodeck offers the crew rest and recreation, escape and entertainment on long interstellar voyages. (*Metaphysics*, 122)

The scenario of the Holodeck breaks down into the following themes:

1. You enter *(active embodiment)* . . .
2. into a picture *(spatiality of the display)* . . .
3. that represents a complete environment *(sensory diversity).*
4. Though the world of the picture is the product of a digital code, you cannot see the computer *(transparency of the medium).*
5. You can manipulate the objects of the virtual world and interact with its inhabitants just as you would in the real world *(dream of a natural language).*
6. You become a character in the virtual world *(alternative embodiment and role-playing).*
7. Out of your interaction with the virtual world arises a story *(simulation as narrative).*

8. Enacting this plot is a relaxing and pleasurable activity *(VR as a form of art).*

Active Embodiment

Once in a while one hears cultural critics emit the opinion that VR is a disembodying technology, a comment that in our body-obsessed, hedonistic culture amounts to a sweeping dismissal of the project. The chief complaint of these critics (Simon Penny, Anne Balsamo, Arthur Kroker) is that VR replaces the body with a body image, thereby causing a Cartesian mind-body split.[2] This opinion is justifiable if under VR one understands "cyberspace" and the imaginary geography of the Internet, where face-to-face encounters in physical meeting places give way to conversations with strangers lodged in invisible bodies and hiding behind digital avatars. In his period-making novel *Neuromancer,* a work that profoundly shaped the popular conception of VR, William Gibson reinforced the conception of computer technology as hostile to the flesh by insisting on the need to leave the "meat" of the body to reach the Matrix, a global computer network through which the mind enjoys the mystical contemplation of the world translated into a fully intelligible display of digital information.

But as a technology of representation, the Lanier-inspired conception of VR is neither cyberspace, the Internet, nor the product of Gibson's imagination.[3] In the brand of VR that is the concern of this chapter, the participation of the physical body is a primary issue, even when the body is clad in a "smart costume" (i.e., a body image constructed by the system) or when it manipulates a distant puppet through teleoperations. "Our body is our interface," claims William Bricken in a VR manifesto (quoted in Pimentel and Teixeira, *Virtual Reality,* 160). Or as Brenda Laurel argues, VR offers the rare opportunity to "[take] your body with you into worlds of imagination" ("Art and Activism," 14). Compared with walking around town, exploring a virtual world with headset, data gloves, or wired bodysuit may involve a significant loss of corporeal freedom, especially since current systems restrict the reading of the body to head and hand movements, but even in its rudimentary state of development the VR experience allows far more extensive physical action than sitting at a computer terminal and typing on a keyboard.

Spatiality of the Display

For a body to enter into a world, this world must be fully spatial. Lanier, in his pioneering pronouncements about VR, describes the VR experience as follows: "When you put [VR glasses] on you suddenly see a world that surrounds you—you see the virtual world. It's fully three-dimensional and it surrounds you, and as you move your head to look around, the images that you see inside the eyeglasses are shifted in such a way that an illusion is created that while you're moving around the virtual world is standing still" (Zhai, *Get Real*, 176). The idea seems simple enough, but if we take a closer look at this description, we notice that the experience of being inside a computer-generated world involves three distinct components: a sense of being surrounded, a sense of depth, and the possession of a roving point of view. Each of these dimensions improves on, or remediates, a variety of earlier technologies.

The ancestors of the surrounding image are the panorama and the cyclorama, both types of installation that flourished in the nineteenth century. Whereas panoramas were moving pictures that unrolled, like a scroll, between two spindles, so that only a portion of the picture would be displayed at any given time, cycloramas were circular paintings displayed on the walls of an equally circular building, affording the viewer a 360-degree angle of vision (Maloney, "Fly Me," 566). VR combines these two ideas by allowing the body to turn around and inspect various parts of the image, as in cycloramas, and by constantly updating the image, as in mechanical panoramas. The sense of depth created by VR displays is the latest development in a series of mathematical or technological innovations that includes the discovery of perspective in the Renaissance, the stereoscopes of the eighteenth and nineteenth centuries, the Cinerama movies of the 1950s, which conveyed a sense of depth when they were viewed with special glasses, and the large-screen IMAX movies of the present. It takes, however, a movable point of view to acquire a full sense of the depth of an image, because it allows objects to slide within the field of vision and to get bigger or smaller to the viewer as their distance from the eye increases or decreases. (This effect is known as motion parallax.)

Here again VR represents the ultimate achievement in the history of what Jay Bolter and Richard Grusin call "point of view technologies" (*Remediation,* 162). First we had flat depictions that did not project a space beyond their surface, and therefore did not assign a point of view to the spectator. We tend to process these representations as "the sign" (or visual icon) of an absent object rather than as its immediate presence in the field of vision. Then we had perspective paintings that extended the pictorial space in front of and behind the canvas. Their two-dimensional projection of a three-dimensional space placed the spectator's body in a fixed location with respect to the depicted object. When a chair in a painting is represented from the right and from above, we will retain this point of view even if we physically move to the left and kneel down to look at the picture from below, though admittedly the effect will not be as dramatic as for a spectator situated at the center of projection.[4] Movies allowed shifts in point of view, as the movements of the camera presented objects from various angles and made them change size for the eye, but the spatial location of the virtual body of the spectator in the movie-world was rigidly determined by the location of the camera. Now imagine that the spectator is able to operate the camera, select the point of view, and maintain a continuous apprehension of the external world. This is exactly what happens when a computer tracks the movements of the user's head and body and updates her vision accordingly. As Frank Biocca and Ben Delaney observe, "With a [headset] the viewer ceases to be a voyeur and comes closer to being an actor in the visual world" ("Immersive Virtual Reality," 68). But it does not take a headset to implement the dimension of roving point of view: any so-called first-person video game on a regular computer screen offers a display that can be navigated with a mouse and that constantly updates itself to reflect the position of the cursor (a substitute for the player's body). In contrast to VR, though, the screen display does not offer three-dimensional stereoscopic effects. VR is the only medium that combines the three properties of 360-degree panoramic picture, three-dimensional display, and a point of view controlled by the user.

Sensory Diversity

Each sense, or faculty, is the target of an art form: literature for the mind, painting for the eye, music for the ear, cuisine for the taste buds, perfume for the nose. Touch is the hardest to link to art, but even touch can be cajoled by "designed experiences": erotic massage techniques, sculptures meant to be stroked, and carnival rides that induce scary but pleasurable shaking or vibrations. The extraordinary development of media in the twentieth century may be in part responsible, together with the influence of such creators as Wagner and Artaud, for the popularity of a conception of total art that insists on the involvement of all the senses in the artistic experience. The closest to this ideal is the opera, with its blend of music, dance, drama, poetry, stage design, costumes, and light effects; but for all its artistic resources, the opera addresses only two of the senses, as do the theater, cinema, and TV.

On a darker note, multisensory experiences have also been cast in the role of antiart. In *Brave New World*, his dystopic novel of anticipation, Aldous Huxley imagined a society stultified by a kind of movie, the Feely, that offered visual, auditive, olfactory, and tactile stimuli. The spectators could feel every hair of the rug on which the protagonists were making love, every jolt in the crash of a helicopter, and they were so fully absorbed in these sensations that they paid no attention to the silliness of the plot. Despite Huxley's warning that multisensory art would extinguish critical sense and render the imagination obsolete, the idea has retained a powerful hold on the modern mind. In *Finnegans Wake*, as Donald Theall has shown, James Joyce attempted to create a syncretic and synaesthetic language that involved the entire sensorium and simulated the effects of all media. The 1950s and 1960s were obsessed with a much more literal but also rather trivial expansion of perceptual dimensions: Cinerama, 3D glasses, movies accompanied by scratch-and-sniff cards, and, to crown it all, Morton Heilig's Sensorama, an arcade-style ride-on machine that simulated a motorcycle ride through New York City in four sensory dimensions: the sight of the Manhattan streets; the roar of the engine and of other traffic; the exhaust fumes of cars and the aroma of pizza cooking in restaurants; and the vibrations of the handlebars (Steuer, "Defining," 43).

Though VR is widely credited with the power to create a richer and more diversified environment than any other medium, its potential contribution to the expansion of the sensory dimensions of an image is really quite limited. It is only through haptic sensations—feeling textures and the resistance of simulated objects—and by enabling the user to grab objects that computers can improve on established technologies of representation. Unfortunately, the simulation of the sense of touch is still in a very primitive stage. The most advanced data gloves provide sensation in only one "finger," though the user is said to adapt quickly to this reconfigured hand because of a phenomenon known in the field as "accommodation to virtual worlds." VR developers have made no serious attempt to include gustatory and olfactory signals, because taste and smell do not lend themselves to computerized simulation: as Biocca and Delaney observe, "Both senses are chemical interfaces with the physical world" ("Immersive Virtual Reality," 96). The addition of rudimentary haptic sensations to the standard repertory of visual and aural data may seem out of proportion with the dream of a complete sensory environment, and in its present stage of development virtual touch does not seem to have much artistic potential, but its significance is perhaps more psychological than purely sensorial. It does not take the simulation of haptic sensations to open virtual worlds, at least virtually, to the sense of touch. The resources of digital imaging make it possible to produce a visual display of such intricate texture and shading that the user feels as if he could reach out and caress the objects. Whether imagined or physically simulated, touch is the sense that conveys the strongest impression of the solidity, otherness, and resistance of an object. As Michael Benedikt has pointed out, the awareness of this resistance is the most fundamental condition of a sense of the real: "What is real always pushes back. Reality always displays a measure of intractability and intransigence. One might even say that 'reality' *is* that which displays intractability and intransigence relative to our will" ("Cyberspace," 160).

Transparency of the Medium

In their book *Remediation* Bolter and Grusin identify the force that inspires cultures to develop new media as a desire for a total lifelikeness that they call transparency: "Our culture wants both to multiply its media and to erase all traces of mediation: ideally, it wants to erase its media in the very act of multiplying them" (5). But if we could develop a medium that provides a perfect copy of the real, or a perfect illusion of reality, would there still be a need for other media? In its ideal implementation, VR is not merely another step toward transparency, to be "remediated" by future media, but a synthesis of all media that will represent the end of media history. Frank Biocca, Taeyong Kim, and Mark Levy suggest that we may be now witnessing "the early stages of the arrival of the ultimate medium" ("Vision," 13). Lanier, predictably, comes up with the most radical pronouncements: "Virtual Reality starts out as a medium just like television or computers or written languages, but once it gets to be used to a certain degree, it ceases to be a medium and simply becomes another reality that we can inhabit" (Zhai, *Get Real,* 184). Or: "Virtual Reality, by creating a technology that's general enough to be rather like reality was before there was technology, sort of completes a cycle" (187).

In this final chapter of media history, transparency is not an end in itself but the precondition for total immersion in a medium-created world. This explains why Pimentel and Teixeira title the first chapter of their book on VR "The Disappearing Computer." The "virtual reality effect" is the denial of the role of hardware and software (bits, pixels, and binary codes) in the production of what the user experiences as unmediated presence.[5] VR represents in this respect a radical change of direction from the conception of the computer that prevailed when artificial intelligence was the most publicized application of digital technology. In the age of VR, and even more in the age of the World Wide Web, computers are no longer credited with an autonomous mind but serve as pure media—as largely hollow channels for the circulation of information. As Brenda Laurel declares, "Throughout this book *[Computers as Theatre]* I have not argued for the personification of the computer but for its invisibility" (143). Jaron Lanier echoes: "With a VR system you don't see the computer anymore—it's gone. All that's there is you" (Lanier and Biocca, "Insider's View," 166).

This disappearance of the computer represents the culmination of the trend toward more user-friendly interfaces in computer design. Binary coded machine instruction once gave way to the mnemonic letter codes of assembly languages; assembly languages were in turn translated into high-level languages with a syntax resembling that of natural languages. Then arbitrary words were supplanted by the motivated signs of icons on the screen. One of the articles of faith of the art of interface design is that the computer is a forbidding object that intimidates the user. Whenever possible, the electronic way of doing things should therefore be explained by metaphors that assimilate the new to the familiar—a strategy reminiscent of the process by which natural languages encode abstract ideas through the transposition of concrete categories. The most famous of these metaphors was the desktop and its assortment of tools represented by icons: pages, files, folders, scissors, glue, erasers, and trash cans. But the icons of the desktop metaphor merely surround the part of the screen where everything happens, and the screen is very much a part of the computer's visible body. For immersion to be complete, visual displays should occupy the entire field of the user's vision rather than forming a world-within-a-world, separated from reality by the frame of the monitor. As Gabriel D. Ofeisch observes, "As long as you can see the screen, you're not in VR. When the screen disappears, and you can see an imaginary scene . . . then you are in VR" (quoted in Pimentel and Teixeira, *Virtual Reality*, 7). In the perfect VR system the disappearance of the computer should be achieved on two levels, the physical and the metaphorical. Physically the computer will be made invisible to the user by being worn on the surface of the skin as what Lanier calls "virtual reality clothing." (Dystopic science fiction warns us of a far more frightening practice, the direct implantation of a computer inside the human body.) Metaphorically the computer will turn into a space that embraces far more than the desktop and the chat room: this space will be a world for the user to inhabit. "Virtual reality" is not just the ultimate medium, it is the ultimate interface metaphor.

Dream of a Natural Language

The dream of an optimal interface is the dream of a command language naturally fitted to the task at hand. This means that in VR symbolic code must disappear, at least in those areas in which it can be more efficiently replaced by physical actions. According to Jaron Lanier, "There's also the ability of communicating without codes. . . . I'm talking about people using their hands and their mouth, whatever, to create virtual tools to change the content of a virtual world very quickly and in an improvisational way" (Lanier and Biocca, "Insider's View," 160). "So, if you make a house in virtual reality, and there's another person there in the virtual space with you, you have not created a symbol for a house or a code for a house. You've actually made a house. It's that direct creation of reality; that's what I call post-symbolic communication" (161). For Michael Benedikt, this postsymbolic communication signals the beginning of a "postliterate" era in which "language-bound descriptions and semantic games will no longer be required to communicate personal viewpoints, historical events, or technical information. . . . We will become again 'as children' but this time with the power of summoning worlds at will and impressing speedily upon others the particulars of our experience" ("Introduction," 12). Through this language without symbols, people will build a shared reality and minds will become transparent to each other: "Simply, virtual reality, like writing and mathematics, is a way to represent and communicate what you can imagine with your mind. But it can be more powerful because it doesn't require you to convert your ideas into abstract symbols with restrictive semantic and syntactic rules, and it can be shared by other people" (Pimentel and Teixeira, *Virtual Reality*, 17).

The mystics of ages past—such as Swedenborg, the esoteric philosopher of the eighteenth century—had a term for this radically antisemiotic mode of communication. They called it the "language of the angels." It would be easy to dismiss the whole project as vaporous New Age mysticism, but the idea of communication without symbols appears much less angelic if we regard it as a *supplement* to and not as a *replacement* for symbolic expression, and if we interpret *symbol* in the narrow sense proposed by Charles Sanders Peirce: a sign whose meaning is based on a social convention that must be learned by the user. Nonsymbolic does not necessarily mean *post*symbolic, *pace* Lanier and Benedikt. In many situations symbolic expression is indeed what comes naturally to human agents. We may, for instance, meet people in the virtual world—people real or virtual—and wish to talk to them in French or in English. Worse even, the Home Reality Engine could fall into the hands of a perverse hacker who might insist on furnishing his homemade world with very visible virtual computers, and on programming them in their native machine language. If, as Susan Brennan observes, "certain actions are more easily done gesturally/spatially (as direct manipulation enthusiasts have noticed)" while "others are more easily done with language" ("Conversation," 403), it would be absurd to exclude symbol-based codes such as language from virtual worlds. Biocca and Delaney give us a more realistic idea of the place of nonsymbolic expression in VR systems than the authors quoted above by offering concrete examples of its advantages:

> The input devices of highly immersive virtual environments try to conform to the way we interact with the physical world by making use of things such as the movement of our limbs, head, eyes, and other motions in physical space. The difference is best illustrated by an example. Say you want to move a computer graphic representation of a cube. In a nongraphic system you might type: Move cube, Location $x = 10$, $y = 55$, $z = 42$. In virtual reality you simply bend down and pick up the computer graphic cube with your hand and place it on a computer graphic table. The floor, the cube, the table, and the graphic representation of your hand are all data entities in a program, as is the computer's representation of your movement. To you it appears as a naturalistic perceptual event. ("Immersive Virtual Reality," 97)

This ambition to develop natural modes of interaction within virtual worlds represents a complete turnaround from the philosophy of early structuralism, in which the arbitrary system of signs of what are ironically called "natural" languages was regarded both as the metalanguage into which all other semiotic codes could be translated and as a universal medium whose categories entirely determine how we think and what can be thought. We are now much more open to the idea that thought is not always verbal, and that some types of thought are better served by expressive resources that do not involve discrete and arbitrary symbols. Partisans of this view include Jaron Lanier, who named his now defunct company VPL, for Visual Programming Language; Pierre Lévy, who believes that the expressive potential of the computer will be better served by a graphic language that he calls "dynamic ideography" than by alphanumeric symbols (*L'Idéographie dynamique*); and Brian Rotman, who argues that mathematics should accept diagrams as proofs rather than relying exclusively on reasoning formalized in the traditional symbols of the field. In an advanced VR system there will be no need for *ekphrasis*—the verbal description of a visual artwork—because the system will encompass all forms of representation, action, and signification. The multisensory will also be the omnisemiotic.

Alternative Embodiment and Role-Playing

The convenience of virtual environments as a workshop for do-it-yourself bodies serving as manifestations of do-it-yourself identities has been widely extolled, and the legitimacy of these identities endlessly debated by cultural critics. Some critics view the virtual bodies of cyberspace—MOOs, chat rooms, computer games, and VR—as the liberating expression of culturally repressed desires; others insist that we have only one body, located in the real world, and that all this play with virtual bodies and virtual personae does not alter the fact that the only body that really matters is the material body; the self cannot be pried loose from the flesh. Halfway in between are those who maintain that the self is multiple and that digital identities actualize its potential, but that all these identities are ultimately supported, held together, or "warranted" (Allucquère Rosanne Stone's expression; see "Will the Real Body Please Stand Up") by the physical body. I am not going to offer here my own solution to these problems, and indeed I have none, because I believe that even in postmodern society, selves are not uniformly diverse, and no theory can speak for the many ways in which we may relate to the virtual bodies and virtual personae that we adopt in cyberspace. It will be sufficient for the present purpose to sketch briefly the importance that role-playing and corporeal involvement have been accorded from the very beginning in the conception of VR.

The possibility of redesigning our bodies and becoming something or somebody else is indeed a central theme in Jaron Lanier's 1988 description of the Home Reality Engine: "The computer that's running the Virtual Reality will use your body's movements to control whatever body you choose to have in Virtual Reality, which might be human or might be something quite different. You might very well be a mountain range or a galaxy or a pebble on the floor. A piano . . ." (Zhai, *Get Real*, 177). What does it mean to become a mountain, pebble, or galaxy, all entities devoid of consciousness? The metamorphoses that Lanier has in mind are not a loss of mental faculties or even a change of personality but primarily a change of point of view and physical abilities: our virtual bodies may fly or creep on the ground, see everything from high above or put up with the limitations of a terrestrial vision, embrace the whole universe or shrink down to the size of a Lilliputian. Some media theorists who do not have much positive to say about VR praise the technology for teaching the relativity of point of view, a lesson that should inspire empathy since it enables users to experience "what it is like to be something or somebody else" (Bolter and Grusin, *Remediation*, 246).

If we inhabit virtual bodies as a point of view, how will we know what these bodies look like and how will we relate to them? To strengthen the bond between the self and the new body, VR systems may paradoxically dissociate the two, so that users will be able to see an image of their virtual embodiment in a combination of first- and third-person point of view that suggests an out-of-the-body experience. According to Ann Lasko-Harvill, former collaborator with Lanier at VPL, "In virtual reality we can, with disconcerting ease, exchange eyes with another person and see ourselves and the world from their vantage point" ("Identity and Mask," 227). But a mere play with point of view falls short of the free design of identities that Lanier describes above. It is only in MUDs and MOOs, the so-called text-based virtual realities, that users can entirely fabricate their own personae, because all it takes to create a virtual individual in these environments is to post its verbal description on the network. When the reconfiguration of the body depends on technological means, such as headset and data gloves, these means determine the range of possible forms of embodiment available to the user. In the present state of development of VR technology, virtual identities must be selected from a menu of ready-made avatars. Playing a role, in these systems, is not a matter of becoming whomever you want to be but a matter of stepping into what Brenda Laurel has aptly called a "smart costume"—smart because it does not merely alter appearance but implements a change of body dynamics.

Simulation as Narrative

VR is not the static image of something or of nothing that Baudrillard calls a simulacrum but an active system of simulation. For Baudrillard, the essence of simulation is deception: "To simulate is to feign to have what one doesn't have" ("Precession," 3). All of Baudrillard's examples of simulacra are images that deceive by virtue of hiding an absence: the Byzantine icon hides the fact that there is no God; Disneyland hides the fact that the city and country surrounding it are just as unreal as a theme park (though Baudrillard does not bother to explain in what sense Los Angeles and the rest of America are unreal); and by the same reasoning VR hides the fact that all reality is virtual. Though these simulacra are fully formed objects, they don't seem to be the product of a creative process, and they don't seem to fulfill a specific purpose. Baudrillard's simulacra are not made, they just are, and they are not *used* to deceive (this would presuppose an agent and an intent), they *embody* deception as a fundamental cultural and epistemological condition. If they have a function at all, it is to satisfy our need for this condition. Computer simulations differ from this conception of the simulacrum on several essential points: they are processes and not objects; they possess a function, and this function has nothing to do with deception; they are not supposed to re-present what is but to explore what could be; and they are usually produced for the sake of their heuristic value with respect to what they simulate. To simulate, in this case, is to test a model of the world. When the simulated world does not exist, as is the case with the projected uses of Lanier's Home Reality Engine, simulation becomes an autotelic activity, but this does not preclude a heuristic value, since the creation and exploration of imaginary worlds can be an instrument of self-discovery.

The essence of computer simulation, whether in VR or in less sophisticated environments, resides in its dynamic character. Ted Friedman calls simulation a "map-in-time" with a narrative dimension ("Making Sense," 86). A typical simulation consists of a number of agents that are given an environment to live in and some rules to follow. The sum of these elements constitutes a narrative world, complete with characters, setting, and principles of action. Because of its power to model the interaction of many forces, and to follow the evolution of a world over a lengthy period of time, computerized simulation is an invaluable tool for the study of complex systems, such as those that form the concern of chaos theory. The simplest simulations comprise only one type of agent; for instance, the computer museum in Boston exhibits a simulation of how termites build heaps of wood by picking up scattered pieces. The system starts by distributing termites and pieces of wood randomly on the screen. The termites are given three rules of behavior:

1. Move randomly.
2. If you bump into a piece of wood, pick it up.
3. If you bump into a piece of wood while carrying another one, drop this piece nearby.

After a number of repetitions of this pattern, the wood starts gathering into distinct piles, but since the termites keep gnawing at the edges, the contours of the piles are never definitely outlined, and the piles themselves never solidify into perfect self-contained shapes. Throw several agents into the system with competing goals—for instance, one species of fish that want to eat other fish and another species that want to swim peacefully in schools—and the narrativity of the system takes on a dramatic shape; if the rules are written in such a way that a goal can be fully attained, the system may even reach a state of equilibrium, the simulative equivalent of narrative closure.

When the system revolves around human input, as is the case in VR and computer games, the simulation becomes the life story of the user, or rather the story of one of the user's virtual lives in the pursuit of a more or less specific goal. Every action taken by the user is an event in the virtual world. The sum of these events may not present proper dramatic form—an Aristotelian rise and fall of tension—but because all events involve the same participant they automatically satisfy the looser pattern of the epic or serial (episodic) narrative. A smart system, as we will see in chapters 8 and 10, may even steer the user's choices toward Aristotelian structures. There are admittedly no courtroom scribes or sports broadcasters who verbalize everything that happens in the virtual world, so the built-in narrativity of VR is strictly a matter of potentiality. The same can be said of the narrativity of life, or even of the theater, and this is why the expression "untold story," so dear to tabloids, is not necessarily an oxymoron. Drama, life, and VR create narrative material with characters, setting, and actions but without narrators. In contrast to narrated narratives, simulation systems do not re-present lives retrospectively, fashioning a plot when all events are in the book and all the potential narrative material is available to the storyteller, but generate events from a prospective point of view, without knowledge of their outcome. The user lives the story as she writes it through her actions, in the real time of a continuously moving present. Taken as a whole, however, a VR system is not merely a nonnarrated narrative but a matrix of doubly possible stories: stories that could be lived, and stories that could be told. Like a "Garden of Forking Paths"—to parody the title of a short story by Borges that has achieved cult status among theorists of interactive literature—the virtual world is open to all the histories that could develop out of a given situation, and every visit to the system actualizes a different narrative path.

VR as a Form of Art

There is no need to dwell at length on the artistic dimension of VR, since it follows from the successful implementation of its other features. VR technology has a number of practical applications, from flight simulators to remote-control surgery and the exploration of the terrain of distant planets, but from the very beginning it has been the potential of the medium as a tool for creative self-expression that has fascinated its advocates. Through its immersive dimension VR inaugurates a new relation between computers and art. Computers have always been interactive; but until now the power to create a sense of immersion was a prerogative of art. Michael Heim has called VR the "Holy Grail" of the artistic quest: "Rather than control or escape or [merely] entertain or communicate, the ultimate promise of VR may be to transform, to redeem our awareness of reality" (*Metaphysics*, 124). In a pure Platonic spirit, the fulfillment of all the senses will stimulate intellectual faculties and offer an experience that blends the aesthetic with the mystical and metaphysical. Lanier's vision is no less exalted, but rather than drawing on philosophical sources, it is inspired by the intellectual current that runs from romanticism and symbolism to dadaism, surrealism, and the drug culture of the 1960s. The Home Reality Engine is nothing less than a technological support for the surrealist/dadaist ideal of an artwork that transforms daily existence into an aesthetic experience, liberates the creative power of the user, and turns poetry into a way of life: "What's exciting are the frontiers of the imagination, the waves of creativity as people make up new things. . . . I want to make tools for VR that are like musical instruments. You could pick them up and gracefully 'play' reality. You might 'blow' a distant mountain range with an imaginary saxophone" (quoted in Zhai, *Get Real,* 49–50).[6]

PRESENCE, IMMERSION, AND INTERACTIVITY

The relative importance of immersion and interactivity in a VR system depends on the system's function. In practical applications, immersion is a means to guarantee the authenticity of the environment and the educational value of the actions taken by the user. In a flight simulator, for instance, the usefulness of the system as a test of what a pilot will do with an actual airplane depends on its power to reproduce the complexity and stressful demands of real flight situations. In artistic applications, by contrast, interactivity tends to be subordinated to immersive ideals. It is because they can act upon the virtual world, and because this world reacts to their input, that users acquire a sense of its presence. This concept of presence is often used in the technical and semitechnical literature on VR to describe the experience that forms the goal of research in the field: "A virtual reality is defined as a real or simulated environment in which a perceiver experiences telepresence" (Steuer, "Defining," 76). Telepresence—or simply presence, in the VR world—relates to physical presence as virtual reality relates to reality: "Telepresence is the extent to which one feels present in the mediated environment, rather than in the immediate physical environment. . . . This [mediated environment] can be either a temporally or spatially distant *real* environment . . . or an animated but nonexistent *virtual world* synthesized by a computer" (ibid.).

The issue of presence involves two conceptually distinct, though practically related, problems: How do we experience what is *there* as being here (telepresence proper), and how do we experience what is made of information as being material? The answer to these questions breaks down into a technological and a psychological, or phenomenological, problem. Jonathan Steuer observes that on the level of hardware, a system's ability to establish presence is a matter of the depth and breadth of information that it can handle (81). Depth is a function of the resolution of the display, while breadth is dependent on the number of senses addressed by this information. One must assume that in order to create presence, a significant amount of information must be devoted to the production of three-dimensional representation. Presence requires a photorealistic display, with detailed effects of texture and shading, but it does not require a real-world content. Another factor of presence involves the mobility of the user's body with respect to the "present" object. In the real world, an object seen through a window may be just as real as an object that we can touch, but we experience it as far less "present" because the sense of presence of an object arises from the possibility of physical contact with it. The object and the body of the perceiver must be part of the same space.

A theory of presence must therefore incorporate a theory of interactivity. Thomas Sheridan ("Musings," 122) acknowledges this dependency when he lists the following three items as the variables that control the experience of presence:

— Extent of sensory information (a category that covers both depth and breadth)
— Control of relation of sensors to environment (e.g., the "ability of the observer to modify his viewpoint for visual parallax or visual field, or to reposition his head to modify binaural hearing, or ability to perform haptic search")
— Ability to modify physical environment (e.g., "the extent of motor control to actually change objects")

The first of these factors is responsible for the lifelikeness and three-dimensionality of the display, while the other two represent two distinct modes of interactivity: the ability to explore an environment, and the ability to change it. At this point we face two terminological choices: label the product of the first factor immersivity, and the sum of the three presence; or label factor 1 realism and call the total effect either immersion or presence. I prefer the second choice, because the sense of belonging to a world cannot be complete without the possibility of interacting with it. As for the terms *immersion* and *presence*, they capture two different but ultimately inseparable aspects of the total effect: *im*mersion insists on being *inside* a mass substance, *pre*sence on being *in front of* a well-delineated entity. Immersion thus describes the world as a living space and sustaining environment for the embodied subject while presence confronts the perceiving subject with individual objects. But we could not feel immersed in a world without a sense of the presence of the objects that furnish it, and objects could not be present to us if they weren't part of the same space as our bodies. This approach means that the factors that determine a system's degree of interactivity also contribute to its performance as immersive system.

Steuer lists the following factors of immersive interaction, without claiming that the list is exhaustive:

> *Speed,* which refers to the rate at which input can be assimilated into the mediated environment; *range,* which refers to the number of possibilities for action at any given time; and *mapping,* which refers to the ability of a system to map its controls to changes in the mediated environment in a natural and predictable manner. ("Defining," 86)

The first of these items requires little explanation. The speed of a system is what enables it to respond in real time to the user's actions. Faster response means more actions, and more actions mean more changes. (Existing systems, because of hardware limitations, are somewhat deficient in this domain. With currently available head-mounted displays, the generation of visual data is said to lag annoyingly behind the movements of the head.) The second factor is equally obvious: the choice of actions is like a set of tools; the larger the set, the more malleable the environment. The factor of mapping imposes constraints on the behavior of the system. The user must be able to foresee to some extent the result of his gestures, otherwise they would be pure movements and not intent-driven actions. If the user of a virtual golf system hits a golf ball, he wants it to land on the ground, and not to turn into a bird and disappear in the sky.[7] On the other hand, the predictability of moves should be relative, otherwise there would be no challenge in using the system. Even in real life we cannot calculate all the consequences of our actions. Moreover, predictability conflicts with the range requirement: if the user could choose from a repertory of actions as vast as that of real life, the system would be unable to respond intelligently to most forms of input. The coherence of flight-simulation programs stems, for instance, from the fact that they exclude any choice of activity unrelated to flying. Meaningful interactivity requires a compromise between range and mapping and between discovery and predictability. Like a good narrative plot, VR systems should instill an element of surprise in the fulfillment of expectations.

Technical features such as these explain how digital information systems can connect the user to a virtual world, but if we want to understand in its subjective meaning the experience of "being there," we need a phenomenological approach to the question of virtual presence. As a philosophy of the first-person point of view, of the "being-for" of things rather than of being in itself, phenomenology is uniquely suited to analyze the sense of presence to a world that arises from the inscription of the body in the VR system. In what follows I propose to read VR in the light of the insights of Maurice Merleau-Ponty, the most forceful advocate of the embodied nature of cognition, on the phenomenology of perception.

THE PHENOMENOLOGICAL DIMENSION OF THE VR EXPERIENCE

The "there" of VR may not be anywhere, objectively speaking, but since we are supposed to relate to virtual worlds as if they were real, the phenomenological investigation of immersion begins with the investigation of the corresponding experience of "belonging to a world" in real environments. This experience forms the major concern of Merleau-Ponty's major work, *The Phenomenology of Perception*. In this book Merleau-Ponty seeks a compromise between an objectivist ontology that attempts to capture the being of things independently of the observer and a subjectivist stance by which my perception creates objects and endows them with properties. Far from denying the mind-independent existence of the world, Merleau-Ponty focuses on the coming together and mutual determination of the world and consciousness. For the perceiving subject, the world is phenomenal; consciousness assumes its existence because it *appears* to the senses.

Moreover, since consciousness is intentional, it apprehends itself as directed toward the world; self-consciousness is thus inseparable from consciousness of the world. *Emergence,* a term made popular by recent cognitive science and the theory of complex systems, describes for Merleau-Ponty the apprehension of things in their alterity: "We must discover the origin of the object at the very center of our experience; we must describe the emergence of being and we must understand how, paradoxically, there *is for us an in-itself*" (*Phenomenology,* 71; italics original). *For us* suggests a subjective stance, but what is for us is an *in itself,* a sense of objective existence.

The conception of consciousness as an intentional act directed toward the world is common to all philosophies affiliated with the phenomenological project; what singles out Merleau-Ponty's thought and makes it particularly relevant to the case of VR is the emphasis he places on the embodied nature of consciousness:

> The perceiving mind is an incarnated mind. I have tried, first of all, to reestablish the roots of the mind in its body and in its world, going against the doctrines which treat perception as a simple result of the action of things on our body as well as against those which insist on the autonomy of consciousness. These philosophies commonly forget—in favor of a pure exteriority or of a pure interiority—the insertion of the mind in corporeality. (*Primacy,* 3–4)

If consciousness is both incarnate and directed toward the world, the body functions as "point-of-view on the world" (*Phenomenology,* 70) and constitutes "our general medium for having a world" (147). It is by imagining ourselves physically reaching out toward things that we acquire a sense of their presence:

> We grasp external space through our bodily situation. A "corporeal or postural schema" gives us at every moment a global, practical, and implicit notion of the relation between our bodies and things, of our hold in them. A system of possible movements, or "motor projects," radiates from us to our environment. Our body is not in space like things; it inhabits or haunts space. (*Primacy,* 5)

The difference between "being in space," like things, and "inhabiting" or "haunting space," like the embodied consciousness, is a matter of both mobility and virtuality. Whereas inert objects, entirely contained in their material bodies, are bound to a fixed location, consciousness can occupy multiple points and points of view, either through the actual movements of its corporeal support or by projecting itself into virtual bodies. The ultimate test of the material existence of things is the ability to perceive them under many angles, to manipulate them, and to feel their resistance. When my actual body cannot walk around an object or grab and lift it, it is the knowledge that my virtual body could do so that gives me a sense of its shape, volume, and materiality. Whether actual or virtual, objects are thus present to me because my actual or virtual body can interact with them. In the case of an image, for instance, effects of texture and shading invite the viewer to touch the picture in imagination, thereby creating the corporeal relation that tells her, This is a real, solid, three-dimensional object that belongs to my world. Perspective creates a similar effect, by suggesting that the depicted objects have a hidden side that could be inspected by a mobile body. This sense of presence can only increase when the technology of representation makes it possible for the physical body to walk around or touch the virtual object, as is the case in VR. For the psychologists Pavel Zahorik and Rick Jenison, the presence of objects in VR is a function of their "possible action relationship to the user, or affordance" (a term coined by the psychologist J. J. Gibson). "Successfully supported action in the environment is a necessary and sufficient condition for presence" ("Presence," 86–87). The ideal VR system is conceived here as an ecology, in which every object is a tool that extends the user's body and enables her to participate in the ongoing creation of the virtual world.

In this VR ecology, it is not just individual objects that extend the user's body; the same can be said of the virtual world as a whole. In stark contrast to the extensive family of frozen metaphors that describe space as a container,[8] VR turns space into data that literally flow out of the body. The computer creates the virtual world dynamically by tracking the movements of the head of the user and by generating in real time the display that corresponds to his current point of view. The flesh and blood body of the user is bound to the virtual world by a feedback loop that reads the position of the body as binary data and uses this input to produce the sensory display. Writing about the centrality of the actor's body in modern theater, the drama theorist Stanton B. Garner observes that the field of performance is an environmental space "subjectified and intersubjectified by the physical actors who body forth the space they inhabit" (*Bodied Space*, 3). This metaphor becomes almost literal for the user of VR. In the virtual environment, as in certain shamanistic rituals described by Mircea Eliade, the body stands at the center of the world, and the world irradiates from it. The "lag" that separates the user's movements from the updating of the display in today's imperfect VR systems should act as a reminder of the productive implication of the body in the phenomenal world. Through this generation of space in response to the movements of the body, VR technology offers a dramatization of phenomenological doctrine. As Merleau-Ponty writes, "Far from my body's being for me no more than a fragment of space, there would be no space at all for me if I had no body" (*Phenomenology*, 102). Or again: "By considering the body in movement, we can see better how it inhabits space—and, moreover, time—because movement is not limited to submitting passively to space and time, it actively assumes them" (ibid.). This active engagement of the mobile body with space and time produces a succession of points of view through which the spectacle of the world smoothly unfolds to perception:

> Our own body is in the world as the heart is in the organism: it keeps the visible spectacle constantly alive, it breathes life into it and sustains it inwardly, and with it forms a system. When I walk round my flat, the various aspects in which it presents itself to me could not possibly appear as views of one and the same thing if I did not know that each of them represents the flat seen from one spot or another, and if I were unaware of my own movements, and of my body as retaining its identity through the stages of those movements. (203)

There could be no better evocation of the emergent quality of the VR experience of space than this description of an architectural walk-through.[9] It is no coincidence that one of the major applications of VR, both in its full-body implementation and in its downsized, mouse-operated screen versions, has been the simulation of tours through man-made or natural landscapes such as cities, buildings, campuses, gardens, or imaginary geographies. As David Herman observes in *Story Logic*, the tour offers a dynamic experience of space that contrasts with the static representation of the map. Whereas the map captures a disembodied "god's-eye view" that embraces the entire territory at once, the tour temporalizes the experience of space by revealing it one visual frame at a time. Whereas the map is an abstract model of space, the walk-through is a lived experience. Whereas the map has no direction, the tour traces an oriented path through space.

The same contrast operates between the landscaping philosophy of the formal French gardens of the seventeenth and eighteenth centuries and the nature-imitating (and -improving) design of the English gardens of the romantic age. With its symmetrical patterns of alleys bordered by meticulously sculptured bushes, the French garden must be seen from an elevated point and contemplated in its totality (both features symbolic of the king's political power); with its meandering walkways, diverse features (temples, ponds, grottoes), and seemingly random grouping of trees, the English garden must be walked through, and every turn of the path reveals a different landscape. As a static spectacle offered to an omniscient gaze and meant to be apprehended from a fixed perspective, the French garden is the horticultural equivalent of a framed painting; as an emergent landscape choreographed for a wandering eye and a moving body, the English garden is a metaphor for the space management and representation that we find in VR.[10]

This VR relation to space is totally different from what we experience in the "cyberspace" of the Internet. Cyberspace projects not a continuous territory but a relatively loose net made of links and nodes, of routes and destinations, with nothing in between. The destinations, or sites, may be centers of interest, but the connecting routes are not. Travel from site to site is not a voyage through a developing landscape but an instantaneous jump that negates the body, since material bodies can move through space only by traversing it one point at a time. The standard metaphor for cyberspace travel, surfing, gives a false impression of continuity. Rather than riding the crest of a swelling wave, the cybernaut is teletransported to more or less random destinations—the faster, the better—by clicking on hyperlinks. In the nonspace of cyberspace, travel time is wasted time, since there is nothing to see between the nodes. In the simulated space of VR, on the contrary, moving around the virtual world is a self-rewarding activity. If surfing were performed in a fully implemented VR system, we would feel the contour of the wave, rise with it above the water plane, and tumble down at the end of the ride. It would not matter where we ended; the pleasure would be the ride itself, the experience of being carried away by a smooth but mighty force. Even in a system that falls short of full-body immersion, the user can find delight in the sensation of bodily movement that results from the changing perspective on the environment, the growing and shrinking of objects, the pursuit of the horizon.[11] To label VR "cyberspace," or cyberspace "virtual reality," is to confuse kinetics with mere transportation, and the making-present of space to the body with its disappearance.

Virtual Realities of the Mind

Baudelaire, Huysmans, Coover

The French literary movement known as *décadence,* an off-shoot of symbolism that "blossomed," so to speak, during the nineteenth-century *fin de siècle,* shares with the culture of the recent turn of the century a powerful attraction toward artificial realities. This attraction takes center stage in two classic texts of *décadence:* Baudelaire's *Artificial Paradises,* a description of drug experiences, and J.-K. Huysmans's *A Rebours (Against Nature),* a one-character novel about an aristocrat, Des Esseintes, who literally attempts to recreate reality through art.

The late-nineteenth-century obsession with the artificial is much more escapist than the present involvement with VR.[1] Nowadays we are drawn to simulacra out of infatuation with the power of our own technologies to produce near-perfect copies of the real. Judging by the importance of nature themes in digital installation art, many artists also seek compensation in computer simulations for the disappearance of natural environments. We hope to recapture through technology the pristine world that technological culture took away from us.

For symbolist and *décadence* authors, on the contrary, Nature is the archenemy, and it needs to be corrected by art. Baudelaire regards the artificial not as a copy that should make up for a lost original but as a way to overcome the terrifying chaos of organic life. In his tellingly titled prose poem "Any where out of the world" (English in the French text), he visits in imagination several landscapes in the hope of finding one that will soothe a hypersensitive soul unable to find a home in physical reality. Though none of these potential travel destinations ultimately tempts the soul—they are still too much anchored in *this* world—it is revealing of Baudelaire's aesthetic preferences that one of them is

INTERLUDE

Lisbon, a city made of marble and water whose inhabitants, we are told, harbor such a hatred of vegetation that they cut down all the trees. Anguished by any kind of uncontrolled proliferation, Baudelaire's soul needs the dynamic symmetry, rhythmic movement, rule-governed metamorphosis, and structured multiplicity of art-made realities in which everything is "ordre et beauté, luxe, calme, et volupté" (order and beauty, luxury, calm and voluptuousness).[2]

Yet to reach the worlds where his vision of patterned beauty will be fulfilled, Baudelaire is willing to trust the forces of randomness. These forces will take him on journeys of uncertain destination (*anywhere out of the world*), to the bottom of the unknown ("au fond de l'inconnu pour trouver du nouveau"),[3] or, in the case of *Artificial Paradises*, toward the unpredictability of drug-induced hallucinations: "You are now sufficiently bolstered for a long, remarkable voyage. The steam whistle blows, the sails are set, and you, among all the other travelers, are a privileged exception, for you alone are unaware of your destination. You wished it to be so; long live destiny!" (40).

The drug trip unfolds in three stages, but it is the second that truly matters in terms of artistic vision. In the first stage—a warm-up for the second—the drug user rediscovers ordinary reality and ordinary language, the first through an increased acuity of the senses, and the second through the enhancement of the mind's analogical faculties. In the third stage, the mind is overtaken by a mystical feeling of peace and love. Situated beyond physical reality and beyond sensory perception, this "calm and immobile beatitude," this "glorious resignation" (57), is a purely spiritual experience that evades poetic language, since the poetic way of expressing the spiritual is to capture it in a metaphorical body. The truly poetic moment occurs, therefore, in the second stage, when thoughts are invaded by a tumult of images, when analogy let loose ties together perceptions with ideas and perceptions between themselves (the famous synaesthesia), when the abstract becomes concrete and the concrete intelligible, when the mind's altered state of consciousness reveals the laws that produce the sensible world.

Baudelaire's description of this second stage anticipates many of the themes of cyberculture. His predilection for images of water and fluidity prefigures Marcos Novak's vision of cyberspace as "liquid architecture":

I have before suggested that for those who are artistically inclined, water takes on a disturbing charm when illuminated by hashish. Waterfalls, babbling jets of water, harmonious cascades, and the blue expanses of the sea will sing, flow, and sleep in the innermost depth of your mind. It would be, perhaps, less than wise to permit a man in such condition to linger on the banks of some still pool; like the fisherman in the ballad, he might allow himself to be carried off by the undine. (*Artificial Paradises*, 23)

Though nothing could be stylistically further removed from Baudelaire's polished lyrical prose than the oral brainstorming of a child of contemporary pop culture, the theme of fluidity is no less prominent in Lanier's dreams of a digital paradise than in Baudelaire's evocation of the chemical version:

Our egos are very important to us and we really separate ourselves off from the environment and from the overall flow of life. What'll happen is that in Virtual Reality we'll recreate the flow. The flow anywhere is the same flow, so the flow that we create in Virtual Reality will be a new flow but it's also a part of the same eternal flow and we'll become all of a sudden . . . [sentence unfinished]. (Zhai, *Get Real*, 187)

In the liquid architecture of Baudelaire's artificial paradises, shapes morph into each other as easily as in computer graphics—

Then the hallucinations begin. External things, forms and images, swell to monstrous proportions, revealing themselves in fantastic shapes as yet unimagined. Instantly passing through a variety of transformations, they enter your being, or rather you enter theirs. The most singular ambiguities, the most inexplicable transpositions of ideas take place in your sensations. (*Artificial Paradises*, 19)

—and the self undergoes alternative experiences of embodiment:

You stare at a tree that harmoniously rocks in the breeze; in a few seconds what would be for a poet a natural comparison becomes a reality for you. You endow the tree with your passions and desires; its capriciously swaying limbs become your own, so that soon you yourself *are* that tree. Thus, when looking skyward, you behold a bird soaring in the deep azure. At first, the bird seems to represent the immortal yearning to soar above earthly concerns. But you have already become that bird. . . . The idea of evaporation—slow, uninterrupted and obsessive—grips your mind and soon you will apply the idea to your own thoughts, to your own thinking process. Through some odd misunderstanding, through a type of transposition of intellectual quip, you feel yourself vanishing into thin air. (51)

This passage is worth comparing with Lanier's statement: "You could become a comet in the sky one moment and then gradually unfold into a spider that's bigger than the planet that looks down at all your friends from high above" (Zhai, *Get Real*, 177). Both Lanier and Baudelaire envision their artificial realities as the site of a corporeal participation in a work, or rather in a state of art. In keeping with the idea that VR flows out of the body, Lanier imagines himself becoming a musical instrument and "blowing"—bodying forth—an entire landscape out of this virtual body:

You can have musical instruments that play reality in all kinds of ways aside from making sound in Virtual Reality. That's another way of describing arbitrary physics. With a saxophone you'll be able to play cities and dancing lights, and you'll be able to play the herding of buffalo's plains made of crystal, and you'll be able to play your own body and change yourself as you play the saxophone. (ibid.)

Meanwhile, in Baudelaire's hallucinations, music "enters within you, and you mingle with it" (*Artificial Paradises*, 64). Paintings take on a life of their own and open their space to the spectator: "You take your place and play your part in the most wretched paintings" (21). The frenetic activity of the drug-stimulated imagination, enhanced by the intensification of sensory perception, writes novels in the mind about a self that Baudelaire conceives, long before postmodernism came up with the notion of the decentered and multiple subject, as living many lives under different avatars: "For the proportions of time and being are thoroughly disrupted by the multiform variety of your feeling and the intensity of your ideas. You could say that many lives are crowded in the compass of one hour. Do you not, then, bear resemblance to a fantastic novel, which will come to life rather than be written?" (52).

Through synaesthetic relations—a phenomenon that inspires many of Baudelaire's poetic metaphors—the sensations provided by these various art forms blend together into an experience that foreshadows the VR dream of total art. If sounds have colors, colors have smells, smells have taste, and words acquire a body of their own ("words are reborn, clothed of flesh and blood" [63–64]), there are many doorways into this complete experience, but as in Lanier's imagination, the surest access is by way of music: "At other times, music recites you infinite poems, or places you within frightening or fantastic dramas. Harmony and melody become inextricably linked with the objects around you" (20–21). Total art, for Baudelaire, is not merely the fusion through music of poetry, drama, and visual representation but above all the contemplation of the spiritual essence of things. As a sympathizer with esoteric doctrines—neo-Platonism, cabalism, and Swedenborgian mysticism—Baudelaire conceives this essence as the mathematical laws that govern the spectacle of the physical world. To the hallucinating mind, the language of music is equally adept at creating horizontal correspondences between the senses and at revealing[4] vertical correspondences between numbers and things:

> Musical notes become numbers, and if you are gifted with any mathematical aptitude, the melody and the harmony, while retaining their sensuous and voluptuous qualities, are transformed into a vast arithmetical operation: numbers engender numbers, the phases of generation of which you follow with inexplicable facility and an agility equal only to that of their execution. (51)

This view of the world as governed by numbers will be literalized in the sensory displays of VR, since they are generated by binary code. Through the multiple correspondences of musical language, Baudelaire is thus able to enjoy the exuberant show of the artificial world and to access at the same time the underlying program. This vision offers a synthesis of unity and multiplicity, lawfulness and movement, that borrows the dynamic growth and varied spectacle of organic life but protects it from the chaotic proliferation that terrifies the poet's soul.

Or at least so it does for a while. By the law of probabilities, a trip taken *anywhere* out of the world can lead to hell as likely as to paradise. The order of the artificial paradises is built on the forces of randomness, and sooner or later this order returns to chaos. Of a fellow drug user Baudelaire writes:

> He told me that amid the intermingling of pleasure and delight, that supreme delight in which one feels so full of life and so possessed of genius, he had suddenly encountered an object of terror. The beauty of his sensations, which had at first dazzled him, were quickly displaced by horror. . . . "I was," he said, "like a frightened horse that flies off into a gallop toward the edge of a precipice, wishing to stop, yet knowing that he cannot. Certainly, that was a terrifying gallop. My thoughts, bound to my circumstances and surroundings, to the accidental and all that the word chance implies, had taken an absolute and purely rhapsodic turn. It's too late! I continually repeated to myself in despair." (44)

Baudelaire must have sensed that secondhand narratives of a bad trip[5] weigh little against the temptation constituted by his poetically inspired first-person evocations of the drug experience. To acquit himself of his ethical responsibility toward the reader he dwells forcefully on two additional dangers: hashish drains the will, a faculty indispensable to artistic creation, and it affords no opportunity for growth and transformation, no escape from the monster within, for hashish is nothing more than a magnifying mirror of our inner being; the drug user "is subjugated, but much to his displeasure, only by himself—that is to say, by the part of himself that is already dominant" (39). The instant satisfaction granted by artificial paradises is a prodigal expense of mental energy that ends up in pure waste, because this energy cannot replace itself, as does willpower—a faculty that for Baudelaire actually accrues through use—and because it does not produce anything that can be shared. In the controlling power of the mind, and in the potential communality of the experience, lies the main difference between Baudelaire's artificial paradises and Lanier's vision. Explicitly conceived as a remedy for the drug experience, VR "is like having shared hallucinations, except that you can compose them like works of art; you can compose the external world in any way at all as an act of communication" (Zhai, *Get Real,* 182).

The danger of experimenting with the forces of randomness is not lost on Huysmans's hero, Des Esseintes, when he embarks on a quest for his own artificial paradise. Whereas Baudelaire chooses to be passively taken toward unknown destinations, Des Esseintes builds his world against nature through an act of pure will, following with steely mental discipline a precise blueprint that leaves nothing to chance and a minimum to improvisation: "The secret is to know how to go about it, to know how to concentrate the mind on one single detail, to know how to dissociate oneself sufficiently to produce the hallucination and thus to substitute the vision of reality for reality itself" (*Against Nature,* 20).

After a youth spent trying out all worldly pleasures and finding nothing but boredom in this pursuit, Des Esseintes, the last scion of a once illustrious family, decides to withdraw from society and to seclude himself in a world of his own making where he will devote himself entirely to the satisfaction of his elitist tastes. He sells the ancestral chateau and buys a modest house in a lonely suburb of Paris, indifferent to its external appearance, because he intends to banish everything natural from his life, including the light of day. His resentment of nature and his worship of the artificial are even more strident than Baudelaire's: "Des Esseintes considered, furthermore, that artifice was the distinctive characteristic of human genius. As he was wont to remark, Nature has had her days; she has finally exhausted, through the nauseating uniformity of her landscapes and her skies, the sedulous patience of men of refined taste" (20). Des Esseintes's life will be spent indoors, in a decor designed to offer aesthetic gratification to all of the senses, especially to the "lower" and most artistically deprived of them: smell, touch, and taste. The book is almost entirely occupied by lengthy discussions of such topics as the furnishing of the house, Des Esseintes's selection of artworks and color schemes, his experiments with new perfumes, his ruminations on painting, music, and literature, the development, through what we would call today bioengineering, of strange new species of flowers, and erotic fantasies synaesthetically induced by the taste and smell of a purple bonbon.

This compendium of decadent refinements anticipates many of the cultural *topoi* of the most recent *fin de siècle*. Foremost among them is Des Esseintes's Baudrillardian preference for copies over originals. His fishtank contains only mechanical fish (virtual pets), the color schemes in his house are designed to be seen in artificial light only, and he feeds himself with pills that simulate the taste of haute cuisine. But a perfect copy of nature is still too natural for Des Esseintes; the ultimate triumph of art is to denaturize nature itself: "After having artificial flowers that imitated real ones, he now wanted real flowers that mimicked artificial ones" (73). From a philosophical point of view, this pursuit of the artificial is much more sophisticated than the purely consumerist obsession of Baudrillard's subject with the hyper-real; or at least so it appears to the hermit, who justifies his hedonistic project as a spiritual quest:

> Thus, were not his propensities towards artifice, his need for ec-
> centricity, the result, in short, of the specious subjects he had
> studied, of otherworldly subtle distinctions, of quasi-theological
> speculations? They were, essentially, outbursts of feeling, im-
> pulses towards an ideal, towards an unknown universe, towards
> a far-off blessedness as desirable as that which we are promised
> by the Holy Scripture. (66)

This reminds us of Baudelaire's pursuit of the Ideal in "Any where out of the world"—a poem prominently displayed in a reliquary on the mantelpiece of Des Esseintes's fireplace—and of the mystical over-tones so often found in cybercultural discourse.

When he furnishes his home—expression of his inner self—Des Esseintes approaches the task with the same maniac attention to detail that today's hackers bring to the design of their Web pages, or MOO players to the construction of a private room that truly reflects the personality of their on-line character:

> And then during a period when Des Esseintes had felt the need
> to draw attention to himself, he had devised sumptuous, pecu-

liar schemes of decoration, dividing his salon into a series of variously carpeted alcoves, which could be related by subtle analogies, by indeterminate correlations of tone, either cheerful or gloomy, delicate or flamboyant, to the character of the Latin or French works he loved. (11)

Like windowed computer displays and MOO architecture,[6] Des Esseintes's living space is structured as a series of subspaces with different themes and functions, between which he divides his days according to a rigid hourly schedule.

Also very contemporary is Des Esseintes's fascination with forms of sexuality that pass as unnatural because they challenge the traditional binarism of gender categories. One of the erotic fantasies induced by the above-mentioned bonbon is an encounter with Miss Urania, an American woman and circus acrobat who gradually turns into a man. Des Esseintes takes particular delight in imagining the figure of the Roman high priest Elagabalus, spending his time "surrounded by his eunuchs, at woman's work, giving himself the title of Empress, and every night bedding a different Emperor" (28). Anticipating the libidinal fixation of many contemporary writers on the modern myth of the cyborg, Des Esseintes voices his lust for a sexual coupling of man and machine in a rapturous celebration of the erotic appeal of the locomotive: "Does there, in this world, exist a being conceived in the joys of fornication and born of the birth pangs of a womb, of which the model and the type is more dazzling or more splendid than those two locomotives [the Crampton and the Engerth] now in service on the railroads of Northern France?" (20). For Des Esseintes, the machine is to man's artificial recreation of the world what woman is to God's natural work: its supreme achievement.

In the literary domain, Des Esseintes's taste elaborates on Baudelaire's conception of poetry as the "mathematically exact" art of choosing the right word, and of expanding its meaning through the "evocative witchcraft" (sorcellerie évocatoire) of a network of correspondences:

The words chosen would then be so inevitable that they would render all other words superfluous; the adjective, positioned in so ingenious and so definitive a manner that it could not legitimately be displaced, would open such vistas that for days on end the reader would ponder over its meaning, at once precise and manifold, would know the present, reconstruct the past, and make conjectures about the souls of the characters, as these were revealed by the light of a single epithet. (162)

This description suggests the twentieth-century aesthetics of the open work, but it differs from the postmodern ideal of the self-renewing "work in movement" in that infinite signification is generated not through a physical permutation of textual elements, as will happen in hypertext and other literary combinatory games (see chap. 6), but by assigning the right place to every word in a frozen structure that reflects, on the whole, the static architecture of Des Esseintes's artificial reality.

This static architecture is what will eventually doom the project to failure. Des Esseintes's "refined Thebaid" ends up as a prison for both mind and body for a very simple reason: there is nothing to do in it. As Doležel has observed (*Heterocosmica*, 48), the novel is entirely devoid of physical action. Whereas the chaos that overtakes Baudelaire's paradise is allegorized as a powerful animal body galloping out of control, the rigid order that stifles Des Esseintes's creation works itself into the flesh of the recluse through various maladies. While the body languishes, the mind is overwhelmed by "currents of emotions," "torrents of anguish," and "hurricanes of rage" as wild and uncontrollable as the forces of nature that the decadent aesthete sought to expel (53).

At the end of the novel, Des Esseintes abandons his retreat and returns to the world to rekindle whatever life is left in his exhausted body. As long as he is engaged in the design of his private space he finds respite from his inner demons, but as soon as a project is completed he is overcome by the same boredom that drove him out of the world. Typical of his overall feelings is the state of mind that follows his horticultural experiments: "He felt somewhat weary, and found this atmosphere of hothouse plants suffocatingly oppressive" (*Against Nature*, 78). For a while Des Esseintes is able to postpone the terminal stage of boredom by throwing himself into new projects or by revisiting his art collections, but he eventually runs out of new senses to explore, and even the art experience loses the self-renewing quality that sustains mental activity. Randomness and the other have been so completely excluded from the design of Des Esseintes's artificial paradise that once its order is in place it affords no discovery, no potential for growth and metamorphosis, no relief from the mind's obsessions. It is a totally lifeless environment. Though they take opposite routes, Baudelaire's and Des Esseintes's quests for total gratification lead to the same state of morbid self-contemplation.

It was left to an author who later developed a keen appreciation for the literary potential of electronic technology to propose a virtual reality of the mind that sustains life and achieves autonomy by instilling a seed of randomness in an otherwise carefully designed alternative world. In *The Universal Baseball Association, Inc.*, Robert Coover tells the story of a middle-aged, in appearance utterly ordinary, office worker, J. Henry Waugh, who spends all of his free time managing the colorful cast of players of a fantasy baseball league. Let us not be blinded by the apparent triviality of this project: though Coover chooses the humorous approach to reality recreation, there is as much at stake in J. Henry's humble game as in Baudelaire's and Des Esseintes's ambitious artistic and spiritual quests. J. Henry is the inventor of a baseball-simulation system in which every play is determined by a throw of the dice. While he creates the players, specifies their athletic abilities and human personalities, assembles the teams, arranges trades, makes the schedule, selects the daily lineups, plays the games, and keeps the statistics, he lets the outcome of the games be determined by a combination of rules and chance that brings life and suspense to the unfolding of baseball history. Thanks to the randomness of the dice throw, the fantasy league becomes the site of emergent behaviors that bring to its creator both heartbreak (when his favorite player is killed by a pitch) and delight (when the players literally come to life and undo the tragic accident). By permitting events that subject him to this emotional roller coaster, J. Henry Waugh not only works out a compromise between Baudelaire's terrifying chaos and Des Esseintes's sterile order, he also gives a human dimension to his fantasy world that makes it livable.

PART II The Poetics of Immersion

The Text as World

Theories of Immersion

> One's memory is apparently made up of millions of [sets of images],
> which work together on the Identikit principle. The most gifted writers
> are those who manipulate the memory sets of the reader in such a rich
> fashion that they create within the mind of the reader an entire world that
> resonates with the reader's own real emotions. The events are merely tak-
> ing place on the page, in print, but the emotions are real. Hence the
> unique feeling when one is "absorbed" in a certain book, "lost" in it.
> —TOM WOLFE

When VR theorists attempt to describe the phenomenon of immersion in a virtual world, the metaphor that imposes itself with the
greatest insistence is the reading experience:

> As [users] enter the virtual world, their depth of engagement
> gradually meanders away from here until they cross the thresh-
> old of involvement. Now they are absorbed in the virtual world,
> similar to being in an engrossing book.
>
> The question isn't whether the created world is as real as the
> physical world, but whether the created world is real enough for
> you to suspend your disbelief for a period of time. This is the
> same mental shift that happens when you get wrapped up in a
> good novel or become absorbed in playing a computer game.
> (Pimentel and Teixeira, *Virtual Reality*, 15)

Literary authors have not awaited the development of VR technology
to offer their own versions and dramatizations of the phenomenon.
Charlotte Brontë conceives immersion as the projection of the reader's body into the textual world:

> You shall see them, reader. Step into this neat garden-house on
> the skirts of Whinbury, walk forward in the little parlour—they
> are there at dinner. . . . You and I will join the party, see what is
> to be seen, and hear what is to be heard. (*Shirley*, 9)

Joseph Conrad's artistic goal prefigures the emphasis of VR developers on a rich and diversified sensory involvement:

> My task which I am trying to achieve is, by the power of the
> written word, to make you hear, to make you feel—it is, before
> all, to make you see. (Preface to *Nigger of the Narcissus*, xxvi)

For Italo Calvino, the transition from ordinary to textual reality is a
solemn event, and it must be marked with proper ceremony. The
instructions to the reader that open *If on a Winter's Night a Traveler*
suggest the rites of passage through which various cultures mark the
crossing of boundaries between the profane and the sacred, or between the major stages of life. Opening a book is embarking on a
voyage from which one will not return for a very long time:

> You are about to begin reading Italo Calvino's new novel, *If on a
> winter's night a traveler.* Relax. Concentrate. Dispel every other
> thought. Let the world around you fade. . . . Find the most
> comfortable position: seated, stretched out, or lying flat. . . .
> Adjust the light so you won't strain your eyes. Do it now, be-
> cause once you're absorbed in reading there will be no budging
> you. (3–4)

IMMERSION AND THE "WORLD" METAPHOR

The notion of reading as immersive experience is based on a premise
so frequently invoked in literary criticism that we tend to forget its
metaphorical nature. For immersion to take place, the text must offer
an expanse to be immersed within, and this expanse, in a blatantly
mixed metaphor, is not an ocean but a textual world. The recent
emergence of other analogies for the literary text, such as the text as
game (see chap. 6), as network (Landow, *Hypertext*; Bolter, *Writing
Space*), or as machinic assemblage (Deleuze and Guattari, *A Thousand
Plateaus*), should remind us that "the text as world" is only one
possible conceptualization among many others, not a necessary, objective, and literal dimension of literary language, but this relativiza-
tion should be the occasion for a critical assessment of implications
that have too long been taken for granted.

What makes the semantic domain of a text into a world? All texts
have a semantic domain, except perhaps for those that consist exclusively of meaningless sounds or graphemes, but not all of them con-

struct a world. A semantic domain is the nonenumerable, fuzzy-bordered, occasionally chaotic set of meanings that is projected by (or read into) any given sequence of signs. In a textual world these meanings form a cosmos. "How does a world exist as a world?" asks Michael Heim, theorist of virtual reality. "A world is not a collection of fragments, nor even an amalgam of pieces. It is a felt totality or whole." It is "not a collection of things but an active usage that relates things together, that links them. . . . World makes a web-like totality. . . . World is a total environment or surround space" (*Virtual Realism*, 90–91). For Heim, moreover, worlds are existentially centered around a base we call home. "Home is the node from which we link to other places and other things. . . . Home is the point of action and node of linkage that becomes a thread weaving the multitude of things into a world" (92). Let me sum up the concept of world through four features: connected set of objects and individuals; habitable environment; reasonably intelligible totality for external observers; field of activity for its members.

For the purpose of immersive poetics, a crucial implication of the concept of textual world concerns the function of language. In the metaphor of the text as world, the text is apprehended as a window on something that exists outside language and extends in time and space well beyond the window frame. To speak of a textual world means to draw a distinction between a realm of language, made of names, definite descriptions, sentences, and propositions, and an extralinguistic realm of characters, objects, facts, and states of affairs serving as referents to the linguistic expressions. The idea of textual world presupposes that the reader constructs in imagination a set of language-independent objects, using as a guide the textual declarations, but building this always incomplete image into a more vivid representation through the import of information provided by internalized cognitive models, inferential mechanisms, real-life experience, and cultural knowledge, including knowledge derived from other texts. The function of language in this activity is to pick objects in the textual world, to link them with properties, to animate characters and setting—in short, to conjure their presence to the imagination. The world metaphor thus entails a referential or "vertical" conception of meaning that stands in stark contrast to the Saussurian and poststructuralist view of signification as the product of a network of horizontal relations between the terms of a language system. In this vertical conception, language is meant to be traversed toward its referents. Sven Birkerts describes this attitude as follows: "When we are reading a novel we don't, obviously, recall the preceding sentences and paragraphs. In fact we generally don't remember the language at all, unless it's dialogue. For reading is a conversion, a turning of codes into contents" (*Gutenberg Elegies*, 97).

The concrete character of the objects that populate textual worlds limits the applicability of the concept to a category of texts that Félix Martínez-Bonati calls mimetic texts. This term refers to texts devoted to the representation of states of affairs involving individual existents situated in time and space, as opposed to those texts that deal exclusively with universals, abstract ideas, and atemporal categories. We can roughly equate mimetic texts with narrative texts, though their evocation of particular existents does not necessarily fulfill the conditions of closure and coherence that we associate with the notion of plot. Since the class of mimetic texts includes both fiction and nonfiction, the notion of textual world does not distinguish the worlds that actually exist outside the text from those that are created by it. Both fictional and nonfictional mimetic texts invite the reader to imagine a world, and to imagine it as a physical, autonomous reality furnished with palpable objects and populated by flesh and blood individuals. (How could a world be imagined otherwise?) The difference between fiction and nonfiction is not a matter of displaying the image of a world versus displaying this world itself, since both project a world image, but a matter of the function ascribed to the image: in one case, contemplating the textual world is an end in itself, while in the other, the textual world must be evaluated in terms of its accuracy with respect to an external reference world known to the reader through other channels of information.

The idea of textual world provides the foundation of a poetics of immersion, but we need more materials to build up the project. As we saw in the introduction, poststructuralist literary theory is hostile to the phenomenon because it conflicts with its concept of language. (More about this in chap. 6.) Reader-response criticism, which should be more open to immersion than any other recent critical school, does not clearly put its finger on the experience, though it often comes tantalizingly close.[1] The building blocks of the project will therefore have to be found in the quarries of other fields: cognitive psychology (the metaphors of transportation and being "lost in a book"), analytical philosophy (possible worlds), phenomenology (make-believe), and psychology again (mental simulation).

TRANSPORTATION AND BEING "LOST IN A BOOK"

The frozen metaphors of language dramatize the reading experience as an adventure worthy of the most thrilling novel: the reader plunges under the sea (immersion), reaches a foreign land (transportation), is taken prisoner (being caught up in a story, being a *captured* audience), and loses contact with all other realities (being lost in a book). The work of the psychologists Richard Gerrig and Victor Nell follows the thread of these classic metaphors to explore what takes place in the mind of the entranced reader. In his book *Experiencing Narrative Worlds* (10–11), Gerrig develops the metaphor of transportation into a narrative script that could be regarded as a "folk theory" of immersion:

1. *Someone ("the traveler") is transported* . . . For Gerrig, this statement means not only that the reader is taken into a foreign world but also that the text determines his role in this world, thereby shaping his textual identity.
2. *by some means of transportation* . . . If there are any doubts as to the identity of the vehicle, they should be quickly dispelled by these lines from Emily Dickinson: "There is no Frigate like a Book / To take us Lands away" (quoted in Gerrig, 12, and as epigraph to the whole book).
3. *as a result of performing certain actions.* This point corrects the passivity implicit in the metaphor of transportation and introduces another major metaphor developed in Gerrig's book: reading as performance. The goal of the journey is not a preexisting territory that awaits the traveler on the other side of the ocean but a land that emerges in the course of the trip as the reader executes the textual directions into a

"reality model" (Gerrig's term for the mental representation of a textual world). The reader's enjoyment thus depends on his own performance.

4. *The traveler goes some distance from his or her world of origin,* . . . When visiting a textual world, the reader must "do as the Romans do": adapt to the laws of this world, which differ to various degrees from the laws of his native reality. Readers may import knowledge from life experience into the textual world, but the text has the last word in specifying the rules that guide the construction of a valid reality model.

5. *which makes some aspects of the world of origin inaccessible.* This idea can be interpreted in many ways: (a) When the idiosyncratic laws of the textual world take over, we can no longer draw inferences from the real-world principles that were overruled. (b) Our objective knowledge that fictional characters are only linguistic constructs—as structuralism would describe them—does not prevent us from reacting to them as if they were embodied humans. (c) As is the case with any intense mental activity, a deep absorption in the construction/contemplation of the textual world causes our immediate surroundings and everyday concerns to disappear from consciousness.

6. *The traveler returns to the world of origin, somewhat changed by the journey.* There is no need to elaborate here on the educational value of reading, even when we read for pure entertainment. In lieu of a theoretical development let me offer a literary formulation of the same idea: "The reader who returns from the open seas of his feelings is no longer the same reader who embarked on that sea only a short while ago" (Pavić, *Dictionary of the Khazars* [female edition], 294).

The best illustrations of this script come from the realm of fiction, but Gerrig's stated purpose is to describe a type of experience that concerns "narrative worlds"—what I would call the worlds of mimetic texts—not just fictional ones. The metaphor of transportation captures how the textual world becomes present to the mind, not how this world relates to the real one, and this sense of presence can be conveyed by narratives told as truth as well as by stories told as fiction. Victor Nell writes that "although fiction is the usual vehicle for ludic reading, it is not its lack of truth—its 'fictivity'—that renders it pleasurable" (*Lost in a Book*, 50). Similarly, it is not the imaginative origin of fictional worlds per se that creates the experience that Gerrig calls transportation. But if a theory of transportation—and, by extension, of immersion—should be kept distinct from a theory of fiction, the two cannot be entirely dissociated, because imaginative participation in the textual world is much more crucial to the aesthetic purpose of fiction than to the practical orientation of most types of nonfiction. While nonfiction sends the reader on a business trip to the textual world, often not caring too much about the quality of the experience—what matters most is what happens after the return home—fiction treats the visit as vacation and mobilizes all the powers of language to strengthen the bond between the visitor and the textual landscape.

Another entangled issue is the relation between immersion and aesthetics. We tend to label a literary work immersive when we take pleasure in it, and we (normally) take pleasure in reading when the text presents aesthetic qualities. But aesthetic value cannot be reduced to immersive power: poetry is not as immersive as narrative because its relation to a "world" is much more problematic; and among the texts regarded as narrative, some deliberately cultivate a sense of alienation from the textual world, or do not allow a world to solidify in the reader's mind. For Gerrig, transportation into a narrative world is not dependent on narrative skills. If I read the word *Texas* in a story, no matter how good or bad the text, I will think about Texas, which means that I will be mentally transported to the place: "Some core of processes is likely to allow readers to experience narrative worlds even when the stories themselves are poorly crafted" (*Experiencing*, 5). In Gerrig's Texas example, however, imaginative transportation to Texas is a consequence of the speech act of reference rather than a consequence of embedding the speech act in a narrative context. We must therefore distinguish a minimal form of transportation—thinking of a concrete object located in a time and place other than our present spatio-temporal coordinates—from a strong form of the experience, by which "thinking of" means imagining not only an object but the world that surrounds it, and imagining ourselves contained in this world, in the presence of this object. The minimal form of transportation is built into language and the cognitive mechanisms of the mind; we cannot avoid it; but the richer forms depend on the resonance in the reader's mind of the aesthetic features of the text: plot, narrative presentation, images, and style.

For Victor Nell, the experience of immersion—or rather, as he calls it, of reading entrancement—is a major source of pleasure but not necessarily a trademark of "high" literary value. *Lost in a Book*, his investigation of the "psychology of reading for pleasure," takes its title from a family of metaphors that present equivalents in many languages: "For example, in Dutch the phrase is 'om in een boek op te gaan'; in German, 'in einem Buch versunken zu sein'; and in French, 'être pris par un livre' "(50). The passivity of these metaphors suggests a smooth passage from physical reality to the textual world. It is indeed in terms of easiness that Toni Morrison describes the experience of a young girl who listens for the umpteenth time to the wondrous story of her birth: "Easily she stepped into the told story that lay before her eyes" (*Beloved*, 29). For a reader to be caught up in a story, the textual world must be accessible through effortless concentration: "In terms of attention theory . . . the ludic reader's absorption may seem as an extreme case of subjectively effortless arousal which owes its *effortlessness* to the automatized nature of the skilled reader's decoding activity" (Nell, *Lost in a Book*, 77–78). Immersion is hampered by difficult materials because "consciousness is a processing bottleneck, and it is the already comprehended messages . . . that fully engage the receiver's conscious attention" (77). The most immersive texts are therefore often the most familiar ones: "Indeed, the richness of the structure the ludic reader creates in his head may be inversely proportional to the literary power and originality of the reading matter" (ibid.).

But for Nell, the association of immersion with ease of reading is no cause for contempt. Anticipating the objections of elitist literary critics, who tend to judge the greatness of literary works by the standards of the Protestant work ethic—"no pain, no gain"—Nell insists on the importance of immersive reading for both high and low culture. Sophisticated readers learn to appreciate a wide variety of literary experiences, but they never outgrow the simple pleasure of being lost in a book. This pleasure is limiting only if we take it to be the only type of aesthetic gratification. There is no point in denying that the worlds of the stereotyped texts of popular culture are the most favorable to immersion: the reader can bring in more knowledge and sees more expectations fulfilled than in a text that cultivates a sense of estrangement. But immersion can also be the result of a process that involves an element of struggle and discovery. How many of us, after finally turning the last page of a difficult novel, compulsively return to the first page with the exhilarating thought that deciphering is over and the fun can now begin? In literature as in other domains—ballet, music, theater, and sports—it is through hard work that we reach the stage of effortless performance. The most forbidding textual worlds may thus afford the "easy" pleasures of immersion, once the reader has put in the necessary concentration.

To remain pleasurable, the experience of being lost in a book must be temporary and remain distinct from addiction, its harmful relative. Nell describes the difference between immersion and addiction in terms of eating metaphors: addicted readers are "voracious" consumers of books; they devour the text without taking the time to savor it. The story lives entirely in the present, and when the reading is completed, it leaves no residue in memory: "Addictive behavior . . . predicts an underdeveloped capacity for private fantasy" (212). While the addicted reader blocks out reality, the reader capable of pleasurable immersion maintains a split loyalty to the real and the textual world. The ocean is an environment in which we cannot breathe; to survive immersion, we must take oxygen from the surface, stay in touch with reality. The amphibian state of pleasurable entrancement has been compared by J. R. Hilgard to "dreaming when you know you are dreaming" (quoted ibid.). Nell explains:

> [Hilgard] writes that the observing and participating egos coexist, so that the subject is able to maintain "a continued limited awareness . . . that what is perceived as real is in some sense not real." This disjunction, allowing the reader both to be involved and to maintain a safe distance, is neatly captured by her subject Robert, who comments on a movie screen in which a monster

enters a cave, trapping a group of children: "I'm not one of them but I'm trapped with them, and I can feel the fright they feel." (212–13)

On the basis of these observations, we can distinguish four degrees of absorption in the act of reading:

1. *Concentration.* The type of attention devoted to difficult, nonimmersive works. In this mode, the textual world—if the text projects any—offers so much resistance that the reader remains highly vulnerable to the distracting stimuli of external reality.

2. *Imaginative involvement.* The "split subject" attitude of the reader who transports herself into the textual world but remains able to contemplate it with aesthetic or epistemological detachment. In the case of narrative fiction, the split reader is attentive both to the speech act of the narrator in the textual world and to the quality of the performance of the author in the real world. In the case of nonfiction, the reader engages emotionally and imaginatively in the represented situation but retains a critical attitude toward the accuracy of the report and the rhetorical devices through which the author defends his version of the events.

3. *Entrancement.* The nonreflexive reading pleasure of the reader so completely caught up in the textual world that she loses sight of anything external to it, including the aesthetic quality of the author's performance or the truth value of the textual statements. It is in this mode that language truly disappears. As Ockert, one of the subjects interviewed by Nell, describes the experience: "The more interesting it gets, the more you get the feeling you're not reading any more, you're not reading words, you're not reading sentences, it's as if you are completely living inside the situation" (290). Despite the depth of the immersive experience, however, this reader remains aware in the back of his mind that he has nothing to fear, because the textual world is not reality.

4. *Addiction.* This category covers two cases: (a) The attitude of the reader who seeks escape from reality but cannot find a

home in the textual world because she traverses it too fast and too compulsively to enjoy the landscape. (b) The loss of the capacity to distinguish textual worlds, especially those of fiction, from the actual world. (I call this the Don Quixote syndrome.)

POSSIBLE WORLDS

What does it mean, in semantic and logical terms, to be transported into the virtual reality of a textual world? The answers to these questions are tied to an ontological model that acknowledges a plurality of possible worlds. The fictional worlds of literature may not be, technically speaking, the possible worlds of logicians, but drawing an analogy between the two allows a much-needed sharpening of the informal critical concept of textual world.[2] Originally developed by a group of philosophers including David Lewis, Saul Kripke, and Jaakko Hintikka to solve problems in formal semantics, such as the truth value of counterfactuals, the meaning of the modal operators of necessity and possibility, and the distinction between intension and extension (or sense and reference),[3] the concept of possible worlds has been used to describe the logic of fictionality by Lewis himself, and adapted to poetics or narrative semantics by Umberto Eco, Thomas Pavel, Lubomír Doležel, Doreen Maître, Ruth Ronen, Elena Semino, and myself. The applications of possible-world (henceforth, PW) theory to literary criticism have been as diverse as the interpretations given to the concept by philosophers and literary scholars.[4] Since it would be beyond the scope of this section to try to represent the entire movement, I will restrict my presentation of PW theory to an approach that is largely my own, even though it is strongly indebted to the pioneering work of Eco, Pavel, and Doležel.[5]

The basis of PW theory is the set-theoretical idea that reality—the sum total of the imaginable—is a universe composed of a plurality of distinct elements, or worlds, and that it is hierarchically structured by the opposition of one well-designated element, which functions as the center of the system, to all the other members of the set. The central element is commonly interpreted as "the actual world" and the satellites as merely possible worlds. For a world to be possible it must be linked to the center by a so-called accessibility relation. Impossible worlds cluster at the periphery of the system, conceptually part of it—since the possible is defined by contrast with the impossible—and yet unreachable. The boundary between possible and impossible worlds depends on the particular interpretation given to the notion of accessibility relation. The most common interpretation associates possibility with logical laws; every world that respects the principles of noncontradiction and excluded middle is a possible world. Another criterion of possibility is the validity of the physical laws that obtain in real life. On this account, a world in which people can be turned overnight into giant insects is excluded from the realm of the possible. Yet another conceivable interpretation involves the idea of temporal directionality: the actual world is the realm of historical facts, possible worlds are the branches that history could take in the future, and impossible worlds are the branches that history failed to take in the past.

The distinction of the possible from the impossible is a relatively straightforward matter: all it takes is a particular definition of the criteria of accessibility. A much thornier issue is the distinction of the actual from the nonactual within the realm of the possible. Through its centered architecture, PW theory runs into difficulties with postmodern theory. The idea of a world enjoying special status is easily interpreted as hegemonism, logocentrism, negative valorization of the periphery, and a rigid hierarchical organization based on power relations. Another objection frequently heard against the centered model is that even though we all live in the same physical world and share a large number of opinions about its basic furnishing, there is no absolute consensus as to where to draw the boundary between the realm of actually existing objects and the domain of merely thinkable existence. Some of us believe in angels and not UFOs, some of us in UFOs and not angels, some of us in both, and some of us in neither. Moreover, belief is a matter of degree. I may believe weakly in angels, and the borders of my vision of what exists may be fuzzy. According to this argument, it would take a "naive realism" to postulate a singular actual world; for if reality is incompletely accessible to the mind, or not accessible at all, there will be inevitable discrepancies in its representation. Postmodern ideologues may further object that the idea of a unique center ignores the cultural and historical relativity of perceptions of reality. The current emphasis on the value of diversity seems better represented by philosophies that postulate a variety of "world versions" without establishing any hierarchical relations between them, such as the model described by Nelson Goodman in *Ways of Worldmaking,* than by the necessarily centered structure proposed by modal logic.

These objections to the concept of actual world can be circumvented by adopting what David Lewis has called an "indexical" definition of actuality. The opposition between the actual and the possible can be conceived in two ways: absolutely, in terms of origin, or relatively, in terms of point of view. In the absolute characterization, the actual world is the only one that exists independently of the human mind; merely possible worlds are products of mental activities such as dreaming, wishing, forming hypotheses, imagining, and writing down the products of the imagination in the form of fictions. In the relative characterization—the one advocated by Lewis—the actual world is the world from which I speak and in which I am immersed, while the nonactual possible worlds are those at which I look from the outside. These worlds are actual from the point of view of their inhabitants.[6] With an indexical definition, the concept of actual world can easily tolerate historical, cultural, and even personal variations. Without sacrificing the idea of an absolutely existing, mind-independent reality, we can relativize the ontological system by placing at its center individual images of reality, rather than reality itself. Most of us conceive the world system as centered because this reflects our intuition that there is a difference between fact and mere possibility—an egalitarian model such as Goodman's cannot account for these all-important semantic concepts—but we all organize our private systems around personal representations of what is actual.

I represent this model as shown in figure 1:

— At the center, a hypothetical real world, existing independently of the mind.
— Superposed upon this world of uncertain boundaries, the representations of it held by various individuals or collectively by various cultures. These spheres are the different personal versions of the "absolute" center. Their boundaries overlap because they reflect the same physical reality, and despite the current emphasis on relativity and differences,

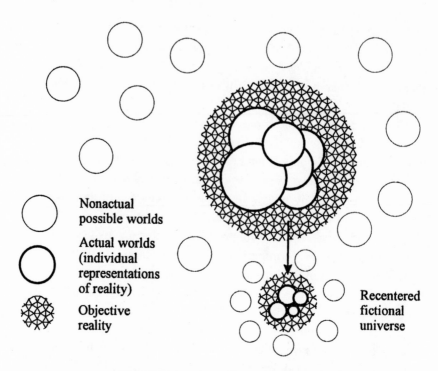

Nonactual
possible worlds

Actual worlds
(individual
representations
of reality)

Objective
reality

Recentered
fictional
universe

FIGURE 1 | A recenterable possible-worlds model

there is a vast area of consensus as to what exists and what
does not.

— Further away, outlined in thinner lines, the worlds that each
of us holds to be possible but nonactual. They stand at various distances from our personal center, depending on how
difficult it would be to enact them, or on what type of accessibility relations link them to the center. If we interpret possible worlds as textual worlds, the model predicts that for
most readers the world of a realistic novel is closer to reality
than the world of a fairy tale, because its actualization does
not require a modification of physical laws. It also predicts
that a modern American reader will see greater discrepancy
between reality and the world of Macbeth than a contemporary of Shakespeare, because belief in witches was more
prevalent in Renaissance England than in the twentieth-
century United States.

The applicability of the model to literary theory is not exhausted with the assimilation of textual worlds to possible worlds. In fact, a straight assimilation would be doubly reductive. First, it would obscure the fact that the distinction actual/possible reappears within the semantic domain projected by the text. In the case of mimetic texts, an essential aspect of reading comprehension consists of distinguishing a domain of autonomous facts—what I call the textual actual world—from the domains created by the mental activity of characters: dreaming, hoping, believing, planning, and so on. Mimetic texts project not a single world but an entire modal system, or universe, centered around its own actual world. Second, if nonactual textual worlds were apprehended as mere statements of possibility, there would be no phenomenological difference between counterfactual statements or expressions of wishes, which embed propositions under predicates of nonfactuality, and fictional statements, which, as Lewis observes, take the form of straight assertions of truth.

The concept of immersion is crucially dependent on this distinction. When I process "Napoleon could have won the battle of Waterloo if Grouchy had arrived before Blücher," I look at this world from the standpoint of a world in which Napoleon loses; but if I read in a novel "Thanks to Grouchy's ability to move quickly and bring his army to the battlefield before Blücher, Napoleon crushed his enemies at Waterloo," I transport myself into the textual world and process the sentence as a statement of fact. Both counterfactuals and fictional statements direct our attention toward nonactual possible worlds, but they do so in different modes: counterfactuals function as telescopes, while fiction functions as a space-travel vehicle. In the telescope mode, consciousness remains anchored in its native reality, and possible worlds are contemplated from the outside. In the space-travel mode, consciousness relocates itself to another world and, taking advantage of the indexical definition of actuality, reorganizes the entire universe of being around this virtual reality. I call this move recentering, and I regard it as constitutive of the fictional mode of reading. Insofar as fictional worlds are, objectively speaking, nonactual possible worlds, it takes recentering to experience them as actual—an experience that forms the basic condition for immersive reading.

Recentered universes reproduce the structure of the primary system, except that in the primary system we see only the white circle of our personal actual world, while in recentered systems the reader has access to at least some areas of the patterned circle. In a fictional universe, objective reality corresponds to fictional truths, and fictional truths are established by textual authority. This authority means that fictional truths are unassailable, whereas the facts of the actually actual world can always be questioned. In figure 1 the boundaries of the textual actual world are not clearly defined because individual readers will complete the picture differently, and because some texts, especially postmodern ones, leave areas of undecidability or present contradictory versions of facts. (These texts could be represented as having two or more actual worlds, in a blatant violation of the classic modal structure.) The individual representations of reality superposed upon the textual actual world correspond to the personal actual world of characters, while the nonactual possible worlds that surround the center stand for the characters' unfulfilled, or partially fulfilled, private worlds. Here again distance from the center stands for degree of fulfillment.

The idea of recentering explains how readers become immersed in a fictional text, but how does the analysis work for texts of nonfiction? It would seem that in this case no recentering is needed, because nonfiction describes the real world and the reader is already there, automatically immersed in this "native reality" by some kind of birthright. But where exactly is the reader of nonfiction imaginatively situated: in a text, or in a world? If, as I have suggested, the world-image projected by the text is conceptually different from the world referred to by the text, the reader-persona is located in the reference world, not in its textual image. In fiction, the reference world is inseparable from the image, since it is created by the text, and the contemplation of the image automatically transports the reader into the world it represents. But in nonfiction we can distinguish two moments: (1) one in which the reader constructs the text (i.e., becomes engaged imaginatively in the representation); and (2) one in which the reader evaluates the text (i.e., distances himself from the image, takes it apart, and assesses the accuracy of its individual statements with respect to the reference world). In the first phase, the reader contemplates the textual world from the inside in, and in the second, from the outside in.[7]

The first phase can be more or less elaborate, the reconstructed image more or less vivid and complete, depending on how badly the user needs the textual information for his own practical purposes, but before we can decide what to believe and disbelieve, remember and forget, we must imagine something, and in this act of imagination we are temporarily centered in the textual world. When the textual and the reference world are indistinguishable, as in fiction, the text must be taken as true, since there is no other mode of access to the reference world, and being centered in the textual world implies recentering into the world it represents. When the two worlds are distinct, the image can be true or false, and the reader evaluates it from the point of view of his native reality. The preliminary operation of imaginative centering in this case does not involve ontological recentering. The distinction of a moment of construction from a moment of evaluation avoids two pitfalls frequently encountered in discourse typology: denying any difference in the mode of reading appropriate to fiction and nonfiction, and treating these two modes as incommensurable experiences. It also explains the phenomenon of subjecting one type of text to the mode of reading appropriate for the other. We read fiction as nonfiction when we extract ourselves from its world and, switching reference worlds, assess its viability as a document of real-world events; conversely, we read nonfiction as fiction when we find the image so compelling that we no longer care about its truth, falsity, or ability to serve practical needs.

MAKE-BELIEVE

Once we are transported into a textual world, how do we bring it to life? Kendall Walton locates the key to immersion in a behavior that we learn very early in life—earlier, arguably, than we learn to recognize the rigidity of the ontological boundary that separates storyworlds from physical reality. The comparison of fiction to games of make-believe is not a particularly new one; it is implicit to Coleridge's characterization of the attitude of poetry readers as a "willing suspension of disbelief" (*Biographia Literaria*, 169), and it has been invoked by other thinkers, including Susanne K. Langer and John Searle (fiction, for Searle, is "pretended speech acts"). But Walton's project is more ambitious than defining fiction: the stated goal of his book *Mimesis as Make-Believe* is to develop a theory of representation and a phenomenology of art appreciation that make the term *representation* interchangeable with *fiction*. The range of the theory includes not only verbal but visual and mixed media:

> In order to understand paintings, plays, films, and novels, we must first look at dolls, hobbyhorses, toy trucks and teddy bears. . . . Indeed, I advocate regarding the activities [that give representational works of art their point] as games of make-believe themselves, and I shall argue that representational works function as props in such games, as dolls and teddy bears serve as props in children's games. (11)

The fictionality of all representations is not demonstrated by the application of the theory to various objects but entailed by the definitions that form the axiomatic basis of the project. Here is my own reconstruction of these definitions:

1. A representation is a prop in a game of make-believe.
2. A prop in a game of make-believe is an object—doll, canvas, text—whose function is to prescribe imaginings by generating fictional truths.
3. A fictional truth is a proposition that is "true in a game of make-believe."

Though Walton proposes no formal definition of "game of make-believe"—apparently taking the concept for granted—a set of rules is easily derived from his analysis:

1. Players select an actual object x_1—the prop—and agree to regard it as a virtual object x_2.
2. Players imagine themselves as members of the virtual world in which x_2 is actual. The actions performed by the players with the prop count as actions performed with x_2.
3. An action is legal when the behavior it entails is appropriate for the class of objects represented by x_2. A legal action generates a fictional truth.

It is easy to see how these rules apply in the case of children's games. In an example proposed by Walton, a group of children decide that stumps are to count as bears. The decision is arbitrary, since any object could be chosen, but once it has been made, the relation between stumps and bears is much stronger than the linguistic relation between the word *bear* and its signified. In the game of make-believe, stumps do not signify absent bears, they are *seen as* present animals. Every time a child sees a stump, she performs an action that counts in make-believe as an encounter with a bear. Players may flee, climb a tree, or shoot the bear, but not pet it, put a saddle on its back, or walk it on a leash. The propositions that describe what the stump stands for and what the players' actions count as are the fictional truths. Participating in the game means stepping into a world in which the real-world proposition "There is a stump" is replaced by the fictional truth "There is a bear." Every time a player performs a legal move, she makes a contribution to the set of fictional truths that describes the game-world: "I am shooting a bear," "I am fleeing from it." In this creative activity resides the pleasure, and the point, of the game.

In visual representation, the stump is the physical image, and the bear is the represented reality. The painting draws the spectator into its world and confers presence to that which it represents. According to Walton, we behave in front of the painting of a windmill as if we were facing the mill. Inspecting the splotches of color on the canvas counts as inspecting a windmill. The generation of fictional truths is the detection of the visual features of the mill. The legitimacy of moves is determined by the visual properties of the prop, by the nature of the represented object, and by the general rule of the game, which restricts participation to acts of visual perception: fondling a painting of a nude does not constitute a legitimate response, no matter how erotic the painting's effect may be.

The question "What does the prop stand for?" is slightly more problematic in verbal representation. Assuming that the prop is simply the text, a naive answer could read, "The prop stands for the world it projects." But as Walton observes (219), we may say "This is a ship" when pointing to the painting *The Shore at Scheveningen* by Willem Van der Velde, but we would never say "This is a ship" when reading *Moby-Dick*. The difference resides in the fact that while paintings depict iconically, words signify conventionally. The only object that a text can reasonably try to pass as is another text made of the same words but uttered by a different speaker and therefore constituting a different speech act. The basic fictional truth generated by a fictional text is that "it is fictional of the words of a narration that someone [other than the author] speaks or writes them" (356). The prop constituted by the authorial text simply stands for the text of a narrator who tells the story as true fact. The game of make-believe performed by the reader involves three mutually dependent operations: (1) imagining himself as a member of this world; (2) pretending that the propositions asserted by the text are true; (3) fulfilling the text's prescription to the imagination by constructing a mental image of this world. The range of legitimate actions corresponds to the various world images that can be produced by following the textual directions.

This analysis implies a sharp distinction between texts of fiction and texts of nonfiction. As Walton observes, "It is not the function of biographies, textbooks, and newspaper articles, as such, to serve as props in a game of make-believe." These works are "used to claim truth for certain propositions rather than to make certain propositions fictional" (70). Through a strange asymmetry, however, the distinction "offered for belief" versus "offered for make-believe" is not found in the visual domain. According to Walton, all representational pictures function as props in a game of make-believe, and there is no such thing as nonfictional depiction: "Pictures are fiction by definition" (351). Even pictures primarily used to convey information, such as anatomical illustrations or passport photos, pass as something else and invite the observer to pretend that she is facing that which they represent. All pictures are make-believe because they convey a sense of virtual presence. (Here Walton obviously rejects the idea of a nonillusionist mode of representation, such as we find in pre-Renaissance and postimpressionist art.) Some pictures, such as Vermeer's interiors, invite the spectator to a rich game of make-believe, one in which many details can be inspected, while other images, such as schematic line drawings, flowers and seashells in decorative patterns, or the silhouettes of children on traffic signs, reduce this game to the basic recognition of shapes. But as soon as recognition takes place, the spectator is engaging in an act of imagining and therefore of make-believe. The propositions considered in this act can only be fictional truths, because they are inspired by a copy and not a real object.

The asymmetry between texts and pictures with respect to the dichotomy fiction/nonfiction suggests that fictionality is an essentially verbal category. Without an other to limit and define it, the concept of fiction loses its identity. The asymmetry is partially explained by the fact that pictures do not literally make propositions, but Walton's categorization is above all the consequence of the reinterpretation to which the concept of make-believe is subjected as it crosses the boundary from textual to visual media. In visual communication, as I noted in the preceding paragraph, make-believe refers to pretended presence: the spectator apprehends the visual features of the depicted object as if she were standing in front of it. In the case of fictional texts, make-believe refers to pretended truth for propositions. This pretended truth presupposes pretended existence. Since pretended presence does not occur in verbal communication—linguistic signs normally refer to absent objects—the diagnosis of fictionality rests on incommensurable criteria for the two media: it is like comparing apples and oranges.

The distinction between fiction and nonfiction in the textual domain creates another difficulty for Walton's theory. The assimilation of representation to fiction and the definition of the latter as a prop in a game of make-believe make the embarrassing prediction that texts designed to elicit belief, rather than make-believe, are not representations. Yet Walton himself admits that "some histories are written in such a vivid, novelistic style that they almost inevitably induce the reader to imagine what is said, regardless of whether or not he believes it. (Indeed, this may be true of Prescott's *History of the Conquest of Peru*.) If we think of the work as prescribing such a reaction, it serves as a prop in a game of make-believe" (71). In this argument, well-written works of history such as *The History of the Conquest of Peru* are rescued from the limbo of nonrepresentations by reinterpreting make-believe as "picturing vividly in one's mind." In other words, these texts are representations because they can be read as fiction. Walton makes a distinction between imagining and considering a proposition, and regards these attitudes as constitutive of the difference between fiction and nonfiction—that is to say, between "representation" and its other, that which is offered for belief. On the contrary, I would like to argue that mentally producing a more or less vivid image of situations is an integral part of the reading of mimetic nonfiction, since it is on the basis of this image that we evaluate the truth of the propositions asserted by the text. The difference between fiction and texts we read for information resides not in the occurrence of an act of imagination but in whether or not it forms the point of the game.

Walton's use of make-believe thus subsumes, and often confuses, two distinct phenomena: (1) regarding texts that describe obviously made-up situations as reports of true facts ("willingly suspending disbelief"); and (2) engaging in an act of imagination, by which depicted objects and their surrounding worlds are made present to the mind. If we disentangle these two aspects of make-believe, the concept is applicable to both the problem of distinguishing fiction from nonfiction (through sense 1), and to the phenomenological description of immersion (through sense 2). While the first sense comes close to being binary, the second is a matter of degree: we can produce sketchy pictures, similar to line drawings, or rich images, similar to a Vermeer picture. When the text restricts itself to abstract ideas and general statements, in other words, when it is nonmimetic in Martínez-Bonati's sense of the term, make-believe as mental picturing reaches its zero degree. I am not saying that all mimetic texts necessarily give rise to a truly immersive experience, but rather that only those texts that are dominated by mimetic statements can be experienced in an immersive manner. The depth of immersion—what Walton calls the richness of the game of make-believe—depends on the style of the representation as well as on the disposition of the reader.

MENTAL SIMULATION

In 1997, when Walton revisits the phenomenology of art appreciation in "Spelunking, Simulation, and Slime," he sharpens his analysis of the mechanics of involvement in a textual world by borrowing from psychology the concept of mental simulation. In its psychological use, the term *mental simulation* is associated with a recent debate concerning the strategies of common-sense reasoning, or "folk psychology." An important aspect of this reasoning is the operations that enable us to imagine the thoughts of others with sufficient accuracy to make efficient decisions in interpersonal relations. In contrast to those psychologists who hold that we are able to make judgments about the psychological state of others by activating "a systematically organized body of information about mental states, their origin, interactions and effects" (Heal, "How to Think," 33), a position known as "theory-theory," simulationists argue that all we need to do to recreate people's thoughts is to use our existing reasoning abilities with different input—what we take to be the beliefs and values of the foreign mind. According to Stephen Stich and Shaun Nichols, we "take our own decision-making 'off-line,' supply it with 'pretend' inputs that have the same content as the beliefs and desires of the person whose behavior we are concerned with, and let it make a decision on what to do" ("Second Thoughts," 91). Simulation theory can thus be described as a form of counterfactual reasoning by which the subject places himself in another person's mind: "If I were such and such, and if I held beliefs p and q, I would do x and y."

Through its implicit shift in point of view, the concept of mental simulation dovetails with the ideas of recentering, transportation, and make-believe, but by locating the reader within the center of consciousness of the characters he tries to understand, it goes further than these concepts in explaining the phenomenon of emotional participation. From a human point of view, one of the most beneficial features of the theory of mental simulation is that it enables us to reason from premises that we normally hold to be false, and to gain more tolerance for the thinking processes of people we fundamentally disagree with: "Here the interesting point is that people can think about, and so explore the consequences of and reflect on the interconnections of, states of affairs that they do not believe to obtain" (Heal, "How to Think," 34). Fiction, similarly, has been hailed (and also decried) for its ability to foster understanding and even attachment for people we normally would condemn, despise, ignore, or never meet in the course of our lives. As we project ourselves into these characters, we may be led to envision actions that we would never face or approve of in real life.

This idea is crucial to Walton's appeal to simulation in support of his theory of mimesis as make-believe. He uses the example of imagining himself participating in a spelunking expedition to demonstrate that simulation can become a means of self-discovery. In the theater of his mind, he crawls for hours in a dark and humid hole until he reaches a shaft so narrow that he must abandon his pack and move forward by wiggling between the hard walls. His headlight goes out, and he lets out a scream of panic as he finds himself in total darkness. Though he does not believe for a moment that he is actually in danger, the simulator undergoes a genuinely upsetting imaginative experience, one that gives him the shivers every time he thinks of it. The act of pretense makes him realize his deep-seated claustrophobia and explains to him his real-life fear of elevators and crowded places. (We cannot, unfortunately, verify this claim, even by replaying the script in our imaginations, because what we would learn in Walton's cave would depend too much on our a priori opinions of his theory.) Through this example—which illustrates not only how we immerse ourselves in the creations of our own minds but also how readers bring textual worlds to life—Walton hopes to answer a criticism that has been frequently raised against his approach to fiction: that if the emotions aroused by fiction are confined to the fictional world and do not engage our real-world selves, reading fiction cannot provide a genuine learning experience. Not so, says Walton: if I can discover my claustrophobia by mentally simulating the cave expedition, I can also discover truths about myself by living in imagination the destiny of fictional characters.

In the spelunking example, mental simulation goes far beyond the attribution of thought to characters; it creates a rich sensory environment, a sense of place, a landscape in the mind. In a reading situation, it executes the incomplete script of the text into an ontologically complete, three-dimensional reality. To the performer of the simulation, the word *cave* does not simply evoke its lexical definition of "natural underground chamber" but awakens all its connotations of darkness, dampness, rough texture, earthy smell, silence occasionally interrupted by the noise of dripping water, and whatever else the simulator may associate with the mental image of the cave. But there is more to simulation than forming a vivid, sensorially diverse representation of a scene or an object; this image must also receive a temporal dimension. Gregory Currie suggests that *mental simulation is simply another name for an act of imagination ("Imagination,"

161), but if the term is to make a significant contribution to the phenomenology of reading, it should be reserved to a special type of imagining: placing oneself in a concrete imaginary situation, living its evolution moment by moment, trying to anticipate possible developments, experiencing the disappearance of possibilities that comes with the passing of time but remaining steadily focused on the hatching of the future.

It is indeed from this prospective orientation, this relentless assessment of the possibilities that still remain open, that simulation derives its heuristic value. Mental simulation should therefore be kept distinct from retrospective and temporally free-floating acts of imagination, such as storymaking, daydreaming, and reminiscing. When we compose a narrative, especially a narrative based on memory, we usually try to represent "how things came to be what they are," and the end is prefigured in the beginning. But when we read a narrative, even one in which the end is presented before the beginning, we adopt the outlook of the characters who are living the plot as their own destiny. Life is lived prospectively and told retrospectively, but its narrative replay is once again lived prospectively. Simulation is the reader's mode of performance of a narrative script.

The term *simulation* may be new, but the idea is an old one. Long before a label was put on the operation, Aristotle recommended its practice to authors of tragedy as a way to ensure the consistency of the plot:

> When constructing plots and working them out complete with their linguistic expression, one should as far as possible visualize what is happening. By envisaging things very vividly in this way, as if one were actually present at the events themselves, one can find out what is appropriate, and inconsistencies are least likely to be overlooked. (*Poetics* 8.3, 27)

This advice is also valid for writers of narrative fiction. In contrast to narratives of personal experience, novels are often conceived from a prospective stance: the author imagines a situation and tries out many possible developments until a good ending imposes itself. As Currie suggests ("Imagination," 163), the process of world construction is only imperfectly under the conscious control of the creator. While simulating the behavior of characters, the novelist comes to imagine them as autonomous human beings who write the plot for her by taking control of their own destinies. There cannot be a more eloquent tribute to the heuristic value of mental simulation than the feeling voiced by many authors that their characters live a life of their own.

The Discipline of Immersion

Ignatius of Loyola

If Aristotle recommended simulation as writing strategy, the credit for developing the technique into a reading discipline should go at least in part to St. Ignatius of Loyola.[1] In the *Spiritual Exercises* the founder of the Jesuits produced a meticulous description of the mental operations that lead to immersion in a textual world. The project may not be viable as a model of reader response—if we imagined textual worlds with the wealth of detail advocated by Ignatius we would never finish a book—but it provides a fascinating document of the utopian dream of a total simulation, and a prefiguration of many of the themes brought to the fore by VR technology.

A program for developing and strengthening faith, the *Exercises* are strangely reminiscent of a program for developing and strengthening muscles. (Ignatius, we are told, was an avid practitioner of the military arts of his time, before a religious conversion occasioned by a physical injury turned his energies toward the salvation of the soul.) Under the coaching of a "director of conscience," the exercitant is led through an elaborate protocol that describes in minute detail a sequence of exercises to be completed over a period of four weeks. The instructions specify how many repetitions of each exercise should be performed, what kind of variations should be introduced with every repetition, how to keep the exercitant interested (by maintaining, as Roland Barthes observes [*Sade, Fourier, Loyola,* 43], a narrative suspense about the next routine to be prescribed), and how to balance spiritual training with everyday life and the demands of the body (the exercitant should coordinate the program with meals and sleep, and the exercises should be made compatible with worldly occupations such as career, civic responsibilities, and what we today would call business interests).

INTERLUDE

The exercises themselves are meditations and contemplations on the biblical narrative, and they are aimed toward a lived participation of the self in the foundational events of the Christian faith. The "self," for Ignatius, is an indivisible "compound of body and soul" (*Exercises*, 136), and both of these components must be involved in the religious experience. But since the body cannot be physically transported to the scenes described in the Gospels, its participation in the sacred events must be mediated by the imagination. The exercitant is enjoined to situate himself or herself spatially with respect to the divinity: "Here [the task] is to see myself as standing before God our Lord, and also before the angels and saints, who are interceding for me" (176). Barthes describes Ignatius's project as a "theater entirely created in order that the exercitant may therein represent himself: his body is what is to occupy it" (*Sade, Fourier, Loyola,* 63). And further: "The body in Ignatius is never conceptual: it is always this body: if I transport myself to a vale of tears, I must imagine, see *this* flesh, *these* members among the bodies of creatures" (62).

But Ignatius is no postmodernist, and if he insists on the importance of the body in religious training, he does not reduce the self to the experience of its embodiment. In accordance with Christian doctrine, Ignatius regards the soul as "imprisoned in this corruptible body" (*Exercises*, 136). The soul remains very much the target of the training program because it is in its power, if it accepts the Redemption, to outlive its prison and receive a new, incorruptible body. The trick here is to put the corporeal part of the self in the service of the soul. Ignatius does not propose an escape from the prison of the body—only death will accomplish that, and the *Exercises* are very much aimed at the living, active members of society—but advocates instead the exploitation of the faculties located in the "smart walls" of the prison.

The originality of the method resides in the idea that the involvement of the senses of the body—sight, hearing, smell, taste, and touch—can be used as stepping stones toward the involvement of the two senses of the soul: the will and the intellect. When he asks the exercitant to contemplate hell, for instance, Ignatius directs his attention from one sense to another, in an order of succession that implies an increasing proximity of the body to the object of contemplation.

This mental picturing of the tortures of hell is not an end in itself but the first step of a three-part exercise that leads from the sensory to the spiritual: (1) realize the gravity of sin and of its consequences; (2) use the intellect to reason against it; and (3) use the will to decide to avoid it (138).

Here is the initial step:

The First Point will be to see with the eyes of the imagination the huge fires and, so to speak, the souls within the bodies full of fire.

The Second Point. In my imagination I will hear the wailing, the shrieking, the cries, and the blasphemies against our Lord and all his saints.

The Third Point. By my sense of smell I will perceive the smoke, the sulphur, the filth, and the rotting things.

The Fourth Point. By my sense of taste I will experience the bitter flavors of hell: tears, sadness, and the worm of conscience.

The Fifth Point. By my sense of touch, I will feel how the flames touch the souls and burn them. (141)

In this exercise, the image brought to life by the involvement of the senses is an apocryphal representation of hell, but the same strategy is proposed to immerse the exercitant in the Holy Scriptures. The candidate is asked not only to apply sight, hearing, smell, and touch to the contemplation of the Nativity—taste this time is omitted—but also to "fill in the blanks" in the biblical text with details of his or her own, until the text projects a world sufficiently vivid and autonomous to open its door to the reader:

The Second Contemplation of the Nativity.

The First Prelude is the history. Here it will be to recall how our Lady and Joseph left Nazareth to go to Bethlehem and pay the tribute which Caesar imposed on all those lands. She was pregnant almost nine months and, as we may piously meditate, seated on a burro; and with her were Joseph and a servant girl, leading an ox.

The Second Prelude. The composition, by imagining the place. Here it will be to see in imagination the road from Nazareth to

Bethlehem. Consider its length and breadth, whether it is level or winds through valleys and hills. Similarly, look at the place or cave of the nativity: How big is it, or small? How low or high? And how is it furnished? (150)

In the third prelude readers are asked to project themselves as a corporeal presence into the textual world and to take up an active role in the narrated events:

The first point [of the third prelude; Ignatius is obsessed with subdivisions]. This is to see the persons; that is, to see Our Lady, Joseph, the maidservant, and the Infant Jesus after his birth. I will make myself a poor, little, and unworthy slave, gazing at them, contemplating them, and serving them in their needs, just as if I were there, with all possible respect and reverence. Then I will reflect upon myself to draw some profit. (ibid.)

In the third point of the same unit, the exercitant is brought to understand that all these events happened "just for me." The spiritual profit to be gained from the biblical narrative is realized by stepping into the story and accepting, not just in imagination but in actuality, the actantial role of beneficiary.

In order to participate fully in the drama of the redemption, the exercitant must not only project a body image into the textual world but also simulate and thereby share the human emotions experienced by the characters. During the contemplation of the Passion, in the third week, "I will try to foster an attitude of sorrow, suffering and heartbreak, by calling to mind often the labors, fatigues, and sufferings which Christ our Lord suffered up to whatever mystery of his Passion I am contemplating at this time" (170), while in the fourth week, when reenacting the Resurrection, "upon awakening, I will think of the contemplation I am about to make, and endeavor to feel joyful and happy over the great joy and happiness of Christ our Lord" (176).

As was the case in VR technology, the condition for immersion in the sacred events is a relative transparency of the medium. Barthes has emphasized the "platitude of style" (*Sade, Fourier, Loyola*, 6) of the text of the exercises: "Purified of any contact with the seductions and illusions of form, Ignatius's text, it is suggested, is barely language; it is the simple, neuter path which assures the transmission of a mental experience" (40). Language, for Ignatius, is not an object of contemplation but a set of freely paraphrasable instructions to the imagination. This philosophy not only shapes his own writing practice but also affects his handling of the biblical text itself. In the fourth and final week of the program, when the exercitant is instructed to retrace and relive the entire narrative of the Passion, Ignatius does not hesitate to substitute his own retelling for the original; the text offered as guide to the imagination is a synthetic summary of the four Gospels. Any story can be retold an infinite number of ways, and as far as the "facts" are concerned, Ignatius's version is as good as any other.

But there is one type of utterance that absolutely resists paraphrase, and that the imagination should not attempt to traverse to get to "the real thing," because these utterances are the real thing itself: the spoken words of God. In his retelling, Ignatius encloses in quotation marks the words directly taken from the four Gospels, and these passages are almost all direct speech acts of Jesus, Mary, the archangel Gabriel, or the disciples witnessing the Resurrection.[2] When the Word is made flesh, it becomes physical presence, and the only traversal of the text needed to experience this presence leads (*pace* Derrida) from the written signs to an original, unique, and yet infinitely reiterable spoken utterance of the very same words: unique because it is inserted into human time and space, but reiterable because it potentially addresses everybody.[3]

During this itinerary through the elaborately designed program of the *Exercises*, the practitioner of Ignatian discipline learns three modes of immersion in the biblical text: imaginative projection of the body into the represented space, participation in the emotions of the characters, and moment-by-moment reenactment of the narrative of the Passion. Each type of experience is associated with one of the basic constituents of narrative grammar: setting, character, and plot. I return in the next two chapters to the poetics and cognitive dynamics of these three dimensions of immersive reading.

FOUR

Presence of the Textual World

Spatial Immersion

It seems plain that the art that speaks most clearly, explicitly, directly, and passionately from its place of origin will remain the longest understood. It is through places that we put up roots. —EUDORA WELTY

I remember the sensation of reading (Freudians can note this) as one of returning to a warm and safe environment, one that I had complete control over. When I picked up a book it was as much to get back to something as it was to set off to the new. —SVEN BIRKERTS

In contemporary culture, moving pictures are the most immersive of all media. Until VR is perfected and becomes widely available, no other form of representation will approximate their ability to combine the spatial extension and fullness of detail of still pictures with the temporality, narrative power, referential mobility (jumping across space and time), and general fluidity of language. This explains why immersion in a book has been compared to "cinema in your head" (Fischlin and Taylor, "Cybertheater," 13). As the reader simulates the story, her mind allegedly becomes the theater of a steady flow of pictures.

How important is the formation of mental images to an immersive reading experience? Do readers construct detailed representations of characters, settings, and actions, something equivalent to a Vermeer painting, or are they satisfied with the schematic outlines created by propositions? The readers who served as subjects in Victor Nell's investigation admit to variable degrees of interest in mental picturing: some describe themselves as "visualizers," some are reading for the plot. "Our imaginings are imprecise and misty," writes William Gass, "and characters in fiction are mostly empty canvas. I have known many who passed through their stories without noses or heads to hold them" (quoted in Nell, *Lost in a Book*, 217). My own experience tells me that novels can occasionally imprint in the mind images of quasi-photographic sharpness, but unlike photographs these images consist of selected features that leave many areas unspecified. The degree of precision and the nature of the immersed reader's mental representation depend in part on his individual disposition, in part on whether the focus of attention is character, plot, or setting. In this chapter and the next one I propose to take a closer look at the textual features and mental operations responsible for three forms of involvement with narratives: spatial immersion, the response to setting; temporal immersion, the response to plot; and emotional immersion, the response to character.

SPATIAL IMMERSION: A SENSE OF PLACE AND A MODEL OF SPACE

From Faulkner's Yoknapatawpha County to Alice's Wonderland and from Joyce's Dublin to C. S. Lewis's Narnia, literature has time and again demonstrated its ability to promote a haunting sense of the presence of a spatial setting and a clear vision of its topography. Whether attractive or repulsive, these mental geographies become home to the reader, and they may for some of us steal the show from the narrative action. A cliché of literary criticism acknowledges this thematic prominence of setting by labeling it "the main character in the novel."[1]

Spatial immersion is often the result of a "madeleine effect" that depends more on the coincidental resonance of the text with the reader's personal memories than on generalizable textual properties. Just as the taste and smell of a piece of madeleine dipped into a cup of tea took Marcel Proust back to the village of his childhood, a single word, a name, or an image is often all the reader needs to be transported into a cherished landscape—or into an initially hated one that grew close to the heart with the passing of time. This phenomenon is documented by the reaction of Gregory Ulmer to this sentence from Michael Joyce's *Twelve Blue*: "Blue isn't anything. Think of lilacs when they are gone." "It so happens," writes Ulmer, "that I never stopped thinking of the lilacs that grew in the backyard of my childhood home, the very scarcity of flowering bushes in Montana making their brief but fragrant appearance all the more impressive. I am hooked" ("Response," para. 2). Such comments are usually judged too impressionistic to be taken seriously by literary theory or literary criticism, but they reveal a dimension of the phenomenology of reading that cannot be ignored.

In the most complete forms of spatial immersion, the reader's private landscapes blend with the textual geography. In those moments of sheer delight, the reader develops an intimate relation to the setting as well as a sense of being present on the scene of the represented events. Since this latter experience involves transportation to a point defined by both spatial and temporal coordinates, I discuss it below in a subsection labeled spatio-temporal immersion. Neither one of these two experiences is easy to convey in language. Unlike pictures, which teletransport the spectator instantly into their space, language can afford only a gradual approach to the textual world. As a temporal medium it discloses its geography detail by detail, bringing it slowly into the reader's mind. And unlike pictures, language is the medium of absence. It does not normally re-present by creating an illusion of presence to the senses, as do visual media, but rather evokes the thought of temporally or spatially distant objects (deictics being a notable exception). To overcome this distance, language must find ways to pull its referents into the theater of the mind, and to coax the imagination into simulating sensory perception.

The philosopher who pioneered the phenomenological study of the experience of space in literature, Gaston Bachelard, conceives spatial immersion in terms of security and rootedness. The titles of the various chapters of his book *The Poetics of Space* are all symbolic expressions of an intimate relation to a closed, enveloping environment: the house; drawers, coffers, and chests; nests; shells; corners; miniatures; and, in a conceptualization of open spaces as cozy habitat, "intimate immensity" and "the universe as house." Lilian Furst (*All Is True*, 99) observes that Victorian novels, which are second to none in creating a sense of place within a narrative structure, were often tales of socially imposed confinement that focused on the emotional bond between a female heroine and the small world of a house, village, or uniform landscape. The boring province, one of the most haunting spatial themes of literature, was a discovery of the nineteenth century. Contemporary "nature writing," such as James Galvin's *The Meadow*, tries to recapture this sense of belonging in a certain place, but usually without the additional immersivity created by a sustained, temporally and emotionally riveting narrative. These sedentary dreams stand in stark contrast to the "deterritorialization" and nomadism that have come to pass, under the influence of Gilles Deleuze and Félix Guattari, as the quintessential postmodern experience of space. Whereas Bachelard reflects on a "sense of place," postmodern literature conceptualizes space in terms of perpetual movement, blind navigation, a gallery of mirrors, being lost in a not-always-so-funhouse, a self-transforming labyrinth, parallel and embedded universes, and discontinuous, non-Cartesian expanses, all experiences that preclude an intimate relation to a specific location. We could say that in Bachelard, space is sensorially experienced by a concrete, bounded body, while in postmodern literature its apprehension presupposes a dismembered, ubiquitous, highly abstract body, since real bodies can be in only one place at one time. The difference is one of a lived versus a conceptualized space: we can conceptualize space as a whole, but we can live it only by developing a relation to some of its specific points.

Yet if the nomadic, alienating space of postmodernism prevents an immersive relation, I would not go as far as to say that spatial immersion precludes travel. Textual space involves not only a set of distinct locations but a network of accesses and relations that binds these sites together into a coherent geography. A sense of place is not the same thing as a mental model of space: through the former, readers inhale an atmosphere; through the latter, they orient themselves on the map of the fictional world, and they picture in imagination the changing landscape along the routes followed by the characters. In the most complete form of spatial immersion, sense of place is complemented by a model of space that J. Hillis Miller has eloquently described:

> A novel is a figurative mapping. The story traces out diachronically the movement of the characters from house to house and from time to time, as the crisscross of their relationships gradually creates an imaginary space. . . . The houses, roads, and walls stand not so much for the individual characters as for the dynamic field of relations among them. (*Topographies*, 19–20)

To create a global and lasting geography, the text must turn in its favor the linearity of its medium. Unable to provide a panoramic glance, the text sends its readers on a narrative trail through the textual world, guiding them from viewpoint to viewpoint and letting them discover one by one the salient features of the landscape. In contrast to virtual realities of the electronic kind, the immersive quality of the representation of space depends not on the pure intensity of the information—which translates in this case as length and detail of the descriptions—but rather on the salience of the highlighted features and on the ability of descriptive passages to project a map of the landscape. A description that merely accumulates details lets its object run through the reader's mind like grains of sand through the fingers, thus creating the sense of being lost in a clutter of data.

In many postmodern texts this effect is deliberately exploited as a way to express the alienation of the subject from the surrounding world. In Alain Robbe-Grillet's *In the Labyrinth*, for instance, setting is painstakingly described through a linear accumulation of details, but space is neither apprehended nor organized by a human consciousness,[2] and details flow by the reader's mind without coalescing into a stable geography. The text does not fail to achieve, it actively *inhibits*, spatial immersion:

> I am alone here now, under cover. Outside it is raining, outside you walk through the rain with your head down, shielding your eyes with one hand while you stare ahead nevertheless, a few yards ahead, at a few yards of wet asphalt; outside it is cold; the wind blows between the bare black branches; the wind blows through the leaves, rocking whole boughs, rocking them, rocking, their shadows swaying across the white roughcast walls. Outside the sun is shining, there is no tree, no bush to cast a shadow, and you walk under the sun shielding your eyes with one hand while you stare ahead, only a few yards in front of you, at a few yards of dusty asphalt where the wind makes patterns of parallel lines, forks, and spirals. (141)

This passage creates a strong sense of atmosphere, but the incantatory tone of the description, its numerous repetitions, and its paratactic accumulation of details have such a dulling effect that some readers may fail to notice the abrupt switch from rain to sun and from winter to summer. If noticed, however, the transformation should lead to an even greater sense of disorientation. In order to support such a discontinuity, the textual universe cannot be a homogeneous Cartesian space with stable reference points but must be something more akin to the space of modern physics: a self-transforming expanse riddled with invisible black holes through which we are unknowingly sucked into parallel worlds. This conception of space is more hostile to immersion than the mental fog that conceals contradictions, because the imagination presupposes the container of a Cartesian space for the shapes of objects to be representable at all.

How can a literary work capture the feel of a place in both its atmosphere and its topography without losing the reader in a descriptive thicket? Balzac's novels also open with meticulous evocations of the setting, but the descriptions never jeopardize the reader's sense of orientation because they trace a precise itinerary through the fictional world. When the novel describes a house, such as the boardinghouse of Mme. Vauquer in *Père Goriot* or the decrepit manor in "La Grande Bretèche," the narrator inspects the building in a systematic manner, approaching it from the street, examining the garden and facade, entering through the main door, and walking from room to room, as would a real estate agent or a prospective tenant. The reader ends up with a precise notion of the configuration of the building, all the way down to the floor plan.

To dramatize the description, Balzac often resorts to the device of figuratively pulling the reader into the scene through a second-person address. The four-page depiction of the provincial town of Saumur that opens *Eugénie Grandet* takes readers on a walk up a narrow cobbled street, lets them peek into the backyards, invites them to browse in the stores ("Entrez"), and finally ushers them into the house where the action is to take place. As it weaves its way through the town of Saumur, the descriptive itinerary creates a narrative thread that facilitates the recalling of the images disposed along the way, thus building a "memory palace" comparable in effect to the mnemonic techniques of the sixteenth century:[3]

> When you have followed the windings of this impressive street whose every turn awakens memories of the past, and whose atmosphere plunges you irresistibly into a kind of dream, you notice a gloomy recess in the middle of which you may dimly

discern the door of Monsieur Grandet's house. Monsieur Grandet's house! You cannot possibly understand what these words
convey to the provincial mind unless you have heard the story of
Monsieur Grandet's life. (37)

The immersive quality of Balzac's descriptions is measured not by
the degree of absorption they arouse in the reader at the time of their
reading but by their lingering effect on the rest of the novel. Many
people find the beginning of *Eugénie Grandet* exasperating rather
than immersive. We may indeed hurry impatiently through the description of Saumur, eager for the real action to begin, but the atmosphere that has been fixed in the first few pages will facilitate the
process of mental simulation and enrich our mental representation of
all the episodes to come. My personal mapping of the topography of
the novel places the house of Grandet at the top and on the left of the
steep street, with the back of the house, where Eugénie's room is
located, overlooking the countryside. When I simulate the various
scenes of the novel, I always look at the house from the perspective of
the bottom of the hill, and I see people enter from right to left and
leave from left to right. This visualization blends text-given information (the house at the top) with a personal filling in of the blanks (the
house on the left), but as I replay the novel in my mind the two types
of detail blend into a seamless picture, and I don't remember what
comes from me and what comes from the text.[4]

Balzac's habit of establishing the setting all at once, at the beginning of every novel, reflects his deterministic belief in the importance of the environment for the development of the individual. In
Emily Brontë's *Wuthering Heights,* by contrast, the setting is constructed throughout the novel, in delicate and brief strokes, and it
seems to emanate from the characters rather than the other way
around. Long after readers have forgotten the details of the plot of
Wuthering Heights they retain the landscape in their minds; yet the
novel hardly ever pauses to give a detailed description of the environment. The sense of place and the model of space are created dynamically by a narration focalized through the character who is being
followed. While the movements of the characters between the two
houses of Thrushcross Grange and Wuthering Heights map the geography, their thoughts and perceptions condense the atmosphere:

[Told as Lockwood, the narrator, enters the court of Heathcliff's house.] Wuthering Heights is the name of Mr. Heathcliff's dwelling. "Wuthering" being a significant provincial adjective, descriptive of the atmospheric tumult to which its station is exposed in stormy weather. Pure, bracing ventilation they must have up here at all times, indeed: one may guess the power of the north wind blowing over the edge, by the excessive slant of the few stunted firs at the end of the house; and by the range of gaunt thorns all stretching their limbs one way, as if craving alms of the sun. (2)

Yesterday afternoon set in misty and cold. I had half a mind to spend it by my study fire, instead of wading through heath and mud to Wuthering Heights. . . . [Lockwood decides to go anyway.] On that black hilltop the earth was hard with a black frost, and the air made me shiver through every limb. (6)

One time I passed the old gate out of my way, on a journey to Gimmerton. It was about the period that my narrative has reached: a bright frosty afternoon; the ground bare, and the road hard and dry. (99)

Through the quasi-instantaneous snapshots of these "narrativized descriptions," as Harold Mosher and others call the technique, the problem of segmentation is minimized, and the experience of space blends with the forward movement of time. The sense of the presence of the environment is out of proportion with the diversity of its features: landscape in *Wuthering Heights* is reduced to a few recurring motifs, such as the wind on the moor, the hard frozen ground in the winter, the soft waves of the grasses in the summer. This economy of detail conveys the vast emptiness of the environment, but it also suggests that textual worlds, like dreamscapes, need only a few mooring points to take hold of the mind, especially when they are already inscribed in the imagination as what Gaston Bachelard calls *rêverie des éléments* (elemental imagination). More than the evocation of a specific English province, landscape in *Wuthering Heights* is a dialogue of earth and wind, an archetypal confrontation of cosmic elements.

A particularly efficient way to create a sense of place without resorting to lengthy descriptions is the use of proper names. From a semantic point of view, proper names contrast with common nouns through their intrinsic lack of sense and the uniqueness of their reference: in a perfect nomenclature, every object in the world would have a different label. The function of names is not to designate the properties of a certain object but to call its existence to the attention of the hearer, to impose it as discourse topic—in short, to conjure a presence to the mind. Through the instantaneous character of the act of reference, the use of a place name teletransports the reader to the corresponding location. For Richard Gerrig, as I noted in chapter 3, the mere mention of the name Texas in a novel lands the reader in Texas, or rather, lands Texas in the mind of the reader. Names may be technically void of sense, but they make up for this emptiness through the richness of their connotations. The name Texas transports the reader not into a barren expanse but into a territory richly landscaped by cultural associations, literary evocations, personal memories, and encyclopedic knowledge. Through this ability to tap into reservoirs of ready-made pictures, place names offer compressed images and descriptive shortcuts that emulate the instantaneous character of immersion in the space of visual media.

From an imaginative and ontological point of view, the place names of fictional worlds fall into several categories. The popularity of regional literature and the predilection of many readers for stories taking place in familiar locations suggest that the most immersive toponyms are the names of real places, either well known or obscure, that we happen to have personally visited, because it is always easier to build mental representations from materials provided by personal experience than by putting together culturally transmitted images—photographs, paintings, movie shots—or by following the instructions of purely textual descriptions. Direct personal memories enable readers to construct a precise map of the textual world and to visualize the changing environments as the characters move from location to location, much in the way the players of the so-called first-person-perspective computer games see the image of the game-world evolve as a result of their movements.

Next on the scale of immersivity are the names of famous real places we have heard of and dreamed about but never visited. (Readers whose imaginations are more oriented toward culture and history than toward space and landscape will probably invert my rankings.) Western culture elevates locations such as Paris, Venice, Vienna, Provence, New York, or California to mythical status, and these names function for most people as catalysts of desire. Proust has eloquently described the magic of such names:

> I need only, to make [these dreams of the Atlantic and of Italy] reappear, pronounce the names Balbec,[5] Venice, Florence, within whose syllables had gradually accumulated the longing inspired in me by the places for which they stood. . . . But if these names permanently absorbed the image I had formed of these towns, it was only by transforming that image, by subordinating its reappearance in me of their own special laws; and in consequence of this they made it more beautiful, but at the same time more different from anything that the towns of Normandy or Tuscany could in reality be, and, by increasing the arbitrary delights of my imagination, aggravated the disenchantment that was in store for me when I set out upon my travels. (*Remembrance,* 420)

These quasi-mythical sites are often surrounded in fictional worlds by obscure real place names that stand for an entire category of nondescript provincial towns. The Paris of Balzac and Flaubert is irreplaceable, but it wouldn't matter much to most French readers if the Saumur of *Eugénie Grandet* were named Troyes or the Rouen of *Madame Bovary* became Nantes or Bayeux. The place name, in this case, represents a stereotype, and readers construct the setting by activating the cognitive frame with which the text associates the name: "provincial town," "fishing village," "slum," "industrial zone," "vacation resort," and so on. If we can use our idea of French provincial towns to imagine the fictional counterpart of the real Saumur, we can similarly activate our conception of American suburbs to visualize an invented Springfield or Glendale, or, to remain in a purely literary domain, we can draw from a standardized "generic landscape," enriched by personal fantasies of idyllic settings, to picture the Arcadia and the *loci amoeni* of pastoral romance. Well-chosen imaginary place names that conform to the toponymy of a certain region are just as efficient at conveying *couleur locale* as the names of actual locations. Proust's invented Combray, Méséglise, and Martinville or the river

Vivonne exude for me the same French *saveur de terroir* (earthy flavor) as Gerard de Nerval's Ermenonville, Châalis, or Loisy—all real names of the province of Valois mentioned in his novella *Sylvie.*

Another space-constructing device that shortcuts the linearity of language is what Tom Wolfe calls the detailing of status life and Roland Barthes ascribes to *l'effet de réel* (the reality effect): the mention of concrete details whose sole purpose is to fix an atmosphere and to jog the reader's memory. For the trivial to exercise its signifying function, it must appear randomly chosen and be deprived of symbolic or plot-functional importance. Intrigued by the mention in Flaubert's tale "Un Coeur simple" of the barometer and pyramid of boxes and cartons in Mme. Aubain's room, Barthes asks the question "Is everything in the narrative meaningful, significant? And if not, if there exist insignificant stretches, what is, so to speak, the ultimate significance of this insignificance?" ("Reality Effect," 12). The ultimate function of such details, according to Barthes, is to tell the reader, "This is the real world." But the device is not merely a convention of realistic fiction. If we read in a fairy tale, "The princess walked into the dragon's lair. Luminescent green scales speckled with ruby-colored dots were scattered on the floor," the mention of the scales fulfills the same reality effect as the barometer and cartons in Mme. Aubain's room: the seemingly random detail conveys a sense of the presence of the setting and facilitates spatial immersion. The reader's sense of being there is independent of the verisimilitude of the textual world.

SPATIO-TEMPORAL IMMERSION: HOW TO TRANSPORT THE READER ONTO THE SCENE

From a logical point of view, the narrator and narratorial audience of a story told as true fact are located in the textual reference world, but this (re)location does not necessarily land them *on the scene* and *at the time* of the narrative window—to the heart of what some narratologists call the story-world. One of the most variable parameters of narrative art is the imaginative distance between the position of narrator and addressee and the time and place of the narrated events. Spatio-temporal immersion takes place when this distance is reduced to near zero.

The following four examples illustrate different degrees of reader proximity to the narrative scene, and different strategies to reduce the distance:

> I say that in the city of Pistoia, there was once a very beautiful widow, of whom, as chance would have it, two of our fellow-Florentines, who were living in Pistoia after being banished from Florence, became deeply enamoured. (Boccaccio, *Decameron*, ninth day, first story, 682)

This passage conforms to what Mary Louise Pratt describes as the standard "natural" (i.e., real-world) storytelling situation: a narrator informs an audience of events that took place at a temporal and spatial distance from the present location, the narrator knows the facts, and he displays their report for the entertainment and/or information of the audience.[6] These parameters are confirmed by the framing tale of the *Decameron*: ten young people locked up in a church during an outbreak of the plague, telling each other stories to entertain themselves during their confinement. While the narrator and his audience are located in the same discursive space—in this case, the storytelling event in the church—neither of them is part of the spatial and temporal window occupied by the narrated events, and neither perceives these events through the senses of the body. This particular passage verifies Seymour Chatman's description of the epistemological foundations of narration: "The narrator can only report events: he does not literally 'see' them at the moment of speaking them. The heterodiegetic narrator never saw the events because he/she/it never occupied the story world" (*Coming to Terms*, 144–45).

Chatman proposes this statement as a general model of narration, but the limits of this account are demonstrated by this passage from *Madame Bovary*:

> The bedroom, as [Homais and Dr. Canivet] entered, was mournful and solemn. On the sewing table, now covered with a white napkin, were five or six small wads of cotton in a silver dish, and nearby a large crucifix between two lighted candelabra. Emma lay with her chin sunk in her breast, her eyelids unnaturally wide apart; and her poor hands picked at the sheets in the ghastly and

> poignant way of the dying, who seem impatient to cover them-
> selves with their shrouds. Pale as a statue, his eyes red as coals,
> but no longer weeping, Charles stood facing her at the foot of
> the bed; the priest, on one knee, mumbled under his breath.
> (367–68)

This episode combines several acts of consciousness: the view offered
to the visitors who enter the room; the sensory perception of an
invisible observer located on the scene; and the general reflections
of an authorial figure about the habits of the dying. These various
perspectives blend so smoothly that it almost seems that the events
inscribe themselves as they occur in a recording mind. The backgrounding of the act of telling annihilates the imaginative distance
between discursive space and story-world, and fuses the consciousness of reader and narrator into the same act of perception. The
virtual body whose perspective determines what is perceived belongs
at the same time to the narrator and the reader—or to be more precise, to the reader's counterpart in the fictional world—just as, in
classical paintings, the eye that contemplates the scene belongs to
both painter and spectator.

In this passage from James Joyce's short story "Eveline," immersion
is made even more complete by the fusion of the virtual body of
narrator and reader with the fictionally real body of a member of the
textual world:

> She sat at the window watching the evening invade the avenue.
> Her head was leaned against the window curtains and in her
> nostrils was the odour of dusty cretonne. She was tired.
>
> Few people passed. The man out of the last house passed on
> his way home; she heard his footsteps clacking along the con-
> crete pavement and afterwards crunching on the cinder path
> before the new red houses. (36)

The reader does not watch a narrator watching Eveline watch the
street through the window, but, by virtue of the transitivity of the
representation of mental processes, she directly perceives Eveline's
perception. Through identification with the body of Eveline, the
reader gains a solid foothold on the scene, as well as a sensory inter-

face to the textual world. The narrative scene becomes as close to her as the smell of dusty cretonne to her nostrils or the texture of the fabric to her cheeks.

The next example presents an equally vivid representation of perceptual phenomena, but these perceptions seem to float around without corporeal support:

> The land was so distant that no shining roof or glittering window could be any longer seen. The tremendous weight of the shadowed earth had engulfed such frail fetters, such snail-shell encumbrances. Now there was only the liquid shadow of the cloud, the buffeting of the rain, a single darting spear of sunshine, or the sudden bruise of the rainstorm. Solitary trees marked distant hills like obelisks. (Virginia Woolf, *The Waves*, 54)[7]

As Monika Fludernik has argued, this description is not attributable to the consciousness of a specific character, and it is too spontaneous, too vivid and live, to represent knowledge kept in memory by a narrator and verbalized after the experience. The floating consciousness must therefore belong to a virtual counterpart of the reader:

> Just as, in figural [i.e., Eveline-type] narrative, the reader is invited to see the fictional world through the eyes of a reflector character, in such a text the reader also reads through a text-internal consciousness, but since no character is available to whom one could attribute such a consciousness, the reader directly identifies with a story-internal position. (*Fictions of Language*, 391)

Fludernik compares this situation with Jonathan Culler's account of the epistemological status of lyric poetry: "The paradigm thus established treats the modern lyric not as patterning of words or as expression of truths (even particular modern truths) but as a *dramatization of consciousness attempting to engage the world*" (quoted in ibid., 394).

The imaginative transportation of the reader's virtual body onto the scene of the events is facilitated by a variety of narrative strategies that often contrast with another device: scene versus summary; internal and variable focalization (representing characters as subjects) versus external focalization (looking at characters as objects); dialogue and free indirect discourse bearing the marks of the characters' idiosyncrasies versus stylistically neutral indirect reports of speech; prospective first-person narration representing the textual world from the point of view of the narrator-then (as hero of the tale) versus retrospective representation informed by the knowledge of the narrator-now (as historian of his own life); totally effaced or aggressively visible "hectoring" narrators versus what Tom Wolfe calls "pale-beige narrators" ("New Journalism," 16); and mimesis ("showing") versus diegesis ("telling"). The most fundamental of these techniques are those that invite the reader to relocate to the inner circle of the narrative action by dissociating the reference of the deictic elements of language, such as adverbs, tense, and pronouns, from the speech situation (i.e., the narrator's spatio-temporal location) and reassigning it from the perspective of a participant in the narrated scene. Let us consider three ways to redirect reference toward the narrative window: adverbial deictic shift, present tense, and second-person narration.

Adverbial Deictic Shift

Literary semantics has described three ways of reporting the speech or thought of characters: direct discourse (DD) ("Eveline thought: 'How can I ever leave my family?' "), indirect discourse (ID) ("Eveline thought that she would never be able to leave her family"), and the predominantly fictional free indirect discourse (FID) ("How could she ever leave her family, thought Eveline"). One of the syntactic trademarks of FID is the combination of a past-tense third-person narration with the adverbials *here, now, today, tomorrow,* rather than the expected *there, then, this day, the next day.* While the reference of the spatial and temporal shifters forces on the reader the perspective of the characters, verb tense and pronouns remain assigned from the point of view of the narrative act: "Even *now,* though she *was* over nineteen, she sometimes felt herself in danger of her father's violence" ("Eveline," 38). Or: "If she *went, to-morrow* she would be on the sea with Frank, steaming towards Buenos Ayres" (40). David Zubin and Lynne Hewitt describe the effect as follows: "The teller seems to fade into the background, and the story world, containing its own deictic center, comes to the fore. This is accomplished by decoupling the linguistic marking of deixis from the speech situation, and reorienting it to the major characters, the locations, and a fictive present time of the story world itself" ("Deictic Center," 131).

The contrast between DD, ID, and FID has been analyzed almost to the point of saturation, but nobody to my knowledge has addressed the issue of their comparative immersive power. The least immersive is clearly ID, not only because it ascribes the reference of all deictics from the point of view of the narrator, but also because of its lack of mimetic properties. While ID paraphrases the quoted discourse in the narrator's vernacular, FID mimics the voice of the quoted character, and DD offers a perfect replica. DD would seem to be the most immersive of the three modes of reporting, but I would like to make a point in favor of FID. In DD, all deictics refer to a center of consciousness located on the scene, but the attributing expression ("Eveline thought") restores the perspective of the narrator and creates a movement of in-and-out between the narrative window and the larger textual world. In FID, by contrast, the reported discourse blends smoothly with the attributing phrase as well as with the rest of the narration because it maintains referential continuity on the level of the most visible and frequent deictic elements, those of tense and person. I leave it to the reader to decide what is more immersive: the form of expression that gives us a complete but temporary relocation to the narrative scene and jogs us in and out of this focal point, or the one that maintains a constant position halfway between the narrator's and the character's spatio-temporal location.

Present Tense

The verbal inflections known grammatically as tense encode many ideas, not all of which are related to time. The present tense, in English, is used for timeless statements ("Two plus two equals four"), for habitual, iterative events ("I run twenty miles per week"), for future events ("Next time I go shopping I will get you some snacks"), for past ones ("There were these teenagers in the park, and I walked past them, and this girl starts screaming at me"), and occasionally to express the (near) coincidence of an event with the time of its verbal description ("I am tired" or "The Babe hits the ball; she is going going gone; home run!"). In conversational storytelling and medieval epics, the so-called historical present is used in alternation with the past to channel the attention of the audience toward certain events and create a profile of mounting and declining tension (Fleischman, *Tense*, 77). The peaks and valleys of this profile correspond to various degrees of imaginative presence of the events, those reported in the present usually forming the peaks. The effect could work the other way around if the present were the standard narrative tense and the past the marked one, but there are good semantic and pragmatic reasons why narratives are usually told in the past: you can only make a story when the events are in the book. Moreover, as the tense of presence, the present is inherently more immersive than the past.

The effect of the contrast is skillfully exploited in these two passages from Marguerite Duras's *L'Amant (The Lover)*:

> Little brother died in December 1942, under Japanese occupation. I had left Saigon after my second baccalaureate, in 1931. He wrote me only once in ten years. . . .

> When he dies it is a gloomy day. I believe it is spring, it is April. Somebody calls me on the phone. Nothing, they don't say anything else, he was found dead, on the floor in his room. (71 and 99; my translation)

Both of these passages—separated by nearly thirty pages—narrate the same event. In the first, the reader is merely informed of the death of the brother; in the second, he shares the narrator's experience of the atmosphere of the day, the breaking of the news, the tragically banal circumstances of the death. Though the narrative use of the present does not literally imply simultaneity between the occurrence of the events and the speech act of their report, as it does in "real-time narration" (such as sports broadcasts and conversations between pilots and control towers), it owes much of its expressive power to the lingering association of the tense with the idea of co-occurrence. We do not naturalize the speech situation of *The Lover* as one in which the narrator tells about her brother's death at the same time she learns about it, but as a prenarrative state of consciousness. The present sends us to a moment when the narrator knows nothing more than what she hears on the phone, a moment in which she is unable to rationalize the event, or even perhaps to realize the finality of its occurrence. As it creates the simulacrum of a real-time "life" (rather than speech) situation, the shift from past to present pulls the reader from the *now* of the storytelling act to the *now* of the story-world and completes the deictic shift toward the narrative window.

Many contemporary texts exploit this pseudo-immediacy of the first-person present-tense report to convey the experience of being swept by the flux of life, overwhelmed by unpredictable waves of events and sensations. Through its insistent use of the narrative present, contemporary narrative casts a resounding vote of nonconfidence in the authenticity of the rational activity of retrospectively emplotting one's destiny;[8] truth, it tries to tell us, lies in the immediacy of experience, not in the artificial form imposed on one's life by narrative activity. Yet if the present enjoys an immersive edge over the past, this edge becomes considerably duller when the present invades the whole text and becomes the standard narrative tense. Continuous presence becomes habit, habit leads to invisibility, and invisibility is as good as absence. For immersion to retain its intensity, it needs a contrast of narrative modes, a constantly renegotiated distance from the narrative scene, a profile made of peaks and valleys.

Second-Person Narration

Until the second half of the twentieth century, narrative came in two forms: first- and third-person, with occasional second-person addresses to the reader (cf. the Brontë and Balzac examples). Now that literature has become a systematic exploration of the expressive potential of language—or is it a systematic exemplification of all the categories of verbal paradigms?—narrative also comes in "you," "we," and "they" form as well as in the past, present, future, and conditional.[9] The reference of the second-person pronoun in a fictional context can be interpreted in many ways, and it can shift in the course of reading. Depending on the text, "you" can be used as a boundary-crossing address from the narrator in the textual world to the reader in the real world (first chapter of Italo Calvino, *If on a Winter's Night a Traveler*); as an intra-textual-world address from the narrator to an anonymous narratee (addresses to the reader by the "engaging narrators" of nineteenth-century novels); as an address from the narrator to a specific individual (= character) in the textual world (Michel Butor, *La Modification*); as a self-address by the narrator (you-form autobiography); and even as an address from an authorial figure to a real-world reader interrupting the textual-world speech act of the narrator (postmodern metafiction).[10]

Despite their different reference, all of these uses play on our instinctive reaction to think *me* when we hear *you*, and to feel personally concerned by the textual utterance. Reading a second-person novel is a little bit like going to the psychoanalyst and wondering what he is going to tell you about yourself that you do not already know. Even when it refers to a well-individuated character in the textual world, the pronoun *you* retains the power to hook the attention of the reader and to force at least a temporary identification with the implied referent. Through this identification, the reader is figuratively pulled into the textual world and embodied on the narrative scene (unless, of course, the I-you communication is of the metafictional type, in which case the effect is a decentering).

The immersive power of the second person is often a short-lived effect. When the shock of the initial identification wears off, second-person fiction tends to be read like a third-person narrative: the reader gradually detaches herself from the pronominal referent, and *you* becomes the identifying label, almost the proper name, of a regular character. I certainly did not experience a closer identification with the second-person protagonist of Butor's *La Modification* than with the first-person narrator of the author's previous novel, *L'Emploi du temps*. As an immersive device, second-person address is the most efficient in small doses, such as Balzac's sudden pulling of the reader into the description of Saumur in an otherwise third-person narration. When it becomes a sustained mode of narration, the second person is often more an allegory of immersion and a programmatic statement than an intrinsically immersive device. This programmatic intent is obvious in the following passage, an advertisement for the Time-Life series of books *The Native Americans*. Through the *you* implicit in the verb form, the narrator makes a conditional promise of immersivity whose fulfillment depends not just on the narrative art of another text but, more importantly, on the purchase of a commodity:

Follow the trail of broken treaties that led to Wounded Knee. Witness raids and battles of terrible intensity—Rosebud and Big Hole, Washita and Battle Butte. Then stand with Crazy Horse and his charging wall of Sioux Warriors at Little Big Horn. . . . Feel the rush of the buffalo hunt, a dawn raid on an enemy camp or the fireside retelling of a rout of bluecoats as you watch two worlds collide—through Indian eyes. . . . Join Chief Joseph on the 1,700-mile trek that ended 30 heartbreaking miles from freedom, when he declared, "I will fight no more forever."

The variability of the distance between the reader's implicit position and the narrated events suggests that narrative phenomenology involves not just one but two acts of recentering, one logical and the other imaginative.[11] The first—described in the previous chapter as the constitutive gesture of fictionality—sends the reader from the real world to the nonactual possible world created by the text; the second, an option available in principle to both fiction and nonfiction, though vastly more developed in the former, relocates the reader from the periphery to the heart of the story-world and from the time of narration to the time of the narrated. This experience of being transported onto the narrative scene is so intense and demanding on the imagination that it cannot be sustained for a very long time; an important aspect of narrative art consists, therefore, of varying the distance, just as a sophisticated movie will vary the focal length of the camera lens.

Immersive Paradoxes
Temporal and Emotional Immersion

> [I found myself] giving more attention and tenderness to characters in books than to people in real life, not always daring to admit how much I loved them . . . those people, for whom I had panted and sobbed, and whom, at the close of the book, I would never see again, and no longer know anything about. . . . I would have wanted so much for these books to continue, and if that were impossible, to have other information on all those characters, to learn now something about their lives, to devote mine to things that might not be entirely foreign to the love they had inspired in me and whose object I was suddenly missing . . . beings who tomorrow would be but names on a forgotten page, in a book having no connection with life.
> — MARCEL PROUST

While spatial and spatio-temporal immersion invite us to slow down the pace of reading, and occasionally to reread a passage so that we may linger on a particularly pleasurable scene, pure temporal immersion incites us to rush through the text toward the blissful state of retrospective omniscience. Temporal immersion is the reader's desire for the knowledge that awaits her at the end of narrative time. Suspense, the technical name for this desire, is one of the most widely appreciated literary effects, but also one of the most neglected by narratologists, in part because of its association with popular literature, but mainly because of its stubborn resistance to theorization. (Most work on suspense has come from empirical approaches to literature, film theory, cognitive psychology, and practical guides on how to write screenplays.)[1] As the label of a certain type of experience—namely, "an emotion or state of mind arising from a partial and anxious uncertainty about the progression or outcome of an action" (Prince, *Dictionary*, 94)—suspense seems so easy to define that it discourages further discussion. But behind its deceptive self-evidence, suspense leaves many questions to be answered: Can one distinguish different types of suspense, what cognitive mechanisms does it involve, what are the narrative devices that favor the experience, does it survive the first reading of a text, and what conception of time does it presuppose?

TEMPORAL IMMERSION

The phenomenological basis of temporal immersion in general, and of suspense in particular, is a "lived" or "human" experience of time, as opposed to what may be called "objective" or "clock" time. If clocks had a philosophical mind, they would describe time as a mechanical, meaningless, nonteleological succession of self-contained moments. Human time, by contrast, is a quasi-musical experience in which the present is not a moving point but a moving window that encompasses memories of the past and premonitions of the future. It is because the preceding notes survive in the present one, and because the present note adumbrates its successor, that we perceive melodic lines; similarly, it is because past events cast a shadow on the future and restrict the range of what can happen next that we perceive narrative lines and experience suspense. Generally speaking, temporal immersion is the reader's involvement in the process by which the progression of narrative time distills the field of the potential, selecting one branch as the actual, confining the others to the realm of the forever virtual, or counterfactual, and as a result of this selection continually generates new ranges of virtualities. The passing of time matters to the reader because it is not a mere accumulation of time particles but a process of disclosure.

As the popularity of sports events suggests, the most suspenseful situations are those that chart the future into a fan of diverging, but reasonably computable, outcomes. Spectator sports such as football or baseball have a lot to teach about temporal immersion in literary works:

— The enjoyment of the spectator is due to the fact that he roots for one of the teams and sees one outcome as vastly preferable to the other.
— Spectators participate in the action through the activity known as "armchair quarterbacking": they imagine scenarios for the action to come and make strategic decisions for the participants. This activity is made possible by the rigidity of the rules that determine the range of the possible.

— Suspense increases as the range of possibilities decreases. It is never greater than in the ninth inning or the last two minutes of the game, when the teams are running out of resources and options are reduced to a sharply profiled alternative: score now and stay alive, or fail to do so and lose the game. At the height of suspense, the ticking of the clock (if the game is limited by time) becomes strategically as important as the actions of the players. When this happens, the spectator reaches a state of complete temporal immersion.

In narrative suspense, similarly, we find the following features:

— Dramatic tension is usually correlated to the reader's interest in the fate of the hero. The prototypical suspense situation occurs when a character is in danger and the reader hopes for a favorable outcome.

— Suspense is dependent on the construction of virtual scripts and events. Though it is tied to uncertainty, it must present what Noël Carroll has called "a structured horizon of anticipation" ("Paradox," 75). This horizon is given shape by potentialities that trace visible roads into the future, such as the processes currently under way, the desires of characters, the goals to which they are committed, and the plans under execution. The reader's ability to project these paths is facilitated by narrative devices that constrain the horizon of possibilities in the same way the rules of games determine what can happen: foreshadowing, predictions, flash forward, or Alfred Hitchcock's strategy of making the reader "aware of all the facts involved" (Truffaut, *Hitchcock,* 72).

— The intensity of suspense is inversely proportional to the range of possibilities. At the beginning of a story, everything can happen, and the forking paths into the future are too numerous to contemplate. The future begins to take shape when a problem arises and confronts the hero with a limited number of possible lines of action. When a line is chosen, the spectrum of possible developments is reduced to the dichotomy of one branch leading to success and another ending in failure, a polarization that marks the beginning of the climax in the action.

This account of suspense as an experience dependent on anticipation may seem to overlook the importance of surprise, but surprise has no part in the building of suspense; it comes into play only at the moment of its resolution. A narrative may create surprise without suspense, and it may conversely relieve suspense without surprise, though a resolution of suspense that reveals a path not foreseen by the reader is aesthetically more satisfying. When a character falls into a manhole on his way to work, the reader may be surprised, but since nothing prepares her for this turn of events, she does not experience suspense. On the other hand, when the hero of a romance novel proposes to the heroine, the reader may be dying to find out whether or not she will accept, but when she finally makes a decision, the plot actualizes a fully foreseeable possibility.[2]

While the reader's curiosity in a suspense situation always concerns the story level of narrative, this curiosity is controlled on the discourse level by the author's strategies of divulging information. The intensity of the reader's temporal immersion depends on the focus of the suspense. In decreasing order of intensity we can distinguish four types.

1. *What suspense.* This brand of suspense is typical of action movies and thrillers. It is epitomized by the classic scene of Western movies in which the heroine is tied to the railroad tracks while a train is approaching. A memorable literary example is found in the last scene of James Joyce's "Eveline": will the heroine follow her lover and board the ship that will take her to a new life in Argentina, or will she remain on shore, tied to the past, to Dublin, and to a joyless life of sacrifice to her family? The focus of attention in this type of suspense is the imminent resolution of a binary alternative: will good or bad happen to the heroine? As Carroll has observed, this concern presupposes an emotional involvement in the fate of a character and a strong desire for an outcome favorable to the good guy. Since the central question is "What will happen next?" this type of suspense is favored by an order of presentation that runs parallel to the chronology of the underlying events (Brewer, "Nature of Narrative Suspense," 113). In order to facilitate the anticipation that creates suspense, the text may allow the reader to know more than the character—for instance, by describing how another character sets a trap while the hero is absent from the scene—but it typically does so without disturbing the chronological order. Through the parallelism of discourse structure and event struc-

ture, the reader or spectator lives the development of the action moment by moment and shares the perspective of the character whose fate is being played out.

2. *How (why) suspense.* This is the suspense involved in finding how Ulysses came to be held prisoner by Calypso, or why the rich de Lanty family treats a mysterious old man with such reverence in Balzac's *Sarrasine.* This type of suspense adopts the format of the enigma and is produced by what Barthes, in *S/Z,* calls the hermeneutic code. The outcome is given in advance through a phenomenon known as prolepsis (or cataphora), and the focus of attention is not the future but the prehistory of a certain state. The reader's experience of time is prospective, but unlike the movement of pure simulation this prospection is oriented, and therefore teleological. While *what* suspense presupposes a choice between two branches leading in opposite directions, *how* suspense involves multiple possibilities converging toward the same point. Since the fate of the hero is known from the very beginning, involvement is less a matter of wishing for a favorable outcome than a matter of curiosity about the solution to a problem. Once the narrative has jumped backward in time and started moving forward toward the known goal, however, the reader may be caught in *what* suspense on the level of the individual episodes. Conversely, a narrative of the *what* type may lead to *how* suspense when the reader's knowledge of generic conventions leaves little doubt about the outcome. It is usually pretty clear to the spectators of Western movies that the heroine tied to the railroad tracks will be saved, but suspense can be maintained by shifting focus from the probability to the circumstances of the rescue.

3. *Who suspense.* This is the suspense of the whodunit, the effect commonly associated with murder mysteries. The reader's interest in the outcome is even more purely epistemic than in the preceding type. Just as there is no sadness for the victim and no moral revulsion for the crime, there is little or no emotional investment in the fate of the characters because all that matters is the intellectual satisfaction of solving the problem. In contrast to the previous type, which leaves open an indeterminate number of paths toward the goal, *who* suspense limits the number of solutions to the number of suspects. This greater structuration of the virtual field enables the author to withhold facts from the reader, such as the motivation of the detective

during individual episodes of the investigation, without creating a sense of blind progress—the situation most hostile to suspense. The action unfolds on two temporal planes, that of the murder and that of its investigation, but the reader is never really caught in the time of the murder sequence. The past is revealed to him piece by piece, like a jigsaw puzzle, by the actions of the detective, and since the pieces do not fall into place in chronological order its experience is more that of a spatial picture than of a linear sequence. If mystery stories allow any kind of temporal immersion, it concerns the investigation sequence, and this immersion is tied to the dynamics of disclosure rather than to the unfolding of human destiny.

4. *Metasuspense,* or critical involvement with the story as verbal artifact. In metasuspense the focus of the reader's concern is not to find out what happens next in the textual world but how the author is going to tie all the strands together and give the text proper narrative form. An example of such suspense for me is the short story "The Assignation" by Edgar Allan Poe. The first scene of the fifteen-page text describes how a stranger rescues a child who has fallen into the canal in Venice. The mother of the child, a beautiful noblewoman, murmurs a few ominous words to the stranger: "Thou hast conquered—one hour after sunrise—we shall meet—so let it be" (197). In the next scene the narrator visits the stranger in his palace in Venice, and they engage in a lengthy (ten-page) discussion about art. On the last page of the story they are still talking about art, and I started to wonder, "How is Poe going to turn this into a story with only half a page of text left?" The author pulls the trick by having the stranger drink a glass of wine and drop dead while a messenger enters the room screaming "My mistress!—my mistress!—poisoned!" (207), thus revealing that the planned meeting was a suicide pact. In my anxious watching of the diminishing textual resources through which the suspense might be resolved, I lived the dynamics of storytelling more intensively than the ticking away of story-time. Because it involves a point of view external to the textual world—a relation reader-author—this last type of suspense does not properly belong to the poetics of immersion but rather anticipates the interactive relation to be discussed in later chapters; I mention it here only for its oppositional value with respect to the other three categories.

For the student of immersion, one of the most intriguing aspects of narrative suspense is what Richard Gerrig has called its resiliency, or anomalous experience. Many researchers have been puzzled by the fact that people can be caught in suspense after multiple readings. When I watched the movie *Apollo 13*, I knew that the crippled spaceship and its crew would make it back safely to Earth, but despite my certainty that everything would turn out for the best I experienced almost unbearable tension during the scene of the return to Earth. Through the representation of a live TV broadcast, the audience is informed that the spaceship has ten seconds to enter the atmosphere; if it does not appear on the screen before the time is up, this will mean that it has disintegrated in outer space. Thus given all the facts, in classic Hitchcockian fashion, I anxiously watched the clock tick away, and my anxiety grew stronger with every passing second. Relief came after twelve seconds that seemed like hours, when a parachute appeared on the TV screen and gracefully descended toward the sea. My involvement in the action was temporal immersion in its purest form—time had become almost tangible—and I experienced it not just once but every time I watched the video of the movie. Repeat suspense is better documented in film than in literature, but the case of children who ask for the same story over and over again, and participate in every retelling with the same intensity, indicates that the experience is not restricted to visual media. A related phenomenon is the ability of stereotyped genres to arouse suspense even though the reader knows that in the end good things always happen to the good guys.

The resiliency of suspense leads to a paradox that poses a serious challenge to its standard definition. Carroll summarizes this paradox as follows: "Conceptually, suspense entails uncertainty. Uncertainty is a necessary condition for suspense. When uncertainty is removed from a situation, suspense evaporates" ("Paradox," 72). Anomalous suspense defies the premise of this reasoning, but the irritation of some readers when they are given the end of a novel suggests that the gripping power of narrative cannot be totally divorced from uncertainty. As William Brewer pointedly observes, if suspense were immune to repetition, people could spend their whole lives rereading one very suspenseful novel, much to the chagrin of publishers ("Nature of Narrative Suspense," 120). The best account of anomalous suspense should therefore be one that tolerates some degree of re-cidivism but predicts a loss of intensity and eventual decay. (Even children, I assume, finally get bored when they hear the same story too many times.) At the same time, however, this account should remain focused on the suspense phenomenon itself, and not invoke a shift of motivation. We must therefore eliminate as a possible expla-nation the otherwise valid observation that multiple readings bring satisfaction because they enable us to notice different features of the text.

For Kendall Walton, the solution to the paradox of repeated sus-pense is encapsulated in the make-believe theory of fiction. If the reader of fiction can pretend that the asserted facts are true when she knows that they are not, and can derive pleasure from this act of pretense, she can just as easily pretend on second reading that she does not know the outcome and experience all over again the thrill of disclosure:

> Although Lauren knows that fictionally Jack will escape from the Giant, as she listens to still another rereading of "Jack and the Beanstalk," it is fictional that she does not know this—until the reading of the passage describing his escape. Fictionally she is genuinely worried about his fate and attentively follows the events as they unfold. It is fictional in her game during a given reading and telling of the story that she learns for the first time about Jack and the Giant. (*Mimesis*, 261)

I find this explanation partly convincing and partly unsatisfactory. The convincing part resides in its implication of a resetting of the narrative clock. When Lauren hears the story for the second or third time, she experiences a temporal relocation that places her at the beginning of narrative time and enables her to share the prospective outlook of the hero. She lives the unfolding of his fate in the real time of a shifted present, rather than being merely informed of what hap-pened in a fictional past. On the negative side, Walton's argument seems too dependent on a fallacious symmetry between suspending disbelief and suspending knowledge altogether. It takes little effort to "make-believe" propositions as true even when one knows that they are not, because making a proposition true of some world is inherent to the act of imagining, but pretending (i.e., "making it fictional") that one learns for the first time about the textual world does not erase propositional content from memory.

Repeated suspense, I would suggest, is not a matter of self-induced amnesia or of pretended ignorance but rather a matter of knowledge being superseded by a more urgent concern: the reader's emotional involvement in the fate of the hero. This suggestion predicts that of the four types of suspense described above, only the first and second (*what* and *how*) will lend themselves to repetition. We may indeed experience concern for the safe return of the crew of Apollo 13 many times over, but we normally do not reread mystery stories, because once we know who committed the crime we cannot relive the purely epistemic excitement of discovering the culprit. (If we do reread them, it is to locate the clues that we missed the first time.) By shifting the phenomenon of anomalous suspense from the epistemic to an emotional, almost existential plane I have not resolved the paradox but have linked it to another problem. To explain how we can repeatedly experience anxiety over destinies that are already written in our memories, we must first understand how we come to invest our desires in the fate of characters who never existed.

EMOTIONAL IMMERSION

Ever since Aristotle defined the effect of tragedy as catharsis, or purification through terror and pity, it has been taken for granted that literary fictions can elicit the same spectrum of emotional reactions in the reader as real-life situations: empathy, sadness, relief, laughter, admiration, spite, fear, and even sexual arousal. The tears amply shed by romantic or Victorian audiences over the fate of such characters as Young Werther and Little Nell throw the testimony of the body and the weight of physical evidence in support of this assumption. Emotional participation in the fate of imaginary characters was accepted as a natural response to literature until textualist approaches overtook realist paradigms and dissolved the human essence of characters into actantial roles or aggregates of textually specified features (or "semes").

In the heyday of structuralism and deconstruction, it became heretical even to mention the phenomenon of emotional response. Who would be moved to tears by the matrix [+ turbulence, + artistic gift, +

independence, + excess, + femininity (though + male)], as Barthes describes in *S/Z* the entity referred to by the name Sarrasine in Balzac's eponymous story?[3] Regarded as utterly unproblematic by one school and as illegitimate by another, the question of emotional response did not emerge as a topic worthy of serious theoretical consideration until the advent of the semantic approaches to fiction inspired by analytic philosophy. (In this class I include make-believe theories and speech-act and possible-worlds approaches.) For the theorists of textual worlds, characters possess among other aspects the mimetic dimension of pseudo–human beings.[4] While the human component of this hybrid dimension justifies emotional responses, the pseudo feature renders them problematic, since in order to attach themselves to imaginary individuals these responses have to cross the ontological boundary between the world of the reader and the fictional world.

As a preamble to the discussion of some of the solutions proposed to the paradox of emotional participation, let me take a closer look at the phenomenon itself. Describing her experience of Dickens, an informant says to Victor Nell, "I get very emotionally involved with people as such. I'm not visualizing [the scene], I'm not seeing it happening, but just the thought that this poor kid is getting into trouble for asking for more . . . upsets me so much that I can't watch it" (*Lost in a Book*, 293). The feeling expressed by this reader—which corresponds to Aristotelian "pity"—is a rather straightforward concept. Directed by the reader toward another person, it expresses empathy for the suffering of favorite characters, regret that their wishes did not come true, and personal revulsion for their fate ("Poor thing, I'd hate to live such a life"). There are good reasons, both pragmatic and stylistic, for the power of fiction to elicit this kind of response. Narratorial omniscience and the techniques of internal focalization allow a greater intimacy with the mental life of fictional characters than with the thoughts or emotions of real-life individuals. Whether or not we like to admit it, voyeurism has a lot to do with the pleasures we take in narrative fiction: where else but in a novel can we penetrate into the most guarded and the most fascinating of realms, the inner workings of a foreign consciousness? Though we are more likely to be moved by real events than by imaginary ones, we are also more likely to be affected by what happens to people we know than by the fate of strangers; and by virtue of the authority of fictional discourse we know certain fictional characters, such as Emma Bovary, better than they know themselves, perhaps even better than we can ever hope to know ourselves.

Another of Nell's informants, reacting to a Hitchcock movie, documents a feeling related to the Aristotelian concept of terror: "I get so frightened . . . I could die of fright. I get heart palpitations. . . . I get so scared I must sit next to someone" (244). This experience is much more problematic than pity because in normal circumstances we are doubly involved in fear: as the experiencer, but also as the object of the feeling. We feel pity for others, and this feeling can stretch from a position of security toward people located on the other side of a glass wall, such as people we read about or see in a picture; but in its literal sense, terror presupposes that we are directly located in the zone of danger. This obviously is not the case with even the most terrifying stories and pictures: there is no chance that a vampire will jump from the pages of a book or the screen of a movie theater and bite us on the neck. The probability would be greater in the theater, but it is still negligible for adult audiences.

What, then, do people mean when they describe a novel or movie as "scary"? It is obvious that different people use the adjective for different experiences. To explore the range of these phenomena I asked the members of an electronic discussion group on narrative—all academics, and therefore professional readers—to give examples of fearful reading or movie-watching experiences and to try to explain the feeling in terms of the relation between the self, the real world, and the fictional world. Here is a sampling of the responses:

1. "Last summer, I took my then-three-year-old son to see a children's performance of 'Little Red Riding Hood,' a story that he had loved until then and had never feared. However, when he was confronted with 'real' people representing the characters, he was confused and frightened. The climax (for us) came when the wolf stepped off the stage as he was singing and sashayed down the central aisle; this proximity was simply too much for my son to handle, and we were forced to leave without ever finding out whether the wolf was truly bad or not! My son does not seem to have that problem with videos or books; it was the 'realness' of the actors/actresses

that created confusion for him, as it forced him to recon-
sider the usual lines between 'fact' and 'fiction.'" (Janet Galligani Casey)

2. "When I was in kindergarten, a foolish parent took all the
   children at her child's birthday party to a horror movie. I
   had recurrent nightmares for two years. I can still remember
   the fear I felt imagining that the monster, which had erupted
   out of the mountain, was about to step on me." (Deborah
   Martinsen)

3. "I like to be made to cry, but I hate being scared. . . . The
   specific fears that I am always trying to avoid are very pat-
   ently death-related (dead bodies, skeletons, buried-alive scenarios, etc.); they seem to be very basic fears and denials of
   mortality." (Antje Schaum Anderson)

4. "[I must also mention] my experience of violent vertiginous
   terror while seated in the Boston IMAX theater watching
   their introductory promo movie, which is shot from a
   small-plane view and features a lot of darting up and down
   between tall buildings, sudden dips toward bridges, etc."
   (Caroline Webb)

5. "What really terrifies me is always something that at first
   seems to be far from everyday life but . . . soon turns out to
   be something that could happen to everyone, myself included." (Edina Szalay)

6. "When reading novels . . . the fear I experience is more often
   anxiety that the characters will jeopardize their futures. . . .
   My fear could best be described as protective to the characters, but it has also quite definitely been fear for myself."
   (Caroline Webb)

7. "I think the key [to fear] is identification—the stronger the
   identification with a character or situation, the more the
   borders between the fiction and reality, self and other, tend
   to collapse." (Charlotte Berkowitz)

8. "As a teenager I remember pleading with the screen while
   watching Zeffirelli's *Romeo and Juliet*—oh please, don't follow the script." (Deborah Martinsen)

9. "Although I haven't read the novel for several years, every
   time I reread *Tess of the d'Urbervilles* I was hoping it would

turn out differently, that Angel Clare wouldn't be such a jerk. I knew the plot, but every time I reached one of the several crucial scenes, I would find myself hoping it would turn out differently this time." (Laura Beard)

10. "In *Native Son* [by Richard Wright], there is a moment when Bigger Thomas is trapped in Mary Dalton's room by her mother (who is blind). The fear he feels leads him to ac-cidentally smother Mary. The scene evokes a powerful sense of fear in readers (myself included). I wonder whether this scene moves us to fear by analogizing to the idea or experi-ence of [a] similar, though not identical, sense of entrap-ment." (Ted Mason)

This last post awakened in me the following literary reminiscence:

11. "One of the passages in fiction that I find the most scary is a scene from Emile Zola's *Thérèse Raquin*. It also plays on the idea of entrapment. Thérèse and her lover have murdered Thérèse's husband and now live together with the mother of the victim. The mother-in-law comes to realize who has murdered her son, but she suffers a stroke and is unable to speak or move. The end of the novel describes the deterio-rating relationship and growing torture of the two lovers as they suffer the unbearable presence of the silent, hostile gaze of that woman trapped in her own body. I read it as a teen-ager, and I still shudder when I replay it in imagination."

Many of these examples can be explained as a collapse of the distinc-tion between the real and the fictional world. In 1 and 2, ontological boundaries are perceived as permeable, and the real-world self be-lieves itself to be exposed to the dangers of the fictional world. As the various examples indicate, this experience affects mostly children. Psychologists have observed that while children learn at a very early age (before four) the difference between make-believe and reality, the consequences of the distinction are slower to sink in. Young children "still remain unsure of the rules that govern transformation between [the realms of fantasy and reality]" (P. L. E. Harris et al., quoted in Gerrig, *Experiencing Narrative Worlds*, 193). In example 3, the reader's or spectator's fear is due to the repulsive character of the object represented. More prominent in visual media, and closely related to the sexual arousal of the consumer of pornography, this type of fear is dependent on the real-life (or illusionist) quality of the representation, on its ability to erase the distinction between image and reality. The reaction described in example 4 is due to a property of the image that creates a very physical experience of dizziness, vertigo, loss of footing in the real world. When a movie camera is mounted on a moving object—roller coaster, racing car, or airplane—the spectator is projected as a passenger of the vehicle and becomes a potential victim of a crash. The purely imaginative possibility of smashing into the ground creates a sensation closely related to the thrill of riding a roller coaster: an adrenaline rush that represents both fear and a very physical form of pleasure. In examples 5 and 6 the fictional world is apprehended as a possible world in the literal sense of the term: if it can happen in the textual world, thinks the reader, it can happen in reality, and if it can happen in reality, it can happen to me.[5] This cause of fear is the most rational of the list, but in 5 it is linked to more personal, subconscious anxieties, and in 6 it combines with the phenomenon of empathy for characters. Fear directed toward others is also illustrated in examples 7, 8, and 9, leading in 8 and 9 to anomalous suspense. In examples 10 and 11, finally, as also in 3 and 5, the text awakens personal phobias (sense of entrapment) and existential anguish (fear of death, of darkness, of the Other, of helplessness).

For philosophically oriented theorists of fiction, the truly problematic cases of emotional immersion are those that do not involve a blurring of ontological boundaries. Logicians tend to dismiss examples 1 and 2 as experiences due to immaturity—though cognitive psychologists might see in them important documents of the functioning of the imagination—and 4 to 6 as a tricking of the senses. Paraphrasing Gregory Currie, who restricts his analysis to the case of empathy for characters, we can state the paradox of emotional participation as follows:

I. We have emotions concerning fictional situations.
II. To have an emotion concerning a situation, we must believe the propositions that describe this situation.
III. We do not believe the propositions that describe the situations represented in fiction.

As Currie observes, these propositions are incompatible: "The problem is to decide which one [should be removed], and what to replace it with" (*Nature of Fiction*, 187). Rejecting III would conflict with the basic tenets of semantically based theories of fiction; and rejecting II would mean that emotion does not require belief in the existence of its object—an outright denial of what Currie calls the cognitivist conception of emotion. Since both of these positions are unacceptable to Currie, as they are to most philosophers, the only alternative left is to reject proposition I. For Currie, emotions are nexuses of relationships between feelings (sadness, happiness), beliefs, and desires: for instance, my belief that a fire destroyed a certain forest, coupled with my desire for the maintenance of the beauty of this forest, will result in sadness. But our propositional attitudes toward fictional characters and situations cannot be beliefs and desires, since we know that they do not exist. Currie proposes to resolve the paradox of emotional reactions by distinguishing two parallel brands of experience: in real life we have beliefs and desires leading to emotions; in fiction we have make-beliefs and make-desires leading to "quasi-emotions." But what do we learn from this seemingly ad hoc sprinkling of prefixes? Given the formulation of proposition I, its rejection in favor of the "quasi" version is a foregone conclusion. If the statement "We have emotions concerning fictional situations" is taken to mean that our feelings regarding fictional situations affect us in exactly the same way as our feelings regarding real-world states of affairs, we can easily dismiss the paradox by observing that the emotions generated by fiction do not prevent pleasure, and do not lead to the same reactions as feelings generated by real situations. As Walton has observed, crying for characters does not make us sad, and a spectator's fear, whatever causes it, does not make him flee the theater.

The phenomenon that really needs to be explained is the conjunction of these two observations:

Ia. We experience emotions regarding fictional situations that can be intense enough to lead to physical symptoms.[6]

Ib. These emotions do not have their normal consequences, and as long as they are not too violent, they do not inhibit pleasure.

In contrast to the philosophical school's loyalty to the cognitivist position—rejecting proposition I in order to maintain II—psychologists are much more inclined to endorse some version of I and to question the validity of a rigid interpretation of II. In the cognitivist approach, Ib is easily explained, but Ia remains problematic, because the reader's attention is split between two worlds, and he is never deeply immersed in the text. Shifting priorities toward the explanation of Ia, at the expense of Ib, the psychological school credits emotional responses to the power of the imagination to push the reader's or spectator's beliefs into the background of consciousness.

Gerrig explains the phenomenon of emotional responses for nonexisting characters through an independently documented psychological principle that he calls the "nonpenetration of belief into emotional experience" (*Experiencing Narrative Worlds*, 181). This principle has been invoked by psychologists to explain the case of phobias: "The beliefs that dominate the appraisal of snakes or spiders at a distance are not able to penetrate the mental processes that produce the extreme emotional responses when the subjects are at hand. Such clinical instances demonstrate a clear capacity for individuals to experience strong emotions that are not ameliorated by beliefs" (183). While the purely cognitive approach regards genuine beliefs as always present to the appreciator's mind, even in the midst of make-believe, the psychological approach suggests that the availability of beliefs is subject to variations. This variability accounts for the difference between the various types of fear documented above. Rather than treating children's reactions of fear, such as the cases illustrated in examples 1 and 2, as qualitatively different from the mental setup of the adult reader who cries for Anna Karenina, this approach explains the difference between pleasurable and painful emotional responses in terms of the intensification of the same process. In pleasurable responses, beliefs are merely backgrounded; in the case of the terrorized child, the force of the imagination makes them temporarily inaccessible.

In a discussion of the related phenomenon of anomalous suspense—in which knowledge, as we have seen, is also backgrounded—Noël Carroll proposes a logical explanation of participatory responses to fiction that complements the psychological approach. He bases his analysis on a distinction between two modes of considering (and offering) propositions. In the mode he calls belief, propositions are held before the mind as asserted; in the mode he calls thought, propositions are contemplated but not asserted. This is the mode of fictional representation: "The author, in presenting his or her novel as a fiction, in effect, says to readers 'hold these propositions before your mind unasserted'—that is, 'suppose p,' or 'entertain p unasserted,' or 'contemplate p unasserted.'" ("Paradox," 85). (Alternatively, the author could be saying, "Contemplate these propositions as asserted by the narrator.") Emotional responses, however, are not sensitive to the distinction between asserted and unasserted propositions: "One can engender emotional states by holding propositions before the mind unasserted. Thus, when I stand near the edge of the roof of a high building and I entertain the thought that I am losing my footing, I can make myself feel a surge of vertigo. I need not believe that I am losing my footing; I merely entertain the thought" (ibid.).

Carroll regards this capacity of the human mind to be emotionally affected by the contemplation of purely imaginary states of affairs as an evolutionary asset that works toward the preservation of the species. As Walton also argues ("Spelunking"), it is on the playground of mental simulation, where emotions happen but do not count, that we learn about our own feelings and prepare ourselves for the trials of life. In the case of fictional suspense, both first-time and repeated, the reader contemplates the unasserted proposition that the hero is in danger, and the intensity of this contemplation generates the emotional response of anxiety. What Carroll is saying here is that to the simulating mind, it does not matter whether the envisioned state of affairs is true or false, and its development known or unknown, because simulation makes it temporarily true and present, and from the point of view of the present, the future has not happened. I can, for instance, relive many times in imagination an accident in which I was almost killed and experience the same terror every time. The power of the imagination to make situations present to the mind is dramatically exemplified by the sexual arousal of consumers of pornographic literature. Readers of pornographic novels can be sexually stimulated by the depiction of a sex act even though they know that it was all made up, because what arouses them is the vividness of their mental representation of the scene, which itself is a response to the vividness of the textual description. In the case of other types of fiction, similarly, the presence of the scene to the mind creates emotional responses that lead to an adrenaline rush, a sensation we enjoy in certain doses, but the backgrounded knowledge that it was all made up and does not affect a human life holds the dosage within the limits of pleasure.

IMMERSION AND REALISM

The features described in this chapter belong, overwhelmingly, to the narrative modes of the nineteenth-century novel, a period style and literary movement widely known as realism. Does this mean that immersion is dependent on a realist configuration of the textual world? It all depends on what we mean by *realism*. At the risk of oversimplifying a complex philosophical and semiotic problem, we can reduce the issue of literary realism to the following options:

1. *The correspondence interpretation.* A realistic text is one that truthfully represents how things are in the real world. This view presupposes the currently discredited stance of philosophical realism: reality is at least partly accessible to the mind, and it can be represented through man-made systems of signs. Since fictional texts obviously cannot satisfy this definition, it must be replaced by definition 2 for the literary domain.

2. *The probabilistic, Aristotelian interpretation.* "It is also clear from what has been said that the function of the poet is not to say what *has* happened, but to say the kind of things that *would* happen, i.e. what is possible in accordance with probability and necessity" (*Poetics* 5.5, 16). In this view, a realistic text is one that depicts situations that could be actualized in the real world. For most readers this requirement means that the textual world does not transgress physical and logical laws, that it respects some basic conceptions of psychological and material causality, and that the plot does not overly rely on events of low probability, such as extraordinary coincidences.

3. *The illocutionary conception.* A literary text is realistic if it reproduces a speech act or discourse genre of real-world communication, and if it is entirely made of sentences whose

felicity conditions could be fulfilled by a real human being. This type of realism is illustrated by the fake diaries, letters, and memoirs of the eighteenth century, but it excludes the "unspeakable sentences," to use Ann Banfield's description, that underlie most of the narrative techniques of nineteenth-century fiction. Illocutionary realism presupposes an embedded act of communication that fulfills literally the schema of real-world communication: the author S_1, located in the real world, becomes in make-believe the narrator S_2 who informs the narratee H_2, fictional counterpart of the reader H_1, of events that took place in the fictional world.

4. *The illusionist conception.* A text is realistic when it creates a credible, seemingly autonomous and language-independent reality, when the style of depiction captures an aura of presence, when the reader is imaginatively part of the textual world and senses that there is more to this world than what the text displays of it: a backside to objects, a mind to characters, and time and space extending beyond the display. This type of realism is indifferent to the type of objects represented—like surrealistic paintings, it can give solidity to dream worlds and existence to impossible objects—and it is entirely dependent on the mode of representation. Its style must offer the verbal equivalent of three-dimensionality, sharpness of lines, and fullness of detail in visual media. It must occasion what Walton calls a "rich" game of make-believe *(Mimesis)*.

It should by now be obvious that if the poetics of immersion presupposes a realist framework, it is a realism of this last type (which in itself does not exclude the others). The nineteenth-century novel was conceived by its authors as realistic in sense 2, and many of its authors supported the philosophical stance mentioned in 1, but it is its endorsement of 4 at the expense of 3 that accounts for its power to call worlds into presence and characters into life, and for the enduring nostalgia of modern readers for its storytelling mode (a nostalgia cunningly exploited by those postmodern authors who parody Victorian style while maintaining an ironic distance toward any mimetic claim).

The "reality effect" of nineteenth-century fiction is achieved by the least natural, most ostentatiously fictional of narrative techniques—omniscient narration, free indirect discourse, and variable focalization, all of which presuppose morphing narrators whose manifestations oscillate between embodied, opinionated human beings and invisible recording devices; but the discrepancy between mode of narration and mimetic claim wasn't noticed until the development of narratology and linguistic pragmatics, because the mode of functioning of these techniques was to pass as spontaneous self-inscription of events. Language has to make itself invisible in order to create immersion. It is this yearning for a fully transparent medium that led the early theorists of the novel, such as Percy Lubbock, to posit "a convenient distinction between 'showing,' which is artistic, and 'telling,' which is inartistic" (Booth, *Rhetoric of Fiction*, 8). Or as Lubbock himself puts it, "The art of fiction does not begin until the novelist thinks of his story as a matter to be shown, to be so exhibited that it will tell itself" (quoted ibid.).

The divorce of the narrative style of the nineteenth-century novel from the modes of expression of factual narrative—historiography in particular—helps explain the current tendency to regard realism as "another set of conventions," in themselves no closer to "the real" than any other period style. But what does contemporary literary theory understand by *convention?* In a weak, rather obvious sense, the term can be taken to mean that most writers of the period known as realism adopted a certain repertory of storytelling strategies (the so-called period style), and that these strategies are relative because stories could be told in a different way (nonimmersively, for instance) or because literature could avoid telling stories altogether. This interpretation is hardly controversial, and as such not particularly interesting.

Another reason why the narrative strategies of realism may be regarded as conventional is that every period has a different view of reality, and every stylistic innovation is made in the name of this interpretation. Along this line of thought, a correlation could be made between the importance of causality in nineteenth-century worldviews and the prominence of motivation in the plots of that period, or between the late-twentieth-century belief that reality does not "present itself to perception in the form of well-made stories, with central subjects, proper beginnings, middles and ends" (White, "Value of Narrativity," 23) and the dismantling of narrative form in the post-modern novel.

In the most radical view of convention, the fact that conventions "could be different" is hyperbolically interpreted as meaning that they are arbitrarily chosen, and that their particular function could be performed just as efficiently by any other sign, technique, or behavior. The designation of the stylistic repertory of realism as a set of fully arbitrary conventions would mean that the power of certain devices to convey the impression "this is real" resides entirely in cultural habit and has nothing to do with the intrinsic effect of these devices on the imagination. In this line of thought, the cultural spread of narrative techniques or stylistic devices would be totally unrelated to their ex-pressive power. This approach to representation has led some phi-losophers to deny any superior lifelikeness to the use of perspective in painting. According to Nelson Goodman ("Reality Remade," 293), the representation of a building in perspective would be as unnatural to a Fifth Dynasty Egyptian as an Egyptian representation of the human body is to a member of modern culture. For Goodman, realism is a matter not of resemblance but of ease of decoding, and this ease is explained by the reader's or spectator's familiarity with a certain set of representational techniques. But once perspective has been intro-duced into art and people have learned to read it, could a return to two-dimensional representation ever pass as truer to life, or is per-spective inherently better suited to create a reality effect?[7] The same question can be asked of narrative techniques. Why, one may want to ask, has the diction of the nineteenth-century novel survived almost intact in today's popular literature, whose appeal depends almost entirely on immersivity, while "high" literature, pursuing more diver-sified ideals and often contemptuous of immersion, dares to use these modes only under the protection of scare quotes and ironic distance?

One author who strongly rejects the conventionalist approach to realism is Tom Wolfe, the vocal advocate of the New Journalism:

> Novelists have made a disastrous miscalculation over the past twenty years about the nature of realism. Their view of the matter is pretty well summed up by the editor of the *Partisan*

Review, William Phillips: "In fact, realism is just another formal device, not a permanent method for dealing with experience." I suspect that precisely the opposite is true. If our friends the cognitive psychologists ever reach the point of knowing for sure, I think they will tell us something of that order: the introduction of realism into literature by people like Richardson, Fielding and Smollett was like the introduction of electricity into machine technology. It was not just another device. It raised the state of the art to a new magnitude. The effect of realism on the emotions was something that had never been conceived before. ("New Journalism," 34)

The literary effect that was pioneered by Richardson, Fielding, and Smollett, and perfected in the next century by the likes of Balzac, Dickens, and Gogol (to remain with Wolfe's examples), is clearly neither the art of revealing "how things are" nor the art of imitating real-world speech acts but the art of getting the reader involved in the narrated events.

When *Madame Bovary* was put on trial in 1857 on charges of outrage to public morality and religion, one of the major points of contention was the nature of Flaubert's realism and its effect on the reader. The defense lawyer, Maître Sénard, argued that the text was not immoral even though it depicted behavior that was considered sinful, because this behavior was justly punished by the painful death of the sinner. For Sénard, "[Flaubert] belongs to the realistic school only in the sense that he is interested in the real nature of things" (*Oeuvres,* 636, trans. in LaCapra, *Madame Bovary on Trial,* 45). The "real" social problem exposed by the novel, according to Sénard, was the danger of providing young women destined for a bourgeois life with an education that would fan dreams of a life reserved for the higher classes.

The prosecution lawyer, Maître Pinard, countered this banalizing interpretation with an immersive conception of realism that makes the novel far more dangerous because it undermines the apparent morality of the ending. The vivid description of the heroine's forbidden desires and "lascivious" satisfactions enables the reader to participate in the pleasure of adultery, and these moments of *jouissance,* no matter how short-lived, cannot be eradicated by any punishment: "When the imagination will have been seduced, when seduction will have descended into the heart, when the heart will have spoken to the senses, do you believe that a very cold reasoning will be very strong against this seduction of the senses and of sentiment?" (*Oeuvres,* 631–32; trans. in ibid., 38).[8] In this argument the prosecution lawyer demonstrated a much keener awareness of the mechanisms of emotional/corporeal participation than his opponent. We may disagree with the purpose of this speech—literary censorship—and with its conception of immorality, but we must give Maître Pinard credit for understanding how the imagination, energized by the evocative power of literary style, can lure the body into its own spectacles.

Virtual Narration as
Allegory of Immersion

Virtual narration, as I propose to define the term, is a way of evoking events that resists the expectation of reality inherent to language in general and to narrative discourse in particular. Philosophy may periodically relativize, destabilize, and even reject the notion of reality, but narrative and expository language know little of these doubts: even in an atmosphere of radical antirealism, such as the contemporary *Zeitgeist,* language remains firmly rooted in truth and reality. The unmarked case of modality is the indicative, and to narrate in the indicative is to present events as true fact. The repertory of semantic categories at the disposal of narrators—or essayists, for that matter—often forces the writer to a firmer commitment to facts than caution would call for. It is through tacked-on modal markers that language defactualizes, relativizes, or switches the reference world from the speaker's reality toward a nonactual possible world. In the type of narration I call real, the narrator presents propositions as true, and the audience imagines the facts (states or events) represented by these propositions. As long as the narrator (or implied speaker) uses the indicative mode, the reference world is identified as the world in which this narrator is located. The real mode of narration is found in both fiction and nonfiction and is independent of the truth value of the discourse: even the false can be told as true fact; otherwise lies would never deceive and errors never mislead.

But the real mode is not the only way to evoke events to the imagination. With appropriate markers of irreality, the reference of sentences can be diverted to the realm of the counterfactual, hypothetical, or merely possible, or to one of the numerous domains of mental life: dreams, beliefs, desires, goals, and commitments. Events may even be called to the imagination as nonexistent. The processing of a negative sentence—for instance, "Mary did not kill her husband"—

involves imagining the world in which Mary actually kills her husband. In the narrative mode I call virtual, the expectation of reality is loosened by an effect that involves the optical interpretation of virtuality. States and events are evoked indirectly, as they are captured in a reflecting device that exists as a material object in the textual world. This reflecting device could be a mirror, story-within-the-story, photograph, movie, or television show. By describing the material support of the representation—a support that functions as primary discourse referent—through such rhetorical devices as ekphrasis, paraphrase, or summary, the speaker conjures images of the reflected world. When these images form a story, the speaker indirectly produces a narration, or the effect of one, even though his discourse is focused on something other than the narrated events. To the extent that the reader of the virtual narration is able to reconstrue these events, virtual narration becomes "as good as" real narration while remaining pragmatically distinct. This phenomenon activates another lexical meaning of the term *virtual:* "for all practical purposes" (as in "virtual dictator").

As an example of virtual narration, consider this description of an engraving titled *The Battle of Reichenfels* in Alain Robbe-Grillet's novel *In the Labyrinth:*

> The picture, in its varnished wood frame, represents a tavern scene. It is a nineteenth century etching, or a good reproduction of one. A large number of people fill the room, a crowd of drinkers sitting or standing, and, on the far left, the bartender standing on a slightly raised platform behind the bar. (150)

As the picture is being read, however, a scene emerges in greater and greater detail; the signs of "pictoriality" gradually disappear, and the scene acquires a life of its own that increases its vividness. Though the following passage still describes the same picture, it does so in a mode that can pass as real:

> The soldier, his eyes wide open, continues to stare into the half-darkness a few yards in front of him, where the child is standing, also motionless and stiff, his arms at his side. But it is as if the soldier did not see the child—or anything else. He looks as if he has fallen asleep from exhaustion, his eyes wide open. (153)

The transition from virtual to real narration is not a return to the primary level of reality, where the picture exists as object, but rather a recentering into the world of the picture. Freed from its flat and static nature, this world now becomes animated, sustains narrative action, and takes over as textual actual world. We know that we have left the primary textual world for good and that the recentering is complete when we read, "It is the child who speaks first. He says: 'Are you asleep?' " (ibid.).

As a technique of representation, virtual narration is rarely maintained for more than a few paragraphs. A case in point is Julio Cortázar's very short (three-page) story "Continuity of Parks." This mind-bending text has been described by Brian McHale as a "strange loop, or metalepsis" (*Postmodernist Fiction*, 119–20; the term *strange loop* is borrowed from Douglas Hofstadter's *Gödel, Escher, Bach,* and *metalepsis* from Gérard Genette). As a strange loop, the story presents the literary equivalent of the Escher lithograph *Print Gallery,* in which a spectator contemplates a painting of a town that loops back upon the gallery and includes the spectator:

> A man reads a novel in which a killer, approaching through a park, enters a house in order to murder his lover's husband—the man reading the novel! The "continuity" in this text is the paradoxical continuity between the nested narrative and the primary narrative, violating and thus foregrounding the hierarchy of ontological levels. (McHale, *Postmodernist Fiction*, 120)

What McHale does not discuss is the technique through which the hierarchy is violated. At the beginning of the story we are told that the character on the primary level—let us call him "the man (or reader) in the armchair"—has been reading a novel for a few days and has permitted himself "a slowly growing interest in the plot, in the characterization" ("Continuity," 63). At this point the book is still an opaque object, described by means of literary metalanguage ("novel"). The real reader does not see the world of the novel until a virtual narration begins. The development of this narration parallels the gradual estrangement of the reader-character from his material surroundings:

> The novel spread its glamour over him almost all at once. He tasted the almost perverse pleasure of disengaging himself line

by line from the things around him. . . . Word by word, licked up
by the sordid dilemma of the hero and heroine, letting himself
be absorbed to the point where the images settled down and
took on color and movement, he was witness to the final encounter in the mountain cabin. (ibid.)

"He was witness": the man is now recentered into the fictional world;
but we real readers have access to this world only through its reflection in the man's consciousness. The virtual narration is less the
description of a book than the description of an act of reading.

Despite his absorption in the plot, the man remains appreciative of
the narrative art: he is able to combine a mimetic and a semiotic
perspective. The fictional world is experienced as both a reality and a
fabrication:

A lustful, panting dialogue raced down through the pages like a
rivulet of snakes, and one felt as if it had all been decided from
eternity. Even to those caresses which writhed about the lover's
body, as though wishing to keep him there, to dissuade him from
it; they sketched abominably the frame of that other body it was
necessary to destroy. Nothing had been forgotten: alibis, unforeseen hazard, possible mistakes. It was beginning to get dark. (64)

As a fabrication, the fictional world is perfectly planned: "Nothing has
been forgotten." Yet as a reality it is full of "unforeseen hazards." The
virtual mode of narration emphasizes this dual perspective through
its unique ability to filter the reflected world through the reflecting
medium. In the last paragraph, however, the narration reverts to the
real mode, and the reader-character loses sight of the text as textuality.
The screen of his mind is now entirely occupied by the characters and
their actions:

Not looking at one another now, rigidly fixed upon the task
which awaited them, they separated at the cabin door. She was
to follow the trail to the north. On the path leading in the
opposite direction, he turned for a moment to watch her running, her hair loosened and flying. (65)

As the future murderer enters an estate and sees a man in an armchair,
the man reading the novel has become so deeply absorbed in his

reading that the world of the novel becomes the only reality: the two-level ontology of Cortázar's story has collapsed into a single level. This collapse becomes evident when the last sentence ends the story rather abruptly:

> The door of the salon, and then, the knife in hand, the light from the great windows, the high back of an armchair covered in green velvet, the head of the man reading the novel. (ibid.)

Why isn't the murder narrated? Because the reflecting object is the consciousness of the man in the armchair, and consciousness is terminated by death. Virtual realities, be they text-created or computer-generated, are normally safe environments for the experiencer, but the parable of "Continuity of Parks" suggests that when they are lived too fully they are no longer protected from death. For those who cannot breathe under water, total immersion leads to a fatal deprivation of oxygen.

In Cortázar's story, immersion comes easily to the fictional reader. When he opens the book, he has already read most of the text, he is familiar with the fictional world, and he nears the climax in the action. If I can judge by my own experience, it takes much more time to read the first fifty than the last fifty pages of a traditional narrative, because during those first fifty pages the reader must construct characters, setting, and motivations, while in the last fifty pages she can harvest the fruit of this labor. Once the fictional world is in place, it seems to evolve by itself. For an account of the struggle that precedes immersion, let us turn to Italo Calvino's *If on a Winter's Night a Traveler.*

There cannot be any better way to introduce this work than to quote a virtual narration that constitutes a *mise-en-abyme* (miniature replica) of the whole plot. Toward the end of the novel one of the characters, the writer Silas Flannery, describes a literary project:

> I have had the idea of writing a novel composed only of beginnings of novels. The protagonist could be a Reader who is continually interrupted. The Reader buys the new novel A by the author Z. But it is a defective copy, he can't go beyond the beginning. . . . He returns to the bookstore to have the vol-

ume exchanged. I could write it all in the second person: you, Reader. . . . (198)

The same scenario is repeated throughout the novel, generating a tale of frustrated immersion: as soon as a fictional world begins to solidify around the reader, he is expelled from it and must start all over again with a new book.

In the first of these embedded novels, whose title is the same as that of the main book, the disorientation of the reader and his efforts to find his way in a strange world are reflected within the fictional world itself through a virtual narration that retraces the formation of images in the mind of the reader. Because its operations are foregrounded, the reading mind is objectified within the text as a visible reflecting surface. We see not only *what* it reflects but also the active process by which it captures textual input and translates it into a representation.

In contrast to the situation in Cortázar's story, the reader is not merely a character in the primary fictional world, reading a book about a secondary reality. Rather, he is a "you" involving what David Herman calls a "double deixis" ("Textual You and Double Deixis"). On one hand, this "you" refers to a character in the main plot reading the book (he will turn out to be male and have a love affair with a female reader, Ludmilla); but on the other hand, the deixis extends to the real reader outside the fictional world. You—and I—are made witnesses of the mental operation through which we form the mental representation of a fictional world. We get to know this world, as we are told how to construct it:

> The novel begins in a railway station, a locomotive huffs, steam from a piston covers the opening of the chapter. . . . All these signs converge to inform us that this is a little provincial station, where anyone is immediately noticed. Stations are all alike; it doesn't matter if the lights cannot illuminate beyond their blurred halo, all of this is a setting you know by heart. (10–11)

The process of world construction involves the activation of familiar frames of knowledge and the import of real-world experience. Despite the text's reluctance to yield information, the reader should be able to form a representation of the setting because "stations are all alike," and this is a typical station. But the text does its best to frustrate the process: "The lights of the station and the sentences you are reading seem to have the job of dissolving more than of indicating the things that surface from a veil of darkness and fog" (11). Ironically, by tracing step by step the emergence of the fictional world in the mind of the reader, the text prevents this emergence. The fictional world remains partially hidden behind the activity that constructs it.

Up to this point the virtual narration could be read as the representation of the act of reading a novel from a point of view external to the world of this novel, as was the case in the Cortázar story. The third-person narrator of "Continuity of Parks" is located in the same world as the man in the armchair—namely, the primary level of reality within the global fictional universe—and he represents the mind of the reader by making use of narratorial omniscience. If the "you" of Calvino's text designates the actual reader, then the implied "I" who describes the text as text should normally refer to the author of the novel. On the other hand, if the "you" is read as a character you, the "I" should be read as referring to a narrator individuated as author-persona. Communication presupposes that sender and receiver be members of the same world. But as an imaginative construct, Calvino's text is free to challenge the ontological basis of communication. The speaker is neither the author nor an author-character but the protagonist of the novel: "I have landed in this station tonight for the first time in my life. . . . I am the man who comes and goes between the bar and the telephone booth. Or, rather, this man is called I and you know nothing about him" (11). The description of the reflection of the fictional world in the reader's mind thus originates in the fictional world itself. The narrator's ability to read the reader's mind creates an ontological paradox, not so much because it transgresses ontological boundaries—after all, authors have access to the minds of their characters—but because it transgresses them in the wrong direction: characters are not supposed to be aware of readers. If the real world is level 0 on the narrative stack and the primary fictional world level 1, then we have a narrator at level 1 who describes the reflection of the world of level 1 in the mind of a member of level 0.[1] In Cortázar's story, by contrast, level 0 is not involved. A narrator on level 1 describes the mind of a reader on level 1, which reflects a fictional world of level 2.[2]

As the text progresses, the passages of virtual narration that present the narrator and the fictional world as reconstructed by the reader ("something must have gone wrong for me" [13]) become sparser, until they are totally displaced by real narration ("I walk along the platform" [24]). This should come as no surprise: as the text yields more information about the setting and characters, whether in the real or the virtual mode, the fictional world becomes less puzzling and the characters better profiled. Finally at home in the fictional world, the reader can follow the action without asking new questions with every new sentence. The fog has been lifted, immersion is complete, reading from now on will become easy, but the challenge may be over. For, as Ludmilla, Calvino's Dream Reader, states in the main plot:

> Reading is going toward something that is about to be, and no one yet knows what it will be. The book I would like to read now is a novel in which you sense the story arriving like still-vague thunder. (72)

> If the first effect is fog, I'm afraid the moment the fog lifts my pleasure in reading will be lost, too. (30)

Thanks to the defect occurring in the material copy of all the embedded novels, however, the pleasure of the story arriving never yields to the disappointment of the story arrived.

In both Cortázar's and Calvino's texts, virtual narration functions as a distancing device that emphasizes the textual nature of the represented reality. It locates the reader on a lower level than the world focused upon, thus preventing recentering into this world. This anti-illusionist, anti-immersive interpretation of virtual narration is justified as long as the mode is maintained. But as we have seen, the device hardly ever dominates a text from beginning to end. Through its very instability, virtual narration subverts its own self-referentiality. The tendency to revert to regular narration is so strong that the reader usually fails to notice the transition to the real mode. We can no more observe the stages of our own immersion than we can watch ourselves falling asleep. It is only retrospectively, like a person who awakens from a dream, that the reader realizes that the virtual image has come to be experienced as primary reality. Immersion cannot be reflected upon from within immersion—this would amount to destroying it—but it can be forcefully enacted by the text from a state of distanced contemplation. In this enactment, virtual narration functions as launching pad, not as destination. Its fate is to fade into real narration, so that immersion can be lived as well as allegorized.

PART III   The Poetics of Interactivity

From Immersion to Interactivity

The Text as World versus the Text as Game

SIX

"There's glory for you!" [said Humpty Dumpty.]
    "I don't know what you mean by 'glory,' " Alice said.
    Humpty Dumpty smiled contemptuously. "Of course you don't—till I
tell you. I meant 'there's a nice knock-down argument for you!' "
    "But 'glory' doesn't mean 'a nice knock-down argument,' " Alice
objected.
    "When I use a word," Humpty Dumpty said in a rather scornful tone,
"it means just what I choose it to mean—neither more nor less."
    "The question is," said Alice, "whether you can make words mean so
many different things."
    "The question is," said Humpty Dumpty, "which is to be master—
that's all."   —LEWIS CARROLL

As we have seen in the introduction, opposition to immersion—in both its literary and technological versions—runs rampant in contemporary criticism. The experience requires a transparency of the medium that makes it incompatible with self-reflectivity, one of the favorite effects of postmodernism. Sven Birkerts captures in no uncertain terms the phenomenon of the disappearance of signs: "We project ourselves at the word and pass through it as through a turnstile. . . . A reader in the full flush of absorption will not be aware of turning words into mental entities" (*Gutenberg Elegies*, 81). In an age that regards signs as the substance of all realities, this traversal of language is a semiotic crime: whatever "freedom from signs" the mind can reach is achieved not through their disappearance but through the awareness of their omnipresence and through the recognition of their conventional and arbitrary character. Signs must be made visible for their role in the construction of reality to be recognized. The idea of transparency tends to be interpreted as a denial of the importance of the medium in what can be expressed and represented. If the medium is transparent, so the argument goes, the medium does not matter.

On the contrary, I would like to argue that the disappearing act of the transparent medium is not a lack of autonomous properties but a hard-won and significant property that plays a crucial role in shaping the experience of the appreciator. It matters crucially that some media, and some representations within a given medium, achieve greater transparency than others. The traversal of signs is to be deplored only when it causes signs to vanish permanently, when immersion is so deep that it precludes a return to the surface. If there is such a thing as a "truth universally acknowledged" by literary theorists, this truth is that attention to the rhetorical devices through which a world emerges out of words is an essential aspect of aesthetic appreciation. To restore contact with the surface, we need an alternative to the metaphor of the text as world that complements, rather than invalidates, the poetics of immersion. This chapter is devoted to the most prominent of the possible candidates.

THE TEXT AS GAME

Textual worlds reached their greatest expansion and maximal consistency in the novels of Balzac, Dickens, Tolstoy, Dostoevsky, and Proust. Then they began to collapse under their own weight. In the second half of the twentieth century, a process of shrinking, fissuring, splitting, and multiplying worlds within a larger textual universe reduced big worlds to little worlds or dismantled them into heterogeneous fragments. Their scattered remnants could no longer build a coherent imaginary space and time, but they provided the perfect material for play. This is how the metaphor of the text as world came to be supplanted by the metaphor of the text as game, not only as the dominant aesthetic guideline for the production of texts but also as a critical paradigm that promoted a rereading of the texts of the past.

The concept of game is one of the most prominent in twentieth-century thought. The list of its manifestations covers such various projects and phenomena as Johan Huizinga's groundbreaking study *Homo Ludens*, Roger Caillois's taxonomy of games, Ludwig Wittgenstein's notion of language games, Hans Vaihinger's philosophy of "as if," John von Neumann's development of the mathematical field of game theory, Jaakko Hintikka's "game-theoretical semantics," Jacques Derrida's doctrine of the play of signs, Vladimir Nabokov's and Italo Calvino's fascination with combinatorics, the Oulipo movement's promotion of games as literary structures, all the way to pop psychologist Eric Berne's description of human behavior as "games people play," the advent of computer culture and electronic games, and the rising prominence of gambling and competitive sports in leisure culture.

Among postmodernists, game enjoys special favor because it exemplifies the elusive character of the signified and the slippery nature of language. In one of the most famous pronouncements in the philosophy of language, Wittgenstein describes the concept of game as a "family resemblance" notion that cannot be held together by a stable core of common features:

> Consider for example the proceedings that we call "games." I mean board-games, card-games, ball-games, Olympic games, and so on. What is common of them all?—Don't say: "There *must* be something common, or they would not be called 'games'"—but *look* and *see* whether there is anything that is common to *all*, but similarities, relationships, and a whole series of them at that. . . . [We] see a complicated network of similarities overlapping and criss-crossing: sometimes overall similarities, sometimes similarities of detail.
>
> I can think of no better expression to characterize these similarities than "family-resemblance"; for various resemblances between members of a family: build, feature, color of eyes, gait, temperament, etc. etc. overlap and criss-cross in the same way.—And I shall say: "games" form a family. (*Philosophical Investigations*, segment 1-32)

What constitutes a family, however, is not resemblance but kinship relations. The set of games may be fuzzy, which means that there is no set of necessary *and* sufficient conditions for an activity to be covered by the word *game*, but there still may be a set of necessary conditions surrounded by nonnuclear but reasonably typical properties whose number determines the degree of "gameness" of an activity.

One feature that comes to mind as a necessary condition is the pleasure dimension: games are freely played, and played for their own sake. If we focus on institutionalized games, as opposed to free play—a distinction made in English by the contrast *game/play* but expressed in neither the French *jeu/jouer* nor the German *spiel/spielen*—we can add a second basic feature: games are constituted by rules, and these rules, as Huizinga observes, are "absolutely binding and allow no doubt" (*Homo Ludens*, 11). This feature distinguishes Plato's concept of *ludus* (rule-governed behavior requiring "effort, adroitness, and ingenuity on the part of the players") from *paidia*, an activity characterized by "fun, turbulence, free improvisation, and fantasy" (Motte, *Playtexts*, 7). Bernard Suits—who strongly disagrees with Wittgenstein's pronouncement regarding the elusive character of the concept of game—proposes a definition entirely based on the notion of rules: "To play a game is to engage in an activity directed towards bringing about a specific state of affairs, using only means permitted by rules, where the rules prohibit more efficient in favor of less efficient means, and where such rules are accepted just because they make possible such activity" (*Grasshopper*, 34). This definition mentions only one of the two core features, but pleasure is not alien to it, since without enjoyment there would be no point in choosing a difficult route over an easy one.

Surrounding the two candidates for nuclear properties is a cluster of widely distributed but nonbinding features. As Huizinga explains:

> Summing up the formal characteristics of play [Huizinga means here the playing of formalized games] we might call it a free activity standing quite consciously outside "ordinary" life as being "not serious," but at the same time absorbing the player intensely and utterly. It is an activity connected with no material interest, and no profit can be gained by it. It proceeds within its own proper boundaries of time and space according to fixed rules and in an orderly manner. It promotes the formation of social groupings which tend to surround themselves with secrecy and to stress their difference from the common world by disguise and other means. (*Homo Ludens*, 13)

The flexible character of the concept explains why there is no specific game that serves as prototype for the textual analogy: literary critics invoke game and play in a loose generic way, and they feel free to borrow features from a number of different games.

Another variable in the literary usage is the nature of the relation between "game" and "text." A text can be called a game literally, metonymically, metaphorically, and the latter in a narrow or a broad way. (As Suits observes ["Detective Story," 200], there is plenty of "loose talk" in the use of the term by literary critics.) The closest thing to literary games in the literal sense of the term—unless computer games are regarded as literature—are folklore or literary genres regulated by fixed formal constraints, such as rhymes, puns, anagrams, acrostics, palindromes, or the secret languages invented by children ("pig Latins"). Whether they concern the graphic, phonic, or semantic substance of words, the rules that define these genres clearly throw "unnecessary obstacles" in the way of message formation. These voluntarily chosen obstacles bracket out the utilitarian, referential function of language and turn words into toys, much in the way that children at play recycle broomsticks as horses and cardboard boxes as fairy-tale castles. In these textual games, the partners in play are language and the writer who completes the fixed pattern in the most ingenious way, but from the point of view of the receiver, the games are more a spectator sport than a participatory activity.

Another way for a text to be a game in a fairly literal sense is by offering a problem to the reader. In genres such as riddle, tongue-twister, crossword puzzle, or even mystery novel, both author and reader are involved as players, and each of them performs exactly one move: the author asks a question, the reader tries to answer, and the author wins if the reader must be given the solution (Suits, "Detective Story," 204). In text-based computer games, a variant of this category, the author's one move is a global design that foresees, or embeds, many moves by the user. Within the text-as-problem category, however, we should distinguish those texts in which the user is supposed to find the answer by following textual directions (crossword puzzles, computer games) from a more metaphorical brand of game in which the pleasure of the reader is actually greater when he cannot solve the problem, because he is more interested in a clever answer than in beating his opponent.

A last way for language to be implicated in a game situation is to participate in a competition in which the skills necessary to win involve a mastery of verbal art. In institutionalized speech events such as verbal dueling, oratory debate, and poetry contests, the production of text counts as a strategic move. A more familiar form of ludic text production is the mad-lib party game, in which funny texts are generated through the random combination of words chosen by the players. It is also in the literal game category that I would place certain uses of language made possible by the Internet, such as those collaborative literary projects in which participants perform moves by adding text to an arborescent or rhizomatic structure, or a project developed by Bonnie Mitchell, *Digital Journey*, that closely resembles the "telephone" children's game: "She and her students create visual or verbal 'starter images' or 'theme pages' and circulate them among other individuals nationally or internationally. Each participant alters, adds to, or comments on whatever he or she receives." At the end of the game a dramatically changed text returns to its original sender (quoted in Gaggi, *Text to Hypertext*, 139).

In the metonymic interpretation of game, the manipulation of the text by the reader involves a mechanism borrowed from a standard type of game, such as throwing dicelike objects (the *I Ching*), shuffling cards (Marc Saporta's *Composition No 1*, a novel written on loose pages that can be read in any sequence), or even clicking with a mouse on the links of a hypertextual network, an activity reminiscent of opening the windows of an Advent calendar to find hidden treasures of candy or pictures.

In the narrow metaphorical sense, the text is not in itself an object to be manipulated in a ludic activity but a verbal transposition of the structure of a nonverbal game. Examples of texts explicitly patterned after a game abound in postmodern literature: Italo Calvino's *Castle of Crossed Destinies* simulates the card game of Tarot; the order of presentation of the lives of the tenants of an apartment building in Georges Perec's *La Vie mode d'emploi* is determined by the move of the knight's tour of the chessboard; and the texts of ∈, a collection of poems by Jacques Roubaud, represent the tokens of the Japanese game of Go (each text includes spatial coordinates that specify what move it will perform when it is placed on a game board). In this narrow analogical implementation, as in the metonymic and literal interpretations, the concept of game functions as discriminatory factor: some texts deliberately simulate games and others do not.

As the scope of the metaphor broadens, the analogical support becomes more and more tenuous. If the concept of game is to capture the essence of the literary, rather than the formal, properties of individual texts, it must be reduced to its bare essentials: a rule-governed activity, undertaken for the sake of enjoyment. The game metaphor becomes the expression of an aesthetic ideal in which rules do not function as unnecessary obstacles thrown in the player's way, as Suits defined their role, or as a tyrannical authority that must be subverted but as a means toward a goal. A successful game is a *global design that warrants an active and pleasurable participation of the player in the game-world*—the term *world* being taken here not as the sum of imagined objects but in a nonfigurative sense, as the delimited space and time in which the game takes place. At this level of generality, however, the metaphor threatens to degenerate into triviality. A good metaphor is supposed to provide an original perspective on its tenor, but we do not learn anything that we did not know before by being told that reading literary texts, like playing games, is for most people a pleasurable, nonutilitarian activity. Nor do we need to invoke the concept of game to realize that literature consists to a great extent of verbal sequences put together according to the rules of a specialized code, such as figures of speech, metric forms, standardized themes, conventional images, narrative patterns, and dramatic structures.

Beyond the superficial similarity of game rules and literary rules, however, the differences in their mode of operation are too important to ignore:[1]

1. Game rules are what John Searle (*Speech Acts*, 33–41) calls constitutive principles (they create the game by defining it), but in literary codes descriptive conventions ("regulative rules," for Searle) are much more dominant. A convention arises when a large number of authors conform to a certain type of behavior, but it is not binding, and it describes an independently existing form of discourse.

2. Game rules must be followed strictly, while modern literature encourages creativity and the transgression of its own conventions. The legacy of Russian formalism has accustomed scholars to regard deviance from the norm as a trademark of literary language.

3. If every text creates its own rules, the reader learns the code in the process of playing. This contrasts with the standard game situation, in which the player must learn the rules before stepping onto the playing field.[2]

Another feature of games that seems inapplicable to the literary text is the specificity of the goal to be achieved. While the attainment of this goal—-beating the game or the opponent—puts an end to the playing, the reading of a literary text is widely regarded as a never-ending activity. According to one of the most sacred dogmas of literary hermeneutics, a discipline whose origin in theology shouldn't be forgotten, the literary text is a holy scripture whose meaning cannot be exhausted.

If the core features of the game concept fail to provide interesting insights into the nature of the literary text, it is because the project of drawing a parallel between what is common to all games, on one hand, and to all literary textual practices, on the other, disregards the heterogeneity of the two sets under consideration. The overall analogy between the two domains is better supported by many concrete local similarities than by a few vague global parallels. A useful point of departure for the exploration of these local similarities is the typology outlined by Roger Caillois in *Men, Play, and Games.* Caillois distinguishes four types of game:

1. *Agon.* Games based on competition, such as sports (soccer, tennis), board games (chess, Go), or TV quiz games (*Jeopardy*).
2. *Alea.* Games of chance, such as roulette and lottery.
3. *Mimicry.* Games of imitation and make-believe, such as children playing house or making mud pies.
4. *Ilinx.* A protean family that involves transgression of boundaries, metamorphosis, reversal of established categories, and temporary chaos. Caillois associates *ilinx* with "those games which are based on the pursuit of vertigo and which consist of an attempt to momentarily destroy the stability of perception and inflict a kind of voluptuous panic on an otherwise lucid mind" (23; quoted in Iser, *The Fictive,* 259). The examples that come to mind are not activities that we easily

associate with games: initiation rituals, drug experiences, cross-dressing and masquerades, and the scary rides of amusement parks.

The usefulness of the first category, agon, is largely restricted to the case of the literal domain.[3] In a computer game, the purpose is clearly to win, and the way to win is to defeat enemies. When all obstacles have been overcome, all riddles solved, and the highest level conquered, the player is said to have "beaten the game." Agon is also a driving force on the thematic level of narrative texts—most, if not all plots involve some sort of competition—but the concept has little to offer on the metatextual level. Wolfgang Iser likens the act of reading to playing a game in which the reader can either win by "achieving meaning" or "maintain freeplay" (a state akin to a tie) by "keeping meaning open-ended ("Play of the Text," 252), but the text certainly does not lose when the reader wins. If maintaining free play is superior to achieving meaning, as postmodern theory tells us, the optimal situation is for the reader and the text to tie. This is hardly a competitive situation.

It is perhaps in some types of hypertext fiction that we find the best example of an antagonism between the reader and the text. Approaching the text as a computer game, some readers—here I speak for myself—experience it as an imprisoning maze of secret pathways devilishly designed by the author to make them run in circles. Their goal is to navigate the system with a purpose, thus escaping the tyranny of the labyrinth master, and the means to this goal is the reconstitution of the underlying map of the network. In Michael Joyce's *Afternoon*, for instance, the reader can reach a key fragment only after having tried a number of possible itineraries. Finding this hidden treasure is the reward for having played long enough with the text—for having conquered the labyrinth.[4]

Another way to conceptualize hypertext as a form of agon is to regard it as a puzzle to be solved. Espen Aarseth suggests the idea of a quest for the plot in his discussion of *Afternoon*: "We might label *Afternoon* a *reluctant narrative*, or an *antinarrative*, or a *sabotaged narrative*, terms typical of modernist poetics. But perhaps the best descriptive term for *Afternoon* is *game of narration*" (*Cybertext*, 94). I take this last characterization to mean that the reader of *Afternoon* is motivated by the desire to unscramble and put back together the narrative body that comes to her in dismembered form. This is certainly the spirit in which I approached the text, though I suspected— rightly, as it turned out—that the puzzle had no definitive solution. Jay Bolter also describes the hypertext experience in mildly agonistic terms, but winning or losing does not really matter because the player will always get another turn:

> Playfulness is a defining quality of this new medium. . . . No matter how competitive, the experience of reading in the electronic medium remains a game, rather than a combat, in the sense that it has no finality. . . . The reader may win one day and lose the next. The computer erases the program and offers the reader a fresh start—all wounds healed. . . . The impermanence of electronic literature cuts both ways: as there is no lasting success, there is also no failure that needs to last. By contrast, there is a solemnity at the center of printed literature—even comedy, romance, and satire—because of the immutability of the printed page. (*Writing Space*, 130)

One wonders, however, how the reader of hypertext can lose, even temporarily, given Bolter's earlier claim that there are no bad choices— no aporia, to use Aarseth's term—in hypertext: "[A] hypertext has no canonical order. Every path defines an equally convincing and appropriate reading" (*Writing Space*, 25). Here the labyrinth is no longer a problem to solve but a playground for the reader. This conceptualization removes any competitive spirit from the reading of hypertext and suggests an experience much closer to what will be described below as *ilinx* or as free play.

The idea of turning the text into *alea*, a game of chance, has both its detractors and supporters among the advocates of the game aesthetics. In what Warren Motte calls "a crusade for the maximal determination of the literary sign" (*Oulipo*, 17), a crusade whose ultimate goal is to purify literary language from the randomness of everyday speech, the members of the French literary movement Oulipo promote the use of exacting formal constraints, such as writing an entire novel without using a certain letter (*e* in Georges Perec's *La Disparition*).

By restricting the choice of words to those that can fill a certain graphic or phonic pattern, however, formal constraints give free rein to chance on the semantic level. The more stringent the formal requirements, the less the meaning of the text will express a preexisting vision, and the more it will be produced by the random encounter of signifieds brought together by the similarity of the signifiers. Oulipo's aversion to chance is also undermined by its predilection for combinatorics. Italo Calvino, a marginal member of the group, defines literature as "a combinatorial game which plays on the possibilities intrinsic to its own material" (quoted in Motte, *Playtexts*, 202). The principle is put to work in Raymond Queneau's *Cent mille milliards de poèmes*, a collection of ten sonnets of fourteen lines each printed on pages cut into so many strips, so that by flipping the strips and combining the lines, the reader can obtain 10^{14} different texts. Other examples of the literary exploitation of aleatory principles include the already mentioned *Composition No 1* by Marc Saporta; the surrealistic technique of *écriture automatique;* the practice of generating texts by computer through the random selection and combination of words from a database (the electronic poetry of John Cayley); and the reliance of literary hypertext on the often blind clicking of the reader to keep the textual machine running. If there is a common message to be read from these various attempts to exploit the creative power of chance, it is that meaning is produced by a force outside human control, a force that emanates from the obdurate substance of language.

The importance of games of mimicry for the theory of fictionality has been discussed in the previous chapters. Since these games presuppose a world, they offer a potential reconciliation of immersion and interactivity, and they transcend the aesthetic ideals that the literary theory of the past twenty years seeks to express through the game metaphor. I do not, therefore, include mimicry in my discussion of the game aesthetics.

With *ilinx,* we leave the realm of rule-governed activity and enter the domain of free play. In the history of philosophy, as Mihai Spariosu has shown *(Dionysus Reborn)*, the concept of play is traditionally associated with the creative power of the imagination. Children at play do not follow established rules but invent their own, in a transgression of real-world identities: "I'll be the salesperson, you'll be the

buyer, this bench will be the store, and these pebbles will be candy."
In literature, *ilinx* and its free play are represented by what Bakhtin
calls the carnivalesque: chaotic structures, creative anarchy, parody,
absurdity, heteroglossia, word invention, subversion of conventional
meanings (*à la* Humpty Dumpty), figural displacements, puns, disruption of syntax, *mélange des genres*, misquotation, masquerade,
the transgression of ontological boundaries (pictures coming to life,
characters interacting with their author), the treatment of identity as
a plural, changeable image—in short, the destabilization of all structures, including those created by the text itself. Rimbaud called this
spirit "déréglement de tous les sens" (disruption of all meanings,
directions, and sensory faculties). More than any other category in
Caillois's typology, *ilinx* expresses the aesthetics, sensibility, and conception of language of the postmodern age.

The prominence of the notions of game and play in contemporary
literary theory cannot be entirely explained by these analogies. A
more pertinent question to ask is how postmodern thought appropri-
ates the game metaphor as the expression of its own concerns. In a
period as contemptuous of tradition and formal constraints as postmodernism—Oulipo is the exception that confirms the rule—the
emergence of a form of behavior constituted by "absolutely binding"
rules as metaphor for the literary text is due not to nostalgia for the
time when literature had to conform to rigid models inherited from
the past but to a crisis of the notion of representation. This crisis is
widely credited to the influence of Saussurian linguistics, more particularly to its emphasis on the arbitrary character of linguistic signs.
For Saussure, arbitrariness concerns not only the relation of individual signifiers to their signified—what could be called vertical arbitrariness—but, more radically, the horizontal organization that language
imposes on phonic and semantic substance, as it divides the perception of what Saussure regards as an intrinsically undifferentiated continuum into discrete elements. This view of arbitrariness has fueled
the postmodernist denial of the power of language to describe an
external reality. Terry Eagleton explains:

> If, as Saussure had argued, the relation between sign and refer-
> ent was an arbitrary one, how could any "correspondence" the-
> ory of knowledge stand? Reality was not reflected by language

but produced by it: it was a particular way of carving up the world which was deeply dependent on the sign-systems we had at our command, or more precisely which had us at theirs. (*Literary Theory*, 107–8)

Recent linguistic theories have adopted a more nuanced approach toward the arbitrariness of linguistic signs and their control of the mind, taking more seriously the idea of semantic universals, investigating the metaphorical motivation of linguistic expressions, acknowledging the possibility of nonverbal thinking, and taking a critical view of the doctrine of linguistic relativism (as formulated in the Sapir-Whorf hypothesis). But for many literary theorists the most radical pronouncements of Saussurian linguistics—such as "In language there are only differences without positive terms" (120)—remain *the* authoritative account of the nature of language, not because of their influence on current linguistics, but because they support what postmodernism wants to believe.

The idea of linguistic relativism is a seductive plaything for a thought that conceives itself as play. In the paradigm that currently dominates literary studies, if literature is a game, it is because language itself is one; and if language is a game, it is because its rules form a self-enclosed system that determines, rather than reflects, our experience of reality. This autonomy of game rules with respect to any kind of external world is exemplified by the game of chess. The identity of the various pieces, such as kings and queens, is defined by the strategic movements they are allowed to perform on the board, and these movements have nothing to do with the properties of their real-world namesakes: the king could be renamed "trout" and the queen "carrot" without affecting the game.

This nonmimetic character of the rules of games can be applied to the case of the literary text on two distinct levels. On the first, literature is a game because, like every specialized genre of discourse, it is governed by its own arbitrary (read: "could be different") conventions, which form what Juri Lotman calls a "secondary modeling system."[5] Of all these second-order literary rules, none is more basic than the convention that allows the language of fiction to create its own world without being held accountable for the truthfulness of its declarations with respect to reality. On the most elementary level, then, it is the declarative power of fictional language that supports the text-game analogy. To express this power, Iser proposes to replace the notion of literary representation with the concept of play. The concept "has two heuristic advantages: (1) play does not have to concern itself with what it might stand for, and (2) play does not have to picture anything outside itself" ("Play of the Text," 250).

On the second level—if we accept the poststructuralist interpretation of Saussure—the nonmimetic character of literature is not a choice but the unavoidable consequence of the arbitrary nature of its medium. As Philip Lewis argues, "Playing with words may . . . involve more than juggling them about, re-orienting their interrelationships; it may compromise their very status as linguistic signs, revealing their inadequacy to relay accurately or completely the 'realities' they purport to designate" (quoted in Motte, *Playtexts*, 21). In this perspective, literary wordplay is not a specialized use of language contrasting with serious utterances but an allegory of the very condition of signification.

During the past twenty years, as structuralism gave way to deconstruction, the focus of interest in ludic metaphors has gradually switched from *ludus* to *paidia*, which means from the notion of game as rule-governed activity to that of play as subversion of rules. For A. J. Greimas, the quintessential structuralist, literary signification manifests itself in spatial configurations of semes, such as oppositional axes, triangles, and especially squares, and the rules of the textual game are revealed in the visual mapping of its semiotic structures. The narrative development of the text brings about permutations of elements, but to remain intelligible, the global pattern must retain at all times an axis or center of symmetry. In a seminal essay, "Structure, Sign, and Play in the Discourse of the Human Sciences," Derrida describes as follows the configuration of the structuralist structure:

> The function of this center [of a structure] was not only to orient, balance, and organize the structure—one cannot in fact conceive of an unorganized structure—but above all to make sure that the organizing principle of the structure would limit what we call the freeplay of the structure. No doubt that by orienting and organizing the coherence of the system, the center

of a structure permits the freeplay of its elements inside the total form. And even today the notion of a structure lacking any center represents the unthinkable itself. Nevertheless, the center also closes off the freeplay it opens up and makes possible. *Qua* center, it is the point at which the substitution of contents, element, or terms is no longer possible. At the center, the permutation or the transformation of elements (which may of course be structures enclosed within a structure) is forbidden. At least this permutation has always remained *interdicted*. (247–48)

To the self-contained structure of games, in which movement is limited to specific slots determined by rigid rules, Derrida opposes a "free play" (*jeu libre*) of elements in a decentered, self-transforming, fluid organization. One of the meanings of *jeu* (play), in French, is the space between two pieces that occurs when one of them is not properly tightened up. This space, which allows movement, can be taken to represent the lack of fit between language and the world, and the instability of linguistic signs. Free play arises because there are no "transcendental signifieds," no rules that fix meaning by anchoring language in an extralinguistic reality. Another conceivable interpretation of *jeu* is the uncertainty of the boundaries that circumscribe the semantic territory—or value—of linguistic expressions. In a purely differential conception of language, the value of word A ends where the value of B begins, but since these values are not backed by positive referential relations to a realm of extralinguistic objects, the exact location of the line of demarcation is undefinable.

It is always hazardous to try to paraphrase Derridean terminology, especially with a concept as broad as free play, but without taking an excessive interpretive risk, we can say that it involves the following ideas:

— The rejection of stable meanings in favor of an emergent conception of signification.
— The impossibility of achieving a totalizing and definitive apprehension of the literary text.
— The rejection of binary oppositions between "serious" and "playful," factual and fictional uses of language. Since *jeu* is inherent to the structure of language, its emphasis in litera-

ture is an allegory of the fundamental conditions of
signification.
— A conception of playing that gives initiative to the multiple
meanings of words and to their ability to activate chains of
reactions, so that playing, for both author and reader, becomes indistinguishable from "being played" by language.

Another leading figure of postmodern thought who associates
writing—all kinds, but especially literary—with the concept of *jeu* is
Michel Foucault. In "What Is an Author?" he writes:

Referring only to itself, but without being restricted to the confines of its interiority, writing is identified with its own unfolded
exteriority. This means that it is an interplay of signs arranged
less according to its signified content than according to the very
nature of the signifier. Writing unfolds like a game *(jeu)* that
invariably goes beyond its own rules and transgresses its limits.
(120)

In this passage the playful nature of writing is associated with three
features:

— Lack of reference to an external reality, which leads to self-
reference (since language must refer to something).
— Priority of the signifier over the signified as a principle of
organization (a claim that privileges phenomena such as
puns, rhymes, and alliteration at the expense of metaphor,
narrative structure, and the recurrence of motifs and
themes).
— The subversion of rules.

While a game necessarily involves players, the interpretations of the
game metaphor surveyed above differ widely in their implicit view of
the player's identity. In the structuralist conception, the forces at play
in the organization of the text are the "elementary structures of signification," which can be seen either as the structure of language itself
or, in a Chomskyan framework, as the hardware of the brain. In
Oulipian games, as well as in the fixed forms of folklore, the player is
the author, the plaything is language, and the reader's mode of in-

volvement is mainly that of a spectator or referee. Derrida's notion of free play emphasizes the agency of language, and it regards users as mediators of play rather than as plenary participants. Through their attention to the horizontal/differential relations that make up the texture of the text and the system of language, author and reader release semantic energies that are "always already" contained as virtualities in the substance of the medium. It is only when the reader is cast in the player role—as he will be literally in computer games, figuratively in texts with variable sequentiality, and programmatically in theories of the reader-as-author—that the game metaphor supports a poetics of interactivity.

COMPARING THE GAME AND THE WORLD METAPHOR

With these general observations in place, we can proceed to the systematic comparison of the game and world metaphors. The entries below, as well as table 2, summarize the main points of the preceding discussion in a contrastive manner.

Function of Language

The text-game analogy borrows features from many games, but if one game stands out as a prototype for postmodern aesthetics, it is the game of construction (a game that, paradoxically, eludes classification in Caillois's categories, unless it is regarded as mimetic, a view that does not sit well with postmodern aesthetics). According to Roland Barthes, reading is a "cubist" exercise in which "the meanings are cubes, piled up, altered, juxtaposed, yet feeding on each other" (*S/Z*, 55). The constructivist metaphor is developed by Milorad Pavic in the introduction to his *Dictionary of the Khazars*, a novel presented in dictionary form: "Each reader will put together the book for himself, as in a game of dominoes, or cards" (13). In this perspective, the text is "open" and reconfigurable, a matrix containing potentially many texts, a network of relations between semiautonomous units, a toolbox rather than an image, a renewable resource rather than a consumable good.

In the regime of the text as world, by contrast, the role of language is best compared to a mirror.[6] (Cf. the epigraph of chap. 13 of *Le Rouge*

TABLE 2 | Feature Comparison for the Metaphors
"The Text as Game" and "The Text as World"

	Game	World
Function of language	Cube, matrix, toolbox	Mirror (virtual image), picture
Substance of language	Opaque, visible	Transparent
Meaning	Relational	Referential
	Horizontal	Vertical
	Fluid, emergent	Textual information needs to
	Entirely contained in text	be supplemented by
	("Il n'y a pas de hors-texte")	imported knowledge
Reader's attitude	Reflexive	Nonreflexive
	Lucid	Willing suspension of disbelief
	Refusing illusion	Accepting illusion
Type of activity	Surf the surface	Exploration
	Construction, permutation, transformation	Voyeurism
Form	Form as exoskeleton	
	Emphasis on arbitrary formal constraints	Organic unity of form and content
Role of chance	Ambiguous:	Negative (words express how
	Positive (words take initiative)	things are in fictional
	Negative (signs must be overdetermined)	world)
Conception of space	Space occupied by text:	Space represented by text:
	Figures, arrangement on page, network of accessibility relations between units	Environment, landscape, geography
Requirements	Specialized "literary" competence	General linguistic and cultural competence
		Basic life experience
Critical analogy (Barthes)	"Writerly"	"Readerly"
Computer analogy (Turkle)	Macintosh operating system	IBM PC (MS-DOS) operating system

et le noir by Stendhal: "Le roman: c'est un miroir qu'on promène le long d'un chemin" [A novel is a mirror that one carries along a road].) Rather than focusing on the flat surface of the mirror, the spectator looks into its depth, where she discovers a three-dimensional reality. The metaphor of the text as world is linked to a phenomenology of "as if," of passing as, of illusion.

Substance of Language

If letters and words are the pieces of a construction game, they are opaque objects, and attention is directed toward the visible or phonic aspects of language, the material substance of the medium. Rhymes, alliteration, acrostics, palindromes, and calligraphic effects are among the language uses that are most commonly described as play. On the other hand, if the text is a mirror, words are transparent signs. Their function is to be a passport to the fictional world, to transport the reader into an alternative reality. Once the fictional world becomes present to the imagination, the language that monitored the mental simulation is partly forgotten.

Meaning

The game metaphor supports a Saussurian conception of language in which signs acquire their meaning not from vertical relations with objects in the world but from horizontal relations with other signs. The literary text, like language itself, is a self-enclosed, self-regulating system[7] in which meaning is determined by a strategic configuration of elements. The conception of meaning associated with the game metaphor can also be described as the product of a field of energies. Meaning is not a preformed representation encoded in words and in need of decipherment but something that emerges out of the text in unpredictable patterns as the reader follows trails of associative connotations or attends to the resonance of words and images with the private contents of memory. This operation is like following links on the Internet: surfing the surface, remaining in perpetual motion.

In the world metaphor, meaning is vertical, since language refers to the objects of the fictional world. As Thomas Pavel has observed, the rise of the possible-worlds approach brought an end to the "structuralist moratorium on representational topics" (*Fictional Worlds*, 6), an important part of which is the concept of reference. As we saw in chapter 3, in a vertical conception of meaning the primary role of language is to direct attention toward objects in the textual world, to link them with properties, to animate characters and setting, and to lure the imagination into narrative simulation.

Reader's Attitude

Playing a game is a lucid activity: the player must see through the deceptive moves attempted by the opponent. The attitude favored by the text-as-game aesthetics is therefore one of critical distance: the reader is not allowed to lose sight of the materiality of language and of the textual origin of the referents. One of the narrative strategies most widely described as play is the self-reflexive, anti-immersive metafictional stance through which authors such as Laurence Sterne, Denis Diderot, Flann O'Brien, or John Fowles remind readers that they are dealing with a constructed plot and not with life itself. This ideal of demystification stands in stark contrast to the suspension of disbelief that describes the attitude typical of the world metaphor. Here the reader pretends that there is a reality existing independently of the language that creates it. The pleasure of the text depends on the reader's willingness to bracket out objective knowledge and surrender to illusion.

Type of Activity

What do you do when you play with blocks? You arrange them into various configurations. You permute, transform, try out all the combinations. You build and you create. Contemporary critical idiom calls this activity the construction of meaning. The blocks can also be taken apart to see how meaning is put together, in a game known as deconstruction.

What do you do when you visit a new world? You explore its territory, and you spy on its inhabitants. You act, in other words, like a tourist and a voyeur—two roles that are generally not associated with intellectual sophistication. This might in part explain the prejudice of many academic critics against immersive reading.

Form

The text as world upholds the classical ideal of an organic unity of form and content. The form is dictated by the content and should not attract attention to itself. It supports the text like an internal skeleton.

The text as game treats form as an exoskeleton: rather than subordinating form to content, it treats content as the filler of the form. As we have seen, games are constituted by arbitrary rules, and literary texts imitate games by subjecting themselves to pragmatically unnecessary formal constraints.

Role of Chance

The ambiguous role of chance in the text as game has already been discussed. In some games it is welcome, in others frowned upon. In the text as world, chance plays a clearly negative role, since the function of language, for those who accept the illusion, is to provide a faithful image of an independently existing reality.

Conception of Space

In the text-as-world metaphor, space is a three-dimensional environment to be lived in, an area for travel, a landscape and a geography to be discovered in time. It is mapped by the bodily movements of characters from location to location. In the regime of the text as game, space is the two-dimensional playing field on which words are arranged (page or screen) or a collection of such fields (book or stack of hypercards), and its map is the network of relations (analogies, oppositions, electronic links) that connects textual units, determines patterns of accessibility, and traces formal figures.

Requirements

In order to play a game, the player must be thoroughly familiar with the rules. The text-as-game metaphor is esoteric and elitist; readers need literary competence to appreciate the text. According to Jonathan Culler *(Structuralist Poetics)*, this competence is so specialized that it must be learned like a second language, preferably in school. The text as world is a much more populist conception. All the reader needs to gain access to the fictional world is a basic knowledge of language, life experience, and reasonable cultural competence. If there are rules to learn in order to navigate the textual world, these rules can be learned on the fly.

Critical Analogy

I borrow from Barthes the concepts of writerly and readerly. It is no secret that Barthes regards the readerly as an inferior category; in his view, the sophisticated reader needs to graduate from the readerly to the writerly to attain higher forms of pleasure. The readerly frames the reader as a passive consumer who devours the text and throws it away: "This reader is . . . plunged into a kind of idleness—he is intransitive; he is, in short, serious: instead of [playing][8] himself, instead of gaining access to the magic of the signifier, to the pleasure of writing, he is left with no more than the poor freedom either to accept or reject the text" *(S/Z,* 4). The writerly, by contrast, is seen as promoting an active and playful participation of the reader in the act of writing: "Why is the writerly our value? Because the goal of literary work (of literature as work) is to make the reader no longer a consumer, but a producer of the text" (ibid.). In a later work, *Le Plaisir du texte*, Barthes contrasts *jouissance* (bliss), the experience provided by the writerly, with *plaisir* (pleasure), the experience of the readerly, and though he considers the two experiences to be qualitatively different and therefore complementary, he also insinuates that *jouissance* is an "extreme," unspeakable *(indicible)* form of pleasure that is diffused throughout the text, while *plaisir*, a "little jouissance," depends on anticipation and occurs only at climactic moments (34–35). This attempt to establish an order of precedence among different types of experience is typical of the prescriptive stance so often found in academic literary theory.

Computer Analogy

In her book *Life on the Screen*, Sherry Turkle discusses two philosophies of computer design that present interesting affinities with the contrast between world and game aesthetics. The world aesthetics is represented by the architecture of the IBM personal computer under the MS-DOS operating system. To operate MS-DOS, the user has to type rather clumsy textual commands in a code that resembles a computer language, but these commands provide a sense of power and control over the world of the machine. MS-DOS commands may be a long way from machine language, but they take the user *inside* the operating system, or at least to the outer circles of this inside, and they convey the sensation of understanding what is going on in the "brain" of the computer. By making its mode of operation relatively transparent to the inquisitive user, IBM architecture offers the computing equivalent of an immersive experience. The reader who goes for immersion may not have the same control as the computer user over the world he penetrates, but he too wants a global overview of the working of the system, which consists in this case of a mental map of the textual space, a sense of purpose and direction, an understanding of narrative causality, an idea of the private worlds of characters, and above all the feeling that the passing of time will bring the solution of enigmas.

While IBM architecture satisfies an attraction for hidden depths, Macintosh design relies on what Turkle calls "the seduction of the interface" (31). With its visually attractive display of icons and its use of the click of a mouse to activate programs, the Macintosh is more a game than a (serious) "business machine." People learn to operate it by trying out options—a task at which children are much better than grownups—not by poring over the tedious prose of the user's manual. The price to pay for the pleasure of operation is that the system keeps its lower levels absolutely hidden from the user. The design philosophy of the Mac is known as WYSIWYG (what you see is what you get): icons can be manipulated, but they cannot be opened; they remain as opaque as the cubes of a construction game. "Unlike the personal computers that had come before, the 'Mac' encouraged users to stay at a surface level of visual representation and gave no hint of inner mechanisms" (34). In so doing, "it introduced a way of thinking that put a premium on surface manipulation and working in ignorance of the underlying mechanisms. . . . It encouraged play and tinkering" (35). Through this design philosophy the Mac represents the triumph of "surface over depth, of simulation over 'the real,' of play over seriousness" (44). As a longtime user of IBM and MS-DOS I still recall my frustration when I first tried a Macintosh and found out that in order to retrieve my diskette from the drive, I had to click on an icon representing the file and drag it to the picture of a trash can, rather than pushing on a knob to open the disk drive. This gesture seemed dangerously close to throwing my text away. Even after I realized that my file would survive the operation, I resented the system for taking away from me the most direct form of control over the machine and for turning honest manual work into a silly game of clicking and dragging through which I was made entirely dependent on invisible software.

In labeling MS-DOS more worldlike and the Mac more gamelike I am adopting the perspective of a hacker type of user for whom the operating system is a world in itself and who does not need to imagine that the computer is something else. But many dedicated Mac users will disagree with this view. For these users the icon-based interface of the Mac is more immersive than a command-line system like MS-DOS because, as I stressed in chapter 2, icons are the first step toward the disappearance of the computer, which allows the emergence of a virtual world. Whereas hacker types find command lines transparent and icons opaque (you cannot really tell from the visual appearance of an icon what operations it actually stands for), Mac users may reply that their system is the more transparent of the two because it places them in a concrete world furnished with familiar objects (the desktop and its tools) and because they can learn to use these objects by trial and error. These divergent perspectives presuppose two different interpretations of transparency: the IBM operating system is transparent in the sense of revealing the inner working of the machine, while the Mac operating system is transparent in the sense of making itself invisible.

Whether users feel more immersed with a transparent operating system or with an invisible computer, the introduction of the Windows operating system paved the way at least temporarily toward a cohabitation (rather than a reconciliation) of what Turkle has labeled the "modern" aesthetics of the IBM with the "postmodern" spirit of the Mac design. Until recent versions (NT and 2000), Windows offered icons and the option to navigate the system with a mouse-operated cursor, but it also let the user return to DOS and control operations in a more direct manner. It did not erase the difference between depth and surface, nor did it allow the user to experience the computer simultaneously as a world and as a game, but it did offer a choice.[9]

Can there be an equivalent of Windows in the textual domain? The polarity of the two columns of table 2 suggests a fundamental incompatibility of the two metaphors. A world is not a game; but as postmodern literature has ingenuously demonstrated, textual worlds can be turned into tokens of play. Brian McHale (*Postmodernist Fiction*, 6–11) defines postmodernism as a movement that foregrounds ontological preoccupations, as opposed to modernism, which was dominated by epistemological questions such as "Who am I?" "Can I know the world?" and "Can I know myself?" Postmodern fiction does not directly ask, "What is the nature of being?" but it thematizes ontological problems by treating worlds as toys, by juggling them in the textual space, by building alternative ontologies, by playing with transworld identity, by transgressing ontological boundaries, by making worlds morph into other worlds, and by merging generic landscapes.

The consequence of this play is that immersion becomes thematized. But since immersion is a state of forgetting language and losing oneself in the textual world, its thematization also means breaking the spell that makes it possible. It takes deprivation for the reader to come to realize the importance of immersion. The many-worlds texts of postmodernism offer glimpses of what it means to settle down in a world, but as soon as the reader develops a sense of belonging, they break the illusion or transport her to another world. By shuttling the reader back and forth between worlds, by constantly shifting perspectives, by proposing multiple realities that relativize each other, by constructing and voiding worlds—as it exposes their language-made nature—the postmodern text keeps the reader in a state of permanent jet lag.

Fortunately for those who prefer to grow roots in a textual world, the frenetic world-play of postmodernism is not the only way to reconcile the two metaphors. The best compromise of all is simply to regard the concepts of game and world as complementary points of view on the same object, much in the way modern physics uses the metaphors of wave and particle as alternative conceptualizations of light. This is not to deny that some texts are inherently more game-like (hypertext, visual poetry, postmodern novels) and others more worldlike (realistic texts). But if the nature of the text usually favors one of the conceptualizations, each of the two metaphors provides a point of view from which we can observe features that remain invisible from the other. If we are unable or unwilling to switch perspective, we will never appreciate the language games that are being played in textual worlds or the worlds that are being manipulated through language games. Yet because an observer cannot simultaneously occupy two different points in space, the complementarity of the two metaphors also means that we cannot experience both dimensions at the same time. We must therefore immerse and deimmerse ourselves periodically in order to fulfill, and fully appreciate, our dual role as members of the textual world and players of the textual game.

The Game-Reader and
the World-Reader

Italo Calvino's *If on a Winter's Night a Traveler*

The rivalry of the world and the game aesthetics has been
charmingly allegorized in Italo Calvino's *If on a Winter's
Night a Traveler* through the characters of Ludmilla and
Ermes Marana. The structure of the novel has already been
described in the interlude to chapter 5. In alternating chapters the text reproduces the beginning of ten incomplete
novels and narrates the ongoing story of the two main characters, a male reader referred to as "you" and a female
reader named Ludmilla. These two characters fall in love
and eventually marry despite the machinations of a villain,
Ermes Marana, who wants to spoil Ludmilla's reading pleasure and keep her separated from the male reader.

Ludmilla is the epitome of the immersion-seeking reader
and the dream soulmate of every author—so much so that
every male writer or reader character in the novel falls in
love with her or attempts to seduce her out of jealousy for
her almost promiscuous passion for books. She cannot re-
sist the call of textual worlds, and textual worlds go out of
their way to accommodate her: whenever she expresses a
wish, the next book she reads "magically" conforms itself to
her expectations. When she says in chapter 2, "I prefer novels . . . that bring me immediately into a world where everything is precise, concrete, specific" (30), the next novel
begins with "an odor of onions being fried" (34); when she
declares in chapter 5, "The novel I would like to read at this
moment . . . should have as its driving force only the desire
to narrate, to pile stories upon stories" (92), the next excerpt
thematizes the "saturation of stories" (109), both told and
untold, that surround every event in the life of the narrator;
when she declares herself most attracted by novels "that
create an illusion of transparency around a knot of human

relationships as obscure, cruel, and perverse as possible" (192), she gets the sexually explicit tale of a young man who loves a young woman and makes love to her promiscuous mother in a triangular relation fully known to the daughter; and finally, when she claims to be looking for a book "that gives a sense of the end of the world after the end of the world" (243), her wish is fulfilled through a novel that describes the narrator's "exercises of making the world around me disappear" (249).

Authors fall in love with Ludmilla because she is open to any kind of text. In an age in which everybody reads to become a writer, she is the last genuine pleasure-reader:

> For this woman . . . reading means stripping herself of every purpose, every foregone conclusion, to be ready to catch a voice that makes itself heard when you least expect it, a voice that comes from an unknown source, from somewhere beyond the book, beyond the author, beyond the conventions of writing: from the unsaid, from what the world has not yet said of itself and does not yet have the words to say. (239)

In Ludmilla's apartment, books are neatly arranged on the shelves that line the walls, and their solid bodies build an interior space that her suitors try in vain to pry open: "To you Ludmilla appears protected by the valves of the open book like an oyster in its shell" (147). The organization of books on the wall mirrors the organization of worlds in her mind. Ludmilla likes to read several books at the same time, but their contents never blend, because "your mind has interior walls that allow you to partition different times in which to stop and flow, to concentrate alternatively on parallel channels" (146). Through books, Ludmilla lives many lives and visits many worlds, but the solidity and boundedness of each volume protects these lives and worlds from contamination. The material body of the book, bound by its spine and bounded by its hard cover, ensures the integrity of the textual world.

For Ermes Marana, on the contrary, texts have no solid bodies that keep them together. Marana, whose first name alludes to Hermes, god of thieves, inventor of writing, and trickster of Greek mythology, deals not with books and works and worlds but with loose collections of pages that can be shuffled, cut, pasted, disseminated, and freely re-combined. While Ludmilla lines her walls with discrete volumes, Ma-rana treats writing as a mass substance, foreshadowing Pierre Lévy's pronouncement on the status of textuality in the electronic age: "Now there is only text, as one might say of water and sand" (*Becoming Virtual*, 62). Whereas Ludmilla's immersive relation to books requires the permanence and solidity of print, Marana's constant juggling of text suggests the volatile nature of electronic writing. His manipula-tions make true this description by R. Howard Bloch and Carla Hesse of the digital future of the text: "Genres until now considered to be discrete suddenly will mingle indiscreetly on the screen; any text will be able to mate electronically with any other text in what looms as the spectre of a great miscegenation of types" ("Introduction," 4).

Marana first appears in Calvino's novel as the translator of a Cim-merian (or is it Cimbrian?) narrative. Because he doesn't know a word of Cimmerian, he plugs Cimmerian-sounding names into what he claims to be a Polish novel, but this novel turns out upon investigation to have been a "trashy French novel" that he "makes pass as Cim-merian, Cimbrian, Polish" (*If on a Winter's Night*, 100). In a later chapter Marana morphs into a globetrotting agent, searching for ma-terials for a publishing house but also trying to satisfy the textual needs of voracious women readers who undergo countless reincarna-tions. He mails letters from all over the world, often from locations he has never visited; invents stories to explain why he cannot deliver translations on time; points out new books that the publisher should not let slip through his fingers; invents a machine to measure immer-sion as a way to predict the commercial success of books; discovers the "universal source of narrative material" (117), an old Indian who spills out stories the way a fountain spills out water; intercepts the narrative flow by putting a tape recorder at the mouth of the cave where the old man hides; programs a computer to finish these stories; and designs a "trap novel" for a Sultana who needs a continuous supply of reading material for fear she might become bored and decide to cheat on her husband.

The trap novel is a mirror-image of *If on a Winter's Night a Trav-eler:* at the moment of greatest suspense, Marana breaks off the trans-lation and inserts in its place a translation from another novel so that the Sultana will never experience the "drop in tension that will follow the end of the novel" (124). Supremely disdainful of intellectual property, Marana subverts the notion of authorship (Gaggi, *From Text to Hypertext*, 56) by attributing manuscripts to wrong names, by pasting together pages from different sources, and by designing a computer program that produces novels by a bestselling author, Silas Flannery, when this author is affected by an acute case of writer's block. Meanwhile his female acolyte Lotharia, the sister and negative image of Ludmilla, promotes an electronic system that does away with the reader in much the same way that Marana's manipulations of text do away with the author: the text is scanned for key words, and the message is reconstructed on the basis of frequencies. Thanks to the computer, both writing and reading are simplified into a manipulation of alphanumeric strings.

The restless travel of Marana stands in stark contrast to the sedentary comfort of Ludmilla's apartment, his multiple personalities—expression of the postmodern "decentered subject"—to her quiet sense of identity, his compulsive need to do things with texts to her intuitive grasp of the voice of a text and of the soul of textual worlds. Jealous of Ludmilla's immersive pleasures, Marana tries to lure her into the territory of the absolutely fake by concocting a literature "made entirely of apocrypha, of false attributions, of imitations and pastiches" (*If on a Winter's Night*, 159), a literature fundamentally hostile to the fictional truths of make-believe: "As for [Marana], he wanted, on the contrary, to show her that behind the written page is the void: the world exists only as artifice, pretense, misunderstanding, falsehood" (239). The entire plot of *If on a Winter's Night a Traveler* is revealed in the end to be another plot designed by Marana to force Ludmilla to stare into this void; but through her limitless receptivity, Ludmilla is able to convert every fragment of text into a world that hides the void, a world whose lack of existence is covered up by the fullness of its presence.

The triumph of Ludmilla over Marana may sound like a crushing defeat for the spirit of play, but if we enjoyed the trickery of *If on a Winter's Night a Traveler*, a novel masterminded by another wizard intent on frustrating immersive reading, we should concede at least a tie to the game aesthetics.

SEVEN

Hypertext

The Functions and Effects of Selective Interactivity

Quain was in the habit of arguing that readers were an almost extinct species.

"Every European," he reasoned, "is a writer, potentially or in fact." He also affirmed that of the various pleasures offered by literature, the greatest is invention. Since not everyone is capable of this pleasure, many must content themselves with shams. For these "imperfect writers," whose name is legion, Quain wrote the eight stories in Statements. Each of them prefigures or promises a good plot, deliberately frustrated by the author. One of them—not the best—insinuates two arguments. The reader, led astray by vanity, thinks he has invented them.   —JORGE LUIS BORGES

With the increasing extension of the press, which kept placing new political, religious, scientific, professional, and local organs before the readers, an increasing number of readers became writers. . . . Thus, the distinction between author and public is about to lose its basic character. . . . At any moment the reader is ready to turn into a writer.   —WALTER BENJAMIN

We tend to think of interactivity as a phenomenon made possible by computer technology, but it is a dimension of face-to-face interaction that was shut off by manuscript and print writing and reintroduced into written messages by the electronic medium, together with several other features of oral communication: features such as real-time (synchronous) exchange, spontaneity of expression, and volatility of inscription.[1] A *New Yorker* cartoon reminds us of the interactive character of the oral storytelling situation by showing a man sitting at his daughter's bedside with an open book in his hands. "Stop asking so many questions," exclaims the exasperated father, "or it's right back to Books on Tape for you" (22–29 June 1998, 93). From the viewpoint of the father in the cartoon, the daughter's many questions are annoying interruptions that lengthen the chore of evening storytelling; but from the viewpoint of the daughter, besides postponing the daily banishment to the land of sleep, questions and requests are a way to customize the text to her own desires. Thanks to the inherently interactive nature of oral storytelling, the child is able to elicit explanations, to get her father to develop or retell favorite passages, to encourage him to embellish the descriptions of certain objects, or to force him to change direction if he embarks on a track that she does not approve of. (Children, notoriously, do not like variations on the "real" story.)

Interactivity appears on two levels: one constituted by the medium, or technological support, the other intrinsic to the work itself. All interactive works require a reasonably interactive medium, but the converse does not hold. Television, for instance, is a mildly interactive medium, since it enables users to switch channels at will—even more conveniently after the invention of the remote-control device—but this feature only allows a choice among noninteractive programs. Interactivity of the medium is detrimental in this case to the appreciation of the individual texts, since by channel-surfing the spectator violates their putative integrity. Similarly, the Internet as a whole is an interactive medium, but many of the documents it makes available are themselves standard linear texts.

Types of interactivity can also be distinguished on the basis of the freedom granted to the user and the degree of intentionality of his interventions. The bottom of the scale is occupied by what one may call, with Söke Dinkla, a "reactive" interaction, which does not involve any kind of deliberate action on the part of the appreciator. An artwork may, for instance, react to the amount of noise in the room and display different images depending on whether the visitors are quiet or speaking ("From Participation," 286). One step higher on the intentional scale is a random selection among many alternatives. When the user takes action deliberately but cannot foresee the consequence of his actions, the purpose of interactivity is to keep the textual machine running so that the text may unfold its potential and actualize its virtuality. Such is the random clicking of many hypertexts. But selective interactivity can also be a purposeful action. In a computer game, for instance, the player may be offered a choice between two paths, one of which leads to success and the other to failure, and the game may cue the player as to which path is the good one. In the fullest type of interactivity, finally, the user's involvement is a productive action that leaves a durable mark on the textual world, either by adding objects to its landscape or by writing its history. This chapter and the next one focus on selective interactivity, more specifically on hypertext, its principal manifestation, while chapters 9 and 10 are devoted to the productive variety.

Thanks to the popularity of the Internet, the idea of hypertext is so widely known these days that its presentation can be limited to a short refresher. In a hypertextual system, text is broken into fragments— "lexias," for George Landow; "textrons," for Espen Aarseth—and stored in a network whose nodes are connected by electronic links. By clicking on a link, usually a highlighted phrase, the reader causes the system to display the contents of a specific node. A fragment typically contains a number of different links, offering the reader a choice of directions to follow. By letting readers determine their own paths of navigation through the database, hypertext promotes what is customarily regarded as a nonlinear mode of reading. The applications of the idea nowadays include the World Wide Web, educational databases, the on-line help files of computer programs, Ph.D. dissertations and other scholarly texts, and multimedia works on CD ROM, as well as poetry and literary fiction.

INTERACTIVE VERSUS ERGODIC DESIGN

Technically speaking, an interactive text is one that makes use of reader input; but we will be much closer to the spirit of interactivity, at least as it is conceived by most hypermedia theorists, if we associate the concept with a second feature. To label this feature I borrow from Espen Aarseth the term *ergodic design*. (The word *ergodic*, borrowed by Aarseth from physics, "derives from the Greek words *ergon* and *holos*, meaning 'work' and 'path'" [*Cybertext*, 1].) Aarseth uses *ergodic texts* and *cybertexts* interchangeably, but I prefer *ergodic*, because *cybertexts* suggests to most people an electronic support, a feature that Aarseth considers optional to the category. According to Aarseth's definition, ergodic literature is a class of works in which "non-trivial effort is required to allow the reader to traverse the text" (ibid.). More specifically, an ergodic design is a built-in reading protocol involving a feedback loop that enables the text to modify itself, so that the reader will encounter different sequences of signs during different reading sessions. This design turns the text into a matrix out of which a plurality of texts can be generated. Each new state of the ergodic text is determined by the previous one, and the total run depends both on initial conditions and on the input that the system absorbs between its states.

This characterization may seem at first sight to be synonymous with the feature of accepting and reacting to user input, and Aarseth's discussion often conflates these two properties, but as I understand the definition, the "non-trivial effort" is not necessarily put up by the reader: some ergodic texts are closed systems, and their feedback loop generates transformations without human intervention. Through their generative power ergodic texts introduce the additional level of textuality that has been described in chapter 1 as a consequence of the virtualization to a second degree of the already virtual nature of the text. The French media theorist Philippe Bootz labels these three levels *texte écrit* (the text as written), *texte-à-voir* (the text as seen by the reader), and *texte lu* (the text as mentally constructed by the reader) ("Gestion du temps," 235–36).

Ergodic design and sensitivity to user input (i.e., interactivity proper) may not be unique to digital texts, but the strongest manifestations of interactive textuality are those that implement these two dimensions in an electronic environment. The relation of the sets defined by these features is shown in figure 2. For each category I offer examples of both selective and productive interactivity.

1. *Nonergodic, nonelectronic, noninteractive texts.* Standard literary texts, in which the dynamic construction of the text that takes place during the act of reading concerns meaning exclusively.

2. *Interactive, nonelectronic, nonergodic texts.* (a) Selective interactivity: The exchange between parent and child during storytelling sessions. Through the dialogue between teller and listener, the text is customized according to the listener's specifications, but since the shaping mechanisms come from the conversational context, not from within the story itself, they do not constitute an ergodic design. (In this example I call "text" the story being told as a result of the dialogue, not the dialogue itself.) (b) Productive interactivity: Conversation. As an open-ended, freely course-switching exchange, conversation does not qualify as ergodic, or arguably as text, because it lacks global design.

3. *Electronic, noninteractive, nonergodic texts.* Texts broadcast on

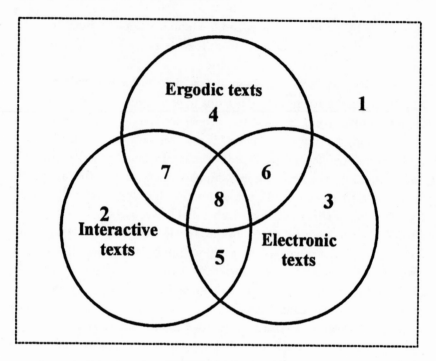

FIGURE 2 | A text typology

TV. In the computer domain, the digital versions of print texts, such as Jane Austen's novels on CD ROM.

4. *Ergodic, nonelectronic, noninteractive texts.* Here we find "texts," or installations, that produce ever-new outputs by simply reacting to their environment without intervention from the appreciator: chimes and aeolian flutes operated by the force of the wind, mobiles and mechanized sculptures (such as Jean Tinguely's machines), or architectural works designed to reflect the light of the sun in different patterns at different times of the day. In the literary (or at least textual) domain the best example I can think of is a work of conceptual art conceived by the Wyoming artist Pip Brant and her artist-team *kunstwaffen*. In this installation words and phrases from a preexisting text, Phyllis Luman Metal's *Cattle King on the Green River,* were inscribed on the bodies of cows, and the cows were let loose on the range. The random movement of their wandering and gathering created ephemeral word combinations that yielded statements (or "poems") on eco-

logical and feminist themes. Not only did the movements of the cows scatter the words and undo the statements, the Wyoming weather did its best to erase the words themselves.

5. *Electronic, interactive, nonergodic texts.* (a) Selective interactivity: A textual database, such as a library catalog, fits this category. The user interacts with it by sending queries, and the text responds by displaying certain parts of itself, but the organization of the text as a whole is static, and the result of a given query is fully predictable. It doesn't matter whether the reader looks up first "Derrida, Jacques" and then "Sade, Donatien, Marquis de" or vice versa; the database will always respond to the same query in the same way. (b) Productive interactivity: Electronic conversations, such as the live exchanges of Internet chat rooms. If oral conversation is not ergodic, neither is its electronic version.

6. *Ergodic, electronic, noninteractive texts.* Purely reactive examples of electronic poetry fit this category—for instance, *The Speaking Clock* by John Cayley, as described by Espen Aarseth ("Aporia," 39–40). *The Speaking Clock* generates an ever-changing display of words selected from a fixed textual database, using the computer clock, as well as aleatory selection devices, to trigger transformations. The rim of the display represents the face of a clock with letters replacing the customary numbers. Each of these letters thus stands for a particular digit, based on its location (the letter at the one o'clock position = 1, etc.). The center of the circle contains a group of words, forming an approximate sentence, with certain letters in boldface. Every second, a letter is replaced by another and a new message is produced, but the letters in boldface offer at all times a coded version of the current time of the computer clock.

7. *Ergodic, nonelectronic, interactive texts.* Here we find the growing class of nonlinear, or multilinear literary print texts that offer to the reader a choice of reading sequence. I propose to reserve the term *multilinear* for the texts that offer a choice among many well-charted sequences, and *nonlinear* for the texts that allow the reader to break her own trail. (Most texts fall somewhere in between.) A classic example of multilinear text is *Hopscotch* by Julio Cortázar, a novel that proposes two fixed reading protocols, one that leads sequentially from chapter 1 to chapter 56, and another that still visits 1 to 56 in numerical

order but inserts additional chapters, labeled 57 and higher, between these units. The fully nonlinear category is illustrated by Marc Saporta's *Composition No 1*, a novel written on a deck of cards that can be shuffled and read in any order. Halfway between these two poles are texts such as *Dictionary of the Khazars*, by Milorad Pavic (a novel written as a series of dictionary entries), *Pale Fire* by Vladimir Nabokov (a story told in the footnotes to a poem), and even, as Peter Stallybrass has shown,[2] the Bible itself (the Christian belief that the New Testament fulfills the Old, as well as the discrete numbering of verses, invites a discontinuous and parallel reading that necessitates the setting of a complex system of bookmarks).

8. *Electronic, ergodic, interactive texts.* (a) Selective interactivity: Literary hypertexts, many types of electronic poetry, surfing the 'Net. (The Internet is a giant database, but I place it in this category, rather than in 5, because of its self-transforming capabilities.) (b) Productive interactivity: Interactive drama, collaborative literary projects on the 'Net, MOOs and MUDs (they may be open-ended exchanges, like conversation, but interaction takes place in a designed environment). VR-based art installations can be either selective or productive, and computer games fall halfway between the two categories: the player does not contribute text, but his shooting, jumping, and riddle-solving moves require more skills, and therefore involve a more active participation, than selecting options from a finite menu.

VARIETIES OF INTERACTIVITY

The interactive text is a machine fueled by the input of the user. Why would the reader want to submit input?

— *To determine the plot.* This type of interactivity is exemplified by the multiple-choice texts in which the reader is asked to decide the destiny of the hero: Do you want Pinocchio to go straight home to Gepetto after school, or do you want him to take the advice of the Fox and the Cat and board the coach to Pleasure Island? The idea of determining *the* outcome of the plot presupposes that the reader gets to make only one pass through the text; otherwise the motivation would belong to one of the next two categories. It is for this reason that Julian Barnes, in the epigraph to chapter 8, advocates burning the envelopes that contain other choices. In an electronic text the links to the branches not taken could be cut off, so that the reader could reread only the plot she has selected.

— *To shift perspective on the textual world.* This can mean: open a new window, follow another plot line, switch focus to a different character, hear a different version of the facts, or enter a new private world.

— *To explore the field of the possible.* If the reader gets a chance to try all the alternatives, the purpose of interactivity is not to determine the plot, since all branches are now part of it, but to gain a global overview of the interweaving of possibilities, and an appreciation of the author's performance as the designer of a "garden of forking paths."

— *To keep the textual machine going.* When the reader's choices are fully blind, as already noted, the purpose of clicking on one link rather than another is simply to get more text to the screen. In this case, interactivity is nothing more than a user-operated random number generator. In the electronic poetry of Jim Rosenberg or John Cayley, for instance, the reader animates certain words, or causes certain sentences to appear, by simply moving the cursor around the screen and randomly clicking. The same principle can be used to animate images: as the cursor moves over certain areas of the picture, it triggers unforeseeable metamorphoses.

— *To retrieve documents.* This is the interactivity granted by hypertextual help files, or by the search engines of the World Wide Web. In this type of motivation the text is treated as a collection of resources that can be brought to the user's screen and used in personal projects. (Cf. Pierre Lévy's approach to reading as discussed in chapter 1.) It is also in this category that I would place electronic textbooks. Here the dialogue between system and user guides the student through the materials, and the system functions as personal tutor.

— *To play games and solve problems.* This category hardly needs explanation. In computer as in other games, winning is a matter of solving problems, and it is by submitting input to the system that the player of computer games performs actions toward her goal.

— *To evaluate the text.* In print literature the process of reviewing, discussing, and more simply buying a book is an interactive mechanism by which professional or amateur readers tell authors what they want to read. The chat groups and bulletin boards of the Internet can similarly be used to comment on the plot and thereby influence the writers of TV series or on-line soap operas.

To these seven varieties of purely selective interactivity we should add two productive types:

— *To participate in the writing of text* by contributing permanent documents to a database or a collective literary project.
— *To engage in dialogue and play roles.* Here the user responds to the system, or to other users of the system, through a written form of conversation. This mode is found in MOOs, chat rooms, e-mail, and interaction with AI-animated "chatterbots" such as Julia and Eliza (presented in chapter 10).

Inverting this inquiry from the point of view of the reader to the point of view of the author, we may also ask, "What is the purpose of the interactive format and the intended effect of clicking on a hyperlink?" The possibilities include both attempts to guide the reader and to turn him loose in a forest of blind options:

— *To control the reader's progress in the discovery of facts.* Even in a nonlinear text, it may be necessary to read certain fragments before others make sense. By creating systems of links, authors establish complex systems of dependencies between lexias. "Guard fields," "multiheaded" and "multitailed" links, may, for instance, ensure that the reader has followed a certain path before reaching a certain node.[3]
— *To propose, and let the reader explore, alternative versions of a reasonably solid core of facts.* This approach interprets the notion of possible world in the epistemic sense: insofar as each path offers to the reader a particular point of view on a

common actual world, it represents the "belief world" of a distinct individual.

— *To propose, and let the reader explore, many possible futures for the textual world.* This use of hyperlinks asks of the reader to adopt the prospective orientation of simulation. Each of the forking paths determined by the links corresponds to a possible world in the alethic/temporal sense: that which could still happen, given the present situation and the laws of the textual world.

— *To suggest analogical relations between segments.* One of the interpretive strategies most characteristic of hypertext fiction (as opposed to print novels) is to ask why the author placed a hyperlink between the preceding and the current lexia. Answering this question often requires a demanding act of memory, since the reader must compare a present with an absent lexia. But this difficulty can be alleviated by split screens that show both lexias.

— *To allow the user to blow up certain screens or passages and get a closer look.* This is common practice in graphics software, but the idea can also be used with text and multimedia works. As Sheldon Renan writes, interactive works "will be *scalable.* They will be forms that shift easily from the epic perspective to infinite detail" ("The Net," 64). In a narrative the system may offer a choice between a "scene" and "summary" representation of an episode, or enable the reader to skip or expand descriptions. (See the interlude to chapter 10 for an imaginary electronic book that implements this idea.)

— *To interrupt the flow of narration,* disrupt, frustrate, puzzle, undermine certainty, subvert or mock the text on the other end of the link, withdraw facts, and place fictional worlds "under erasure."[4]

— *To provide background information,* explanations, supporting material, and intertextual references.

Depending on the particular purpose, the links will either be opaque or involve navigational aides (such as meaningful highlighted key words). Access to a given link will be either context-free or made conditional on the visiting of other links. Links may be either visible or hidden like Easter eggs inside text and pictures. And finally, some links will be permanent while others will self-destruct after a visit. This variety of linking strategies, as well as the many possible functions of branched architectures, underscores the futility of assigning a fixed "message in the medium" to hypertextual design.

PROPERTIES OF THE ELECTRONIC MEDIUM

To understand why interactivity is more fully developed in an electronic than in a print environment, we must turn to the distinctive properties of the electronic support. For those who subscribe to a moderate version of technological determinism, by which new media open wide fields of possibilities rather than forcing their users onto narrowly pre-scripted paths, the expressive potential of selective interactivity and by implication the poetics of hypertext are crucially dependent on the properties of its medium. Listed below are features of either proven or potential relevance for electronic literature.

1. *Mobility of text.* In an electronic environment the reader does not go to the text, the text comes to the screen. Reading hypertext has been widely conceptualized as navigation, but it is because the movement belongs to the text that the reader is much more inclined to explore in an electronic network. With a print text the reader needs to perform a "go to"—targeting a page in a book, a shelf in a library, a store in town—to obtain new materials, but in electronic texts all the reader needs to do is to click on a link.

2. *Kaleidoscopic abilities.* I borrow this concept from Janet Murray (*Hamlet,* 155–62). By this term Murray means the computer's ability to produce "multiple patterns [with] the same elements" (282). This combinatorial power results in a "mosaic organization" (156), a structure that Marshall McLuhan regards as the trademark of twentieth-century media. (Among the other examples of mosaic are cinematic montage, the fragmented appearance of newspapers, and the use of sidebars, captions, columns, text, and pictures in the layout of textbooks or magazines.) In a kaleidoscopic environment the task of the electronic writer is to create a protocol that will ensure pleasurable and meaningful permutations of the basic repertory.

3. *Algorithmic or procedural nature.* The electronic text is the product of a two-leveled semiotic system, by which the signs flashed on the screen are themselves the output of a computer program. Though the signs of this program remain invisible, they regulate the flow of words, determine their appearance, and, more generally, mastermind their choreography. In this respect electronic texts behave like movies: the hiding of the camera produces the illusion of a reality free of technology (Benjamin, "Work of Art," 236), and the hiding of the computer code produces the illusion of a text directly inscribed on the screen. If electronic texts "reconfigure" the notion of author, as many theorists have claimed, it is not by fusing writing and reading into one function but, as Murray has argued, by adding the dimension of algorithmic design to the production of the text: "Authorship in electronic media is procedural. Procedural authorship means writing the rules by which the text appears as well as writing the text themselves" (*Hamlet*, 152). This task requires different abilities than merely writing fiction. As the mathematician John Allen Paulos writes, the main obstacle to successful interactive narrative is "the dearth of writers capable of literary nuance and psychological subtlety and also of the architectural vision and software skills necessary to articulate such a complex branching 'story'" (*Once upon a Number*, 22). It is therefore likely that the electronic interactive texts of the future will be the product of teamwork, as are movies and art CD ROMs.[5]

4. *Hypermediacy.* Through its ability to combine text, sound, and image, electronic technology offers a new twist on the age-old dream of the total language, expressed in the nineteenth century by the opera and in the mid-twentieth century by the dramaturgies of a Brecht or Artaud. This twist is what Jay Bolter and Richard Grusin call hypermediacy. Whereas a genuine multimedia experience fuses various sensory stimuli into a holistic apprehension of the fictional world, hypermediacy "offers a heterogeneous space, in which representation is conceived not as a window onto the world, but rather as 'windowed' itself—with windows that open on to other representations or other media" (Bolter and Grusin, *Remediation*, 34). In a hypermediated display, one window may offer text, another sound, a third pictures or film, but the user experiences these various dimensions strictly one at a time. In contrast to the synthetic multimedia experience of movies,

the theater, the opera, and even VR, hypermediacy is a predominantly analytic assemblage of information bites. Its prototypical manifestation is the fragmented displays of the World Wide Web.

5. *Interrupt structure.* A classic linear text operates according to a protocol of succession known as the queue. The reader is supposed to finish each unit before moving to the next, and the first unit to encounter the reader's eyes is the first to be completely processed. In computer jargon this is known as a FIFO, or first-in, first-out structure. Through the possibility of placing hyperlinks everywhere in a lexia, electronic text transforms this protocol into a stack, or LIFO (last-in, first-out structure). When a reader selects a link located in the middle of a lexia, she interrupts the flow of presentation, argumentation, or narration and jumps to another topic. In a classic stack structure, such as a system of stories embedded in Russian-doll fashion, the reading of the interrupted unit is resumed after the interrupting segment has been processed, but in most types of hypertextual link structures the reader is allowed to drift from segment to segment without ever returning to her point of departure. This open invitation to pursue sidetracks, a strategy widely known as "surfing," explains why hypertext is regarded by its opponents as a mode of reading symptomatic of a generation with a shortened attention span.

6. *Animation and dynamic display.* Print culture was slow to realize the expressive potential of the visual aspect of language. It wasn't until the early twentieth century that poets such as Apollinaire in his *Calligrammes* began to explore the effects that can be derived from the interplay of black and white on the page. In the electronic medium the written word acquires a new, kinetic dimension. What movies have done to the image—putting it in motion—the computer can do to the signs of language: make them dance on the screen, twinkle, flash, change font and color, become three-dimensional, flatten out and disappear, or morph into other words. These effects have been systematically cultivated by electronic poets such as John Cayley, Eduardo Kac, and Jim Rosenberg,[6] and they play an extensive role in Mark Amerika's *Grammatron* (at the cost, some may say, of the more properly verbal quality of the text).

7. *Exploitation of temporality.* Reading a print book takes time, but the book form assumes that time is unlimited. Deadlines come from

teachers and library policies, not from the book itself. In the electronic medium the temporal allowance can be restricted and the passing of time be made significant. Many computer games operate in a simulacrum of real time, as they force the player to act as quickly as possible to defeat a moving opponent. Some electronic texts, such as Stuart Moulthrop's *Hegirascope*, exercise similar control over temporal resources: "In *Hegirascope*," writes Espen Aarseth, "the text fragments are 'pulled,' like a non-interruptible slide-show, after a specific number of seconds, typically fifteen to twenty, and replaced with the next fragment" (*Cybertext*, 80). Temporality could also be exploited by introducing real-time communication between the writer and the reader. The stream of words of a work in progress could, for instance, be displayed on the Internet as the author hits the keys, and the reader would be able to follow live the process of writing.

The linear print book has accustomed readers to an encounter with the text that combines certain duties with certain liberties and guarantees: duty to turn pages in sequential order, but freedom to dwell leisurely on each page, easy return to earlier passages, and security of always knowing how much remains to be read. This mode of processing has become so automatic that we tend to take it for natural and nonsignificant. The properties listed above are the elements of a vocabulary that will enable electronic authors to stage new modes of encounter between the text and the reader, to grant different liberties and to impose different servitudes, an inevitable trade-off that the advocates of hypertext too often overlook. By turning the electronic equivalent of the page into a theater wherein language itself performs in a dialogue with other signs and with a human agent, this exploration of the medium will consolidate the aesthetic significance of the software-design concept of interface.

CONCEPTUALIZATIONS OF INTERACTIVITY

To develop a sense of involvement with a computer program, the user needs a scenario that casts him in a role and projects his actions as the performance of concrete, familiar tasks: writing, editing, drawing, sending mail, building cities, or killing dragons to save a princess. In hypertext this conceptualization is problematic. Clicking on links is not really reading, because reading is something that is done with the eyes and not with the hand. Nor is clicking a matter of turning the pages, because this analogy would revert to the mode of reading characteristic of the codex book. Yet the need for interpretive scenarios that give meaning to the reader's engagement with the text is particularly urgent for a form of textuality as new as hypertext. One of the first theorists of electronic media to recognize the importance of metaphorical mappings for the psychology of computing was Brenda Laurel. In her pioneering book on the strategies of interface design, *Computers as Theatre*, she developed the theater metaphor in the hope of replacing the traditional and rather uninspiring perception of the computer as a tool with "the idea of computer as a representer of a virtual world or system, in which the person may interact more or less directly with the representation" (127). When the metaphor is adapted to the case of interactive textuality, it casts the screen as a stage, the run of the text as a performance, and the user as the director who summons actors to the stage and releases their lines through the click of the mouse.

Another popular interpretive scenario is a development of the spatial metaphor. In this mapping, the text as a whole is a territory, the links are roads, the textual units are destinations, the reader is a traveler or navigator, clicking is a mode of transportation, and the itinerary selected by the traveler is a "story." Since every reader follows a different itinerary, every reading session "writes" a different narrative, and through her own agency the reader becomes the "author" of her own adventures. The spatial metaphor supports different scenarios, depending on whether textual space is conceived as a "smooth" expanse that the reader cruises for the pleasure of the trip or as a "striated" space of freeways whose sole purpose is to lead to a destination. (Here again I rely on Gilles Deleuze and Félix Guattari's famous dichotomy.)[7] In a smooth-space environment, the reader is driven by an obsession to get further, either fortified or dampened in this drive by the thought that her desire to exhaust all the links cannot be satisfied. In a striated space, the reader gives herself a goal, such as reaching the center of a labyrinth, or finding the exit, and her relation to the text is very much that of a player who hopes to beat a computer game.

The reader's role in hypertext can also be conceived along the lines of the supermarket-shopping experience. In this scenario, suggested to me by Christopher Keep ("Disturbing Liveliness," 175), the reader browses along the links, takes a quick look at the commodities displayed on the screens, and either drops them into his shopping basket for careful study or moves on to other screens. This reader does not feel compelled to read the text in its entirety or to pay attention to every screen, because he sees the text not as a work held together by a global design but as a display of resources from which he can freely pick and choose. I was tempted to take this attitude with hypertexts such as Stuart Moulthrop's *Hegirascope* and Mark Amerika's *Grammatron*. In both cases the text gave me the impression of a loose collection of drafts and writing samples whose appeal resides in the imaginative quality, wordplay, or visual effects of the individual segments rather than in a sustained structure of expectations. In such an approach, the links between segments are not regarded as a dimension of meaning but as a mode of transportation to the next screen to be sampled. A reading session is mostly a treasure hunt for individual objects, and it does not build on knowledge acquired during previous sessions. In a variant of the supermarket scenario, the reader puts lexias into his shopping basket not to consume them individually but to use them as material to construct his own stories. This is how George Landow envisions the reader's ability to make narrative sense out of a nonlinear collection of lexias: "In a hypertext environment a lack of linearity does not destroy narrative. In fact, since readers always, but particularly in this environment, fabricate their own structures, sequences, and meanings, they have surprisingly little trouble reading a story or reading for a story" (*Hypertext 2.0*, 197). In this approach the reader creates a linear order that overwrites the system of links.

A third conceptualization relies on the metaphor of the kaleidoscope. In this scenario the text consists of a collection of fragments that can be combined into ever-changing configurations through the random choices of the reader. A variant of this analogy is the image of the construction kit, which attributes purposeful agency to the user and thus leads into the "wreader" theories mentioned in the introduction. In a kaleidoscope, the built-in mirrors create a symmetry that guarantees the aesthetic appeal of all permutations. By analogy, the measure of success for hypertext authors would be to design a system in which, to quote Bolter, "every path defines an equally convincing and appropriate reading" (*Writing Space,* 25). The metaphor of the kaleidoscope presupposes that sequence is significant, otherwise the system would not be able to produce ever-new images. Hypertext, we must remember, is not literally spatial; all segments must be read in succession. The significance of sequence is easily demonstrated in a narrative framework: "Mary had a baby," "Mary married Joseph," "Mary lost her virginity" tells an entirely different story than the same propositions read in the reverse order. (But we must supply a lot of material to make it meaningful, especially when we read it in the sequence listed above!)

Michael Joyce implicitly endorses the metaphor of hypertext as kaleidoscopic storytelling machine when he writes (with a probably broader concept of story in mind than the one illustrated above), "Reordering requires a new text; every reading thus becomes a new text. . . . Hypertext narratives become virtual storytellers" (*Of Two Minds,* 193). But if the system of links is reasonably well developed, the author cannot predict the paths that will be taken beyond the transitions out of, or into, a given segment. This prompts Gunnar Liestøl to write, "In reading hypertext fiction the reader not only recreates narratives but creates and invents new ones not even conceived of by the primary author" ("Wittgenstein," 98). Another implication of the kaleidoscope metaphor is that the reader *cannot* freely pick and choose; every new story consists of the same elements arranged in different configuration. To see a story—or at least, a meaningful juxtaposition—in whatever comes her way, the reader must be a priori willing to supply whatever missing information is needed to give meaning and coherence to the sequence of lexias created by her decisions. For every trip into the text to generate a genuinely new and autonomous story, the reader must erase from memory the knowledge gathered and the expectations created from previous reading sessions.

The kaleidoscope model works better with poetic texts in which the meaning of the sequence is not narrative but lyrical—that is, not logical, causal, and temporal but associative, thematic, and quite tolerant of incongruous juxtapositions. In Raymond Queneau's print text *Cent mille milliards de poèmes*, for instance, 10^{14} different sonnets can be generated by combining the lines of ten different sonnets of fourteen lines each. Here the notion of coherence is much more flexible than in narrative, and it is as easy to see meaning in each sequence of lines as it is to see shapes in the blots of a Rorschach test. But consider what would happen if the various combinations of the kaleidoscopic text were interpreted along strict narrative lines—y read after x meaning that y followed x in time. In certain traversals of Stuart Moulthrop's *Victory Garden*, the reader first learns that a character, Emily Runbird, has been killed by a missile in the Gulf war; then the reader encounters passages that tell of Emily's college days and of her love affair with a professor. If sequence had temporal meaning, this would imply that Emily comes back from the dead, an interpretation that leads to all sorts of unwanted consequences, for *Victory Garden* is on the whole a realistic narrative, not a fantastic tale or a religious myth. Of course, no reader will come to the absurd conclusion that Emily has been resurrected; the most elementary narrative competence tells us to read the passages narrating Emily's college years as a flashback compared with those that presuppose her death.

This strategy may seem obvious, but it has two important consequences for the phenomenology of hypertext reading. First, it means that sequence can be overruled by considerations of global coherence in the construction of meaning. No matter in what order the reader encounters segments, he will assume that Emily first went to college and then died in the Gulf war. And second, if the narrative meaning of sequentiality is neutralized, this means that hypertext does not tell a different story with each reading session but rather creates different modes of presentation for the same underlying plot. For those who subscribe to the classic narratological distinction between story, or *fabula* (what happened in the textual world), and discourse, or *sjuzhet* (the Russian formalists' term for the verbal realization and dynamic disclosure of the *fabula*), what changes from reading to reading in the case of *Victory Garden* is not the *fabula* itself but the *sjuzhet*. The vaunted self-transforming quality of hypertext affects the reading experience much more than it affects the textual world. Moreover, if we assume that hypertext projects a single *fabula,* rather than a radically new story in each reading session, this means that reading is cumulative, and that the construction of the *fabula* can span many sessions.

The interpretive scheme that best expresses the cumulative aspect of reading is the metaphor of the jigsaw puzzle. In this scenario the various lexias are the fragments of an exploded image that the reader tries to put together in a sustained effort. The system of links functions not as a carrier of semantic information but as one of these unnecessary obstacles that game masters put in the way of the player. The player often needs to take several looks at a piece before finding its proper place in the pictorial space. Similarly, the reader of hypertext may need several visits to the same lexia before she can gather sufficient information to place it in a meaningful environment. In the kaleidoscope metaphor, context is created by linear sequence, and every itinerary puts the visited lexia in a different context. This is why hypertext promotes what Liestøl has called a "constant recontextualization" (117). In the jigsaw-puzzle metaphor, by contrast, the order in which the player looks at the pieces of the puzzle has no impact on where he puts them in the global picture. This means that the context of a lexia is formed by those other lexias that shed light upon it, quite independently of whether or not they are electronically connected. The purpose of the act of reading is not to gain an overview of the map of links but to reconstruct a network of thematic, causal, or temporal relations that bears no necessary isotopy to the system of physical connections. The difference between this scheme and the narrative version of the shopping basket is that the reader assumes that there is a specific image to be unscrambled, rather than a collection of recombinant fragments.[8]

Yet another way to deal with the fragmentation and occasional inconsistency of hypertext is the space-travel, possible-worlds approach. Every lexia is regarded as a representation of a different possible world, and every jump to a new lexia as a recentering to another world. If we apply this strategy to Michael Joyce's *Afternoon,* there will be one world in which Peter's ex-wife and son are killed, and one in which they are not; one world in which Peter causes the accident, and one in which he merely witnesses it. Each of these propositions expresses a solid fictional truth in its own reference world. If the reader takes this approach, she will not try to reconstitute a comprehensive world image, and the co-referentiality of the lexias—that is, the fact that they are about the same individuals—will be explained through a semantic model that allows individuals to have counterparts in different possible worlds. This approach is a way to rationalize texts that present a high degree of internal contradiction, since there is no logical problem with p being the case in one world and $-p$ in another, but if every lexia is regarded as describing a self-sufficient monad, narrative development cannot span several lexias, and the only stories that can be told are those that are contained within a node. It seems, therefore, improbable that readers will apply this model in a systematic fashion. Rather, they are likely to assume that certain lexias belong to many possible worlds, so that the world in which Peter's wife and son are dead and the world in which they are alive will share a common history built on the basis of lexias that are compatible with both states of affairs, such as those that assert the existence of an ex-wife and a son for Peter.

Which one of these scenarios will be preferred depends as much on the individual disposition of the reader as on the nature of the text. Easily detectable relations between adjacent segments suggest the metaphors of travel or of the kaleidoscope, because they give meaning to sequence; links that break up semantic continuity tend to redirect the reader's focus of attention toward a global coherence to be reconstituted in jigsaw-puzzle fashion; and relatively autonomous lexias will promote the shopping-cart scheme. The supermarket metaphor is also best suited for readers with a short attention span. Meanwhile, the kaleidoscope metaphor should appeal to those who are fascinated by combinatorics, while the jigsaw-puzzle metaphor will be favored by those who are naturally inclined to approach literary texts as problems to be solved. It is in this spirit that I read Michael Joyce's *Twelve Blue* in the next interlude. The structural fluidity of most hypertexts, *Twelve Blue* included, is usually compatible with several conceptual schemes, but this fluidity also prevents the reader from feeling totally secure in any one of them.

One common feature of all the interpretive scenarios described above is that they present the reader's conceptualization as an attempt to overcome the fragmented appearance of the text and to restore some kind of coherence—either that of a spatial landscape or that of a temporally and causally ordered narrative. Postmodern theorists may object that such strategies try to correct, and therefore implicitly reject, the fundamentally disjointed, fragmented, and untotalizable "nature" of hypertext. The basic dilemma, as I see it, is whether memory and something worth calling "understanding" can live under such conditions of segmentation, or whether linearization and totalization—even if this means only a subset of the text—are elementary cognitive activities.

Adventures in Hypertext

Michael Joyce's *Twelve Blue*

One of the greatest obstacles to the popularization of literary hypertext, both in academia and in the public at large, is the difficulty of finding an appropriate descriptive and critical idiom. As Michel Bernard observes ("Lire"), reading hypertext is a solitary, highly individual experience that is difficult to share. If different readers hardly ever traverse the same material in the same order, if the hypertext novel "changes with every reading" (Joyce, *Of Two Minds,* 35), how could this new literary form create reader communities similar to those that built the reputation of the classics of print literature? It is by exchanging ideas about literary works that readers establish the cultural importance of those works. Will the critical discourse of the electronic age be something on the order of "signal[ing] that by clicking on a certain word on a certain page one can reach a description that nobody has read before" (Bernard, "Lire," 318; my translation)? Or will this discourse retain the stance that has traditionally dominated the criticism of print literature: the perspective of an omniscient Superreader who, having committed every word to memory, and enjoying a panoramic vision of the entire text, authoritatively dissects ideas, themes, style, narrative techniques, and plot (or the lack thereof)? The first alternative reduces criticism to the status of a cheat-guide for a computer game, while the second misses the dynamics of the reading process, whose control forms precisely the point of the interactive framework.

When Michael Joyce writes that "what happens in the work of art *is* the work of art" (*Of Two Minds,* 200), I assume that he has in mind not merely the plot of the novel but mostly the dynamics of its disclosure, as well as the narrative constituted by the reader's quest for meaning. By diversifying the process of discovery and denying complete

INTERLUDE

knowledge of the body of the text—you never know if you have seen all the nodes and followed all the links—hypertext puts to rest all notions of a Model Reader, Ideal Reader, Average Reader, or Super-reader, but it does not completely exclude a shared experience, because all travelers in hyperspace encounter sooner or later the same interpretive challenges. The critical discourse that will secure the place of interactive texts in literary history may still remain to be invented, but it is not too early to derive from the hypertext experience some cognitive lessons about the nuts and bolts of the reading process. It is in the hope that my reading is exemplary at least for the questions it asks and for its quest for coherence that I offer here a narrative based on the diary of my own adventures in *Twelve Blue: Story in Eight Bars*, a hypertext short story by Michael Joyce publicly available on the Internet.

Every hypertext has a fixed entry point—there must be an address to reach first before the system of links can be activated—but in the case of *Twelve Blue* this entry point is not a room but a hallway with many doors: the picture shown in figure 3. The image on the screen consists of twelve largely parallel, occasionally intersecting, lines of different colors—mostly blue, but there are also a striking yellow, a pink, and a purple line—that look like strands of yarn, or like chains of mountains in a hazy landscape. Part of the action indeed takes place in what could be the Blue Ridge Mountains of Virginia, and in a passage buried deep in the text I will find an allusion to the picture on this first screen that associates the colored threads with the destinies (or story lines) of characters:

> She looked out on the creek and measured out the threads like the fates, silk thread in twelve shades of blue. (Is pink blue? Is yellow or purple? She supposed so, she believed in her stories.) ("Fates")

On the right side of the box the strands become fuzzy and disappear before reaching the edge, suggesting unfinished stories. The numbers 1 to 8 divide the picture into eight vertical bars. The beginning of the story is determined not by the line but by the bar on which the reader clicks. In other words, what matters here are the horizontal, not the

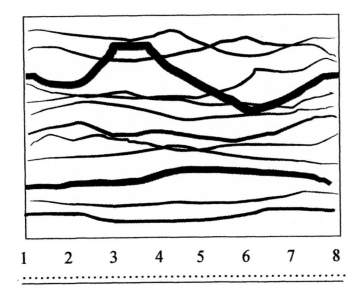

1   2   3   4   5   6   7   8

"So a random set of meanings has softly gathered around the
word the way lint collects. The mind does that."
From *On Being Blue* by William Gass

FIGURE 3 | Title page of Michael Joyce's *Twelve Blue*
Note: Redrawn by author.

vertical, coordinates. There are consequently eight possible beginnings—all different, as I find out through systematic testing. From this
point on, every screen of text includes on the left one of the eight
vertical bars, and the twelve lines that cross the bar function as links to
other fragments. Some of the fragments are too long to appear entirely on the screen, and the reader must scroll to get to the end of the
fragment. This creates the risk of missing part of the text, especially
when a paragraph ends at the bottom of the screen. In addition to the
twelve lines of the bar, certain fragments contain bits of underlined
text that function as internal links, or invisible links hidden in the
spaces between the paragraphs. When the reader clicks in the blank
space, underlined text briefly appears, and then is automatically replaced by a new screen. After a while I become suspicious of the trick,
and I systematically click on all spaces that contain more than one

blank line. Sometimes they hide a link, and sometimes they do not. In contrast to the links on the bar to the left, the text-internal links disappear after they have been followed.

Right from the beginning I face a dilemma: Should I read for the plot (or whatever semblance of plot the text might offer—my acquaintance with Joyce puts a damper on my hopes of finding a well-made, stable story)? Or should I first try to reconstitute the map and the logic of the linking? The strategies appropriate to each goal differ: if I read for the plot I will favor a "depth-first" exploration, venturing further and further along the chains of links, while if I read for the map I will go "breadth-first," performing backtracking operations to try all the paths that lead out of a given node. Fortunately for the pleasure of the investigation, the two operations cannot be kept strictly separate. As I go for the plot I get an idea of the linking strategies; as I go for the map I discover new fragments that fill important gaps in my reconstruction of the plot. My attempts to reconstitute the purely physical map of *Twelve Blue* ultimately produce largely negative results, if by *physical map* one understands a graph that enables the reader to find the path between any two given segments. But I become convinced during the course of my reading that drifting through the text is more rewarding than navigating with a purpose, because the only map that really counts is the one that represents the system of purely *thematic* relations—a system that often overwrites the network of physical links. The thematic logic of the textual space will not reveal itself until I accept the wisdom of the Gass epigraph (fig. 3) and place my trust in random navigation.

This much I am able to establish about the linking logic: that clicking on each of the twelve lines on the bar displayed on the left of every screen will lead to a different segment. By repeatedly clicking on a line of a certain color, one gets for a limited time an impression of continuity in the plot, such as several segments about the same character(s); but after a while the thread breaks down, either by repeating a previous sequence, by remaining stuck on the same screen, or by jumping to another narrative line. Clicking on the internal links is the best way to maintain narrative coherence, but these paths invariably turn out to be very short, since most of the screens do not offer such an option. My final impression of the link structure is that the colorful threads that the text dangles in front of me are mostly deceptive guides—all the more deceptive because they occasionally suggest a semblance of continuity. The effect is that of an amnesiac mind that desperately tries to grasp some chains of association but cannot hold on to them long enough to recapture a coherent picture of the past.

As I begin my exploration, the dilemma between breadth and depth hits me on the level of the individual fragments. A breadth-oriented reader tries to get a general overview of the text before paying attention to detail, and will therefore tend to skim every segment the first time around, while a depth-oriented reader turns the page, or clicks, only after gathering a maximum of information from each segment. The two strategies are comparable to the two modes of transmitting pictures over a computer network: in one mode, the picture fills its frame right away with colored squares of coarse definition, and the grain is progressively refined over subsequent passes; in the other mode, the picture fills its frame pixel by pixel and line by line, slowly eating away the blank part of the screen. All reading is probably a composite of the two strategies—we never get all the information on the first pass, and even if we do not reread segments physically, we revisit some of them mentally—but the depth-first approach seems better suited to linear texts, and the breadth-first approach to hypertext. In a standard print text, each segment appears in a determined context. When I read a passage, I assume that the author knows to what information I have been exposed and expects me to process the new information in the context of the old. Knowing what the reader knows enables the author to plan more effectively the disclosure of new information. But in hypertext, the immediate context of every segment is highly variable. I may reach a segment through various routes, and what is readily understandable to some readers may be totally opaque to others. Under these circumstances it is better to pick and choose and move on than to clutter memory with unclassifiable information, especially since the linking system makes it very likely (though not guaranteed) that the reader will eventually return to the same node.[1]

My first foray into the text is concerned with the inventory of the basic furniture of the fictional world and with the construction of the web of relations that forms its human and spatial geography. To cut

my way through the jungle of data on the screen, I resolve to establish
for every segment a list of characters, identified by name or definite
description, a setting, and possibly a theme or a striking image. These
notations should serve as a mnemonic aid, for nothing is more reluctant to inscribe itself in memory than the volatile signs of the screen. I
also hope that by writing down the name of each segment and its most
salient information I will be able to keep track of the units I have seen.
Here is a sample of what I am able to get out of each screen during my
first steps into the text:

— "Follow Me"
   Setting. The porch of a country house in summer.
   Characters. A fifteen-year-old girl with blueberry cotton
   candy and a gap in her teeth; a woman lying on a couch who
   needs to pee; a man.
   Theme. The girl pours wine for the man, but I don't know if
   this is imagined or real.
— "Each River"
   Setting. Upstate New York (mention is made of the Hudson
   and of Albany).
   Character. A woman scientist who makes biological slides.
   Theme. The woman thinks of a river and of life forms.
— "Look Out"
   Setting. Indefinable. The passage evokes mental processes
   that visit many places. We do not know where the con-
   sciousness is located.
   Characters. Second-person narration. Does "you" count as a
   character?
   Theme. Instructions to "you" on how to keep a mental im-
   age of a scene that contains water and a beach. The last line
   suggests drowning.
— "Cleopatra's Toes"
   Setting. Near a pool.
   Characters. Aurelie (focalizer), Lisa, Tevet.
   Theme. Aurelie watches Lisa swimming, reflects on the similarity of her relations to Lisa and Tevet.

During my first screening of the text I am so overwhelmed with
unclassifiable information (noise?) that I grab the first image that

strikes my fancy and use it as memory aid to identify the segment. During later visits I situate the text in a richer context, and I am able to process more and more information. This incremental mode of reading constitutes the most genuinely nonlinear aspect of the hypertext experience.

The text seems to take a perverse pleasure in frustrating my attempts at classifying information into neat categories. The process of world construction is hampered by the dominant narrative mode of *Twelve Blue,* a kind of lyricized meditation focalized through a third- or second-person character. These meditations do not seem to capture the thoughts of a given individual at a given moment but rather trace the musings of a floating, atemporal and hybrid consciousness that belongs as much to an impersonal narrator thinking for the characters as to the characters themselves. In many cases the center of consciousness is identified by a pronoun, but at first I cannot link these pronouns to the few names I have gathered through my reading. I know what the characters think and perceive but not who they are. The following beginnings of fragments are typical of the referential opacity that permeates the text:

> He tried to look at things from her viewpoint. ("Shipwrecked")

> Think of lilacs when they're gone. She looked out the window to the water and tried to think what they were doing upstream. ("Blue Mountain")

> She says August is a month of Sunday nights. She thinks of an electric fog of blue light in steamy living rooms all along the street. ("Blue Room")

When a standard fiction begins with an unresolved anaphora, such as "She sat by the window watching the evening invade the avenue" (James Joyce, "Eveline," 36), the occurrence of a name after a few sentences usually enables the reader to identify the referent, but in hypertext the process of identification is complicated by the variability of the context. For instance, if I read in succession "Blue Mountain" and "Blue Room," should I apply the rules of discourse coherence and conclude that the referent of the pronoun *she* is the same person on each screen, or should I assume that the reference must be renegotiated with every segment? The first alternative would truly allow the text to change with every reading, since the resolution of the anaphora would depend on which nouns or proper names appear on the preceding screen, while the second approach determines reference on the basis of a stable context—the content of the segment. I decide that the best way to get a coherent picture is to assume that each segment determines its own referents, but throughout the text I feel teased and challenged by the use of pronouns.

As I try to establish a list of dramatis personae by writing down names, pronouns, and definite descriptions of voluntary agents, I gradually realize that there is no absolutely reliable method for deciding who does and does not count as a character among the human referents of the text. "Characterhood," in *Twelve Blue*, is a fuzzy predicate. Some human agents are named (Javier, Lisle, Aurelie, Lisa, Samantha, Tevet, Ed Stanko, Eleanore, and Delores Peters), while others are referred to by definite descriptions (the drowned boy and his girlfriend; the drowned woman in California and her daughter); some seem to exist objectively in the fictional world, while others are merely imagined (the Portuguese sailor, hallucinated lover of the fictionally real Eleanore); some are part of the narrative present, while others are only remembered (Delores Peters, subject of Lisle's childhood recollections); some are native of the fictional world, while others are part of the cultural background (Eleanore of Castile); some possess a unified personality, while others represent the avatars of a multiple identity (Eleanore is also the Elli of Javier's past, as well as a whore and a mad goddess).

The establishment of a stable list of characters is further complicated by phenomena of homonymy (a proper name borne by two different characters) and synonymy (a character bearing two different names). In a self-descriptive statement, the segment "Blue Mountain" tells us, "It's hard to keep the names straight, like a Dickens novel." There are three Eleanores, one a student of George Landow who has gone insane (is this a comment on what hypertext does to the mind?) and the other two queens of England (Eleanore of Castile and Eleanore of Aquitaine). There are also two Lees (one better known as Aurelie and the other as Lisle), an Ed (Stanko) and an Eddie who may or may not be Ed as a boy, and two Javiers, a doctor and a Portuguese sailor, but they are manifestations of the same person in the mind of different characters. Meanwhile, Eleanore is also referred to as Elli, and Tevet changes her name to Beth. A similar doubling of names complicates my effort to keep track of the screens I have seen: there are two "Blue Mountain's," two "Ophelia Falls's," two "White Moth's," two "Anchored," a "Riddle" and a "Riddles," and numerous "Blue" somethings.

Initially I can construct only a list of disconnected names and pronouns associated with occasional properties, but many of my questions are suddenly answered when I hit what Espen Aarseth would call an epiphanic segment that generously discloses a network of interpersonal relations. "Tongues" reveals that Javier Reilly is a doctor who used to be married to Aurelie but is now involved with Lisle, a woman doctor from Canada. Aurelie, after "unmarrying" Javier, has taken a female lover named Lisa, a former competitive swimmer. Javier and Aurelie/Lee have a teenaged daughter, Tevet (or Beth), and Lisle/Lee has a teenaged daughter whose name is Samantha. Now that relationships are straightened out, I feel that I have a hold on the human geography that underlies the plot, and as I continue to circle around old and new fragments I am able to locate most of them in a global picture. This stage of my reading, by far the most pleasurable, reminds me of the moment in the reconstruction of a jigsaw puzzle when pieces suddenly begin to fall into place. The snapshots of narrative action form several branches and an island:

1. *The upstate New York branch.* A woman, Lisle, sits on a porch overlooking a pond with her daughter Samantha. She is making a quilt, an obvious image of the patchwork structure of hypertext.[2] Her thoughts return obsessively to the memory of a deaf boy who drowned while swimming in the pond. The scene of the drowning is replayed in various modes, sometimes as the factual narration of the boy eloping with a girl to go swimming, sometimes as remembered and relived by Lisle ("she willed him to the shore"), sometimes as a story made up by Samantha, but most memorably as the stream of consciousness of the boy in the moment of sinking:

> Once he got used to it it didn't seem so bad, only lonely, far lonelier than he ever had imagined. At first he felt his heart like a falling anchor until it stuck, caught on a bone or rock deep in

the muck and hooked there, steadying him. For awhile he was comforted by the distant light of a woman's eyes scanning over the estuary like the sheriff's searchlight. After a time that too grew familiar and vaguely distant like the ache of the anchor within him. Then he became slowly aware of the damp smells of the shore, lilacs and the metallic smell of blood, musk, clove, a faint odor of fuel oil. Something else, the powdery smell of the girl who haunted it and the sweet, indistinct rot of the log where she waited for him.

Lonely, far lonelier than he imagined.[3]

He pulled the water over him like a blanket and slept, anchored in the gaze of an unknown woman and the girl who loved him. ("Anchored" I)

Another fork of this branch shows Lisa and Aurelie in a garden, Lisa amorously contemplating the body of Aurelie while Aurelie's thoughts move from Lisa to her concerns for her daughter Tevet, who represents the dimension of her life from which Lisa is excluded. It is also in the upstate New York landscape that I place a series of segments recounting the relation between Tevet and Samantha, who are both confronted with a human death during the same summer.

2. *The Canadian branch.* This strand captures memories of Lisle's youth in Canada: rush hour in industrial landscapes, life at a convent school, being excluded by the nuns from the choir, running away with the carnival at age fifteen, sliding in the snow with a boy named Eddie, making love with him on a carnival ride owned by the parents of Delores Peters, and somebody going over Niagara Falls in a barrel.

3. *The Blue Mountains branch.* This is the most elaborate narrative branch. Javier takes his daughter Tevet on a trip to the Blue Mountains to look at the only existing picture of his grandmother, Mary Reilly. The picture has now come into the possession of a coarse beer-drinking man named Ed Stanko who first charges Javier an extravagant amount to borrow the picture for a photocopy and then arranges with the photocopy-shop owner to prevent Javier from getting hold of the reproduction. Through passages narrated from Stanko's point of view, filled with profanities and spelling mistakes, we learn about Stanko's Catholic youth, his failure in marriage, his self-hatred, and

his expectation of burning someday in hell. He inherited a hotel from his dead wife, Flossie (a Reilly relative), and turned it into a dingy apartment house. One of his tenants, Eleanore—presumably the same person as the Landow student, who has sworn to kill Stanko—is believed to be a whore. When Javier and Tevet return to Stanko's house they are greeted by a madwoman who tells them that "señor Stanko is indisposed." Soon after the body of Stanko is taken out of the house in a white sheet, Eleanore is arrested, and Javier and Tevet return, very shaken, to their home in New York.

On later visits to the Blue Mountain action zone, I encounter Eleanore's preparation for a ritual killing of Stanko: she perfumes herself, cleans the tub, fills it with flowers, and stabs him in his bath. (I was surprised to find no textual reference to Marat and Charlotte Corday!) A lyrical passage that starkly contrasts with the vulgar tone of Ed's reminiscences captures the lonely but peaceful experience of surrendering consciousness to the waters of death:

> For awhile he was blinded by the blue rage in the woman's eyes but after a time that too grew familiar and vaguely distant like the anchor within him. Then he became slowly aware of the smells in the killing room, lilacs and the damp, metallic smell of blood, musk perfume, faint clove, and the oily smell of the steel blade, the familiar stench of shit in his drawers. Something else, the smell of a woman in sex.
>
> He pulled the water over him like a blanket and slept, lonelier than he had ever been. ("Anchored" II)

4. *The California island.* Consisting of only two segments, "Naiad" and "Salt Shores," this group evokes the drowning of the wife of a famous scientist in a scuba-diving accident, and the daily visits to the beach of her little girl waiting for her mother to return. I call these two screens an island because the ties to the other narrative clusters are strictly analogical: the theme of drowning, and the mother-daughter relation.

At this point I feel that I have reconstructed a fairly stable narrative foundation, but the motivation of the most salient actions remains obscure. I will never find out why exactly Eleanore wants to kill Stanko or why Stanko is so intent on preventing Javier from getting hold of the picture of his grandmother. As a segment titled "Riddle" suggests, these are probably questions without answers:

> What links the dead man and the murderer, the drowned man and the shore, a once wife and her current lover, dream to memory, November to the new year?
>
> What links daughter to daughter, girl to boy, sky to moon, blue river to blue air?
>
> Why do we think the story is a mystery at heart? Why do we think the heart a mystery?
>
> Who shares one voice?

The text is not a whodunit, and the motivation of its main events is better found in symbolism and textual architecture than in the particular interests of the characters. Eleanore kills Stanko as much for the sake of the symmetry of the two drowning scenes related in "Anchored" I and II as to enact a sexual fantasy. As for Javier's failure to acquire the only existing picture of his grandmother, I interpret it as a warning that the past is not an object that can be owned, framed, and displayed but an interior landscape, a hypertext of the soul whose prominent sites can be reached only through the secret links of memory. While Javier's trip to the Blue Mountains ends in the macabre discovery of a dead body, the past is forcefully brought to life in the reminiscences of a Canadian childhood that play and replay in Lisle's interior monologue. Delores Peters and her father's carnival ride are a presence in the text, but Mary Reilly is nothing more than a dead ancestor.

From this point on my reading mostly wanders through segments that I have already seen, but many of these screens yield new information that tightens the relations between the various strands of the plot. Late in my exploration, for instance, I discover that Javier slept once with a woman named Elli, a "mad goddess," a queen ("Shipwrecked"). This reminds me that Eleanore lives in the mountains "on her wits and a small pension from the king" ("Eleanore Cross"). Javier, meanwhile, is referred to by Stanko as the king of England. Repeated mentions of a mysterious unborn daughter of Eleanore's, fruit of her love for a Portuguese sailor named Javier, lead me to suspect that in the realm of material causality (the textual actual world), she became pregnant by

Javier the doctor, that the child died, and that she now lives on the support that he sends her. But in the various possible worlds of her imagination—real life would diagnose her condition as multiple-personality disorder—Eleanore *is* a queen, a goddess, a whore, the lover of a Portuguese sailor, and the unborn child is alive. Whenever the text penetrates her point of view, the world of her madness acquires reality status.

For all its zones of undecidability, I find that the universe of *Twelve Blue* presents significant areas of ontological stability. The narrative events often read like a dream, especially those that have to do with the murder of Stanko, but even dreams have their actual world. Except for the passages that relate to Eleanore, which can be naturalized by reference to her madness, the text does not assert logically incompatible states of affairs, as does to some extent Joyce's well-known hypertext novel *Afternoon*.[4] Even though all of the segments adopt a highly subjectified perspective—even the impersonally narrated passages avoid the stance of absolute, omniscient narrative authority—we get a reasonable idea of what counts as fictional fact and what is imagined by characters from the mutual comparability of the various private worlds. It is because the text creates a zone of intersubjectivity from the overlapping contents of the minds of Javier, Lisle, Aurelie, Lisa, Tevet, and Samantha that I assume that certain characters exist autonomously, and that some events take place objectively. If *Twelve Blue* challenges classical ontology, as most postmodern texts do, it is not by frustrating the reader's quest for fictional truths or by postulating more than one actual world but by offering a more diversified ontology than the standard binary opposition of actuality and virtuality. Between the realm of the solidly factual and the realm of the hallucinated, the text creates a zone of free-floating, dreamlike existence (though it is nobody's dream), populated by objects and characters that seem to exist mainly as poetic images.[5]

It is in this ontological limbo that I situate three screens that relate to the theme of drowning: the scuba-diving death of the woman in California and the two "Anchored" passages quoted above. The imaginative presence of these episodes of indeterminate ontological status suggests a mode of reading that transcends, or rather supplements, narrative logic and its need to categorize information as either factual or inscribed in a possible world belonging to the domain of a specific character. In this other mode of reading, every screen recenters the textual universe around a subjective world, whether or not the mind that projects it can be identified, and every representation becomes present and actual. Through this recentering, the subjectivity put on display becomes the hero of its own story, and every minor character—minor in other people's stories—gets a turn as major character. As we read in "Fates," "She had taught herself abandon, taught herself to understand that they were not minor characters, she and her daughter, but at the center of something flowing through them." The purpose of traveling around the text is no longer to reconstitute an objective plot but to join a stream of imaginative activity that flows through a network of interconnected subjectivities.

This idea of moving from one perspective to another within what may be called a collective consciousness[6] is reinforced in *Twelve Blue* by the use of pronouns that could be replaced by any of a number of possible referents, as well as by several passages that thematize the blurring of identity boundaries between the self and the other. Lisle and Samantha are so preoccupied with the drowned boy that he lives and dies through them; Samantha sees herself as his girlfriend or as the sister of Tevet, and the reader is invited to enter the consciousness of the drowned boy in a passage narrated in the second person:

> You get used to floating, it is after all only a resumption of what we all once had and lost in the light. Even the sounds are the same: the thump and rush of blood, the dark static of the nerves, the soft cry of silence. ("Bright Balloons")

How does this particular text benefit from the hypertext format? The attitude with which I initially attacked the text—and I mean *attack* to be taken in its full force—had much in common with the frame of mind of the player of a computer game or of the reader of a mystery novel. I was determined to "beat the text" by figuring out what the system of links and the multiple ambiguities were designed to hide from me. The understanding of "what the text is all about" was the hidden treasure at the center of the labyrinth or, to return to the jigsaw-puzzle metaphor, the global picture to be reconstituted from the bits and pieces of narrative information provided by each segment.

Hypertext has been credited with offering an alternative to the Aristotelian curve of dramatic tension—slow rise, climax, and sudden fall—but the pleasure of the problem-solving activity follows its own rhythm of mounting and decreasing intensity. At the beginning, the reader is frustrated by a lot of incoming information and an absence of pattern. Pleasure peaks when a pattern begins to take shape, but this also marks the point at which the rate of new information begins to decline. The more the pattern fills out, the more difficult it is to locate new information to fill in the holes. Reading ends not when the plot is conquered but, as Joyce himself suggests in the introduction to *Afternoon*, when the reader becomes finally tired of circling through the same screens. *Twelve Blue* can be compared to the field in the fable of the plowman and his sons in that it contains no hidden treasure that makes everything fall definitely into place on the narrative level. Yet like the sons who plowed the field and made it more fertile, the reader who has been patient enough to explore the text in depth will find ample rewards in its poetic images and in its complex pattern of recurrent motifs.

When the reader's curiosity about the basic configuration of the fictional world has been reasonably satisfied, she enters a second stage of reading in which the hypertextual format is no longer a means to scramble a plot but a simulation of the dynamics of the imagination. Through the interactive mechanism of the text, the reader is invited not only to attend the projection of the film of the characters' inner lives, as she does in the stream of consciousness of the print novel, but also to run, perhaps even to pretend to *be*, the machine that records and projects images on the screen of the mind. (In this metaphor, consciousness is a camera that records and projects at the same time.) The randomness of the act of clicking figures that which is beyond conscious control—the subliminal, the obsessive, paths to the forgotten—in the mode of production of the imagination. In a collective consciousness patterned on the model of an individual mind, as it is in *Twelve Blue*, reaching a given subjectivity is no more predictable than moving from image to image in the thoughts or dreams of a particular subject. As one of the underlined phrases that serve as internal links self-referentially tells us, "Wake from one dream into another" ("Cornflowers"). Just as we never know what a dream will bring next, we never know into whose dream we will awaken by selecting a link.

The importance of the image of water and the scenes of drowning in *Twelve Blue* can be read as a literalization of the metaphors of flow and fluidity that permeate so much of contemporary thought, from AI to architecture and from New Media to New Age philosophy.[7] Through a pun created by my own preoccupations, the drowning theme also raises the question of the immersive power of the text. We should be careful to distinguish here the immersivity that derives from the plot or images of this particular text from the immersivity of the medium itself. During the phase of the construction of the fictional world I became quite absorbed in this task, but I would not call this a truly immersive experience, because my pleasure, like the thrill of the jigsaw-puzzle solver, had more to do with fitting parts together than with an intrinsic interest in the picture I was reconstructing. The murder of Stanko teased me with a semblance of temporal immersion, but the text did its best to send me on trails that cured me of my hopes of solving the mystery. This suggests, however, that the hypertext format could provide the type of immersivity of the detective novel, as do some computer games, if it were based on a determinate and fully motivated plot.

As far as spatial immersion is concerned, I did encounter some lyrical passages that evoked places and atmospheres in which I was tempted to linger: the porch of the summerhouse where Lisle is making a quilt, the damp shores of the pond where the boy drowns, and the surrounding forest with its earthy smells, buzzing insects, and dense vegetation delicately detailed by botanical names. The evocations of the sylvan flora unlocked the door of some of my favorite childhood memories, just as did the image of the faded lilacs for Gregory Ulmer (cf. quotation in chap. 4). Some screens, such as the two titled "Anchored," were prose poems that I would have enjoyed printed on heavy blue-colored paper and detached from the hypertextual context, even though they do enrich the comprehension of the text as a whole. During my first pass through *Twelve Blue* I was so preoccupied with restoring a semblance of order in its informational chaos that I hardly took the time to slow down and properly savor a passage. The twelve lines on the left of each screen kept urging me to make use of my freedom to click, to move on and find out what lay beyond the screen. When I encountered passages that tempted me to take a deep breath and inhale the flavor of language, I cheated the electronic medium by printing them out. This allowed me to postpone their rereading and to move on in my elusive quest for narrative coherence. (Without printouts, and without a map of the network, who knows if I would ever get another look at a given screen?)

The mosaic of hypertext can contain any kind of writing, including prose of a lyrically haunting quality, as *Twelve Blue* amply demonstrates, but the medium itself does not favor meditative contemplation, and the reader must fight it to pause and take in the presence of a scene. It is only by shutting down the hypertextual machine, by temporarily forgetting about interactivity, that I was able to let myself be carried away, like Ophelia, like Stanko and the drowned boy, by word streams like this one:

> Everything can be read, every surface and silence, every breath and every vacancy, every eddy and current, every body and its absence, every darkness every light, each cloud and knife, each finger and tree, every backwater, every crevice and hollow, each nostril, tendril and crescent, every whisper, every whimper, each laugh and every blue feather, each stone, each nipple, every thread every color, each woman and her lover, every man and his mother, . . . every shadow, every gasp, each glowing silver screen, every web, the smear of starlight, a fingertip, rose whorl, armpit, pearl, every delight and misgiving, every unadorned wish, every daughter, every death, each woven thing, each machine, every ever after ("Each Ever After").

EIGHT

Can Coherence Be Saved?

Selective Interactivity and Narrativity

When the writer provides two different endings to his novel (why two? why not a hundred?), does the reader seriously imagine he is being "offered a choice" and that the work is reflecting life's variable outcomes? Such a "choice" is never real, because the reader is obliged to consume both endings. In life, we make a decision—or a decision makes us—and we go one way; had we made a different decision . . . we would have been elsewhere. The novel with two endings doesn't reproduce this reality: it merely takes us down two diverging paths. It's a form of cubism, I suppose. And that's all right; but let's not deceive ourselves about the artifice involved.

After all, if novelists truly wanted to simulate the delta of life's possibilities, this is what they'd do. At the back of the book would be a series of sealed envelopes in various colors. Each would be clearly marked on the outside: Traditional Happy Ending; Traditional Unhappy Ending; Traditional Half-and-Half Ending; Deus ex Machina; Modernist Arbitrary Ending . . . and so on. You would be allowed only one, and would have to destroy the envelopes you didn't select. That's what I call offering the reader a choice of endings.

—JULIAN BARNES

The future of hypertext as literary form rests to a large extent on its power to generate a type of meaning that is shared by almost all of the popular modes of entertainment, from drama to the novel and from movies to amusement parks. It is the ability to tell stories that will decide whether hypertext will secure a durable and reasonably visible niche on the cultural scene or linger on for a while as a genre consumed mostly by prospective authors and academic critics.

The opinion of theorists, hypertext authors, and game designers on the narrative potential of interactive works reveals two fundamentally different conceptions of narrativity. In the first view, most common among those who build their ideas from complex literary texts, narrative is a form of representation that varies with period and culture; in the second, typical of those who study simple stories (cognitive psychologists, discourse analysts, and folklorists), narrative is a timeless and universal cognitive model by which we make sense of temporal existence and human action. For proponents of the first alternative, the New Novel and postmodern literature have radically altered the conditions of narrativity, while for the partisans of the second, the postmodern subversion of plot, character, causality, closure, linearity, and overall coherence has merely loosened the generic ties of the novel with narrative structure. In the first view, prose fiction that refers to individual existents and relates mental or physical events is automatically narrative; in the second, narrativity requires a certain semantic form, and its realization is a matter of degree. And finally, in the relativist approach, the evolution of the way stories are told affects what counts as a story (if the dichotomy of story and discourse is considered valid at all), while in the universalist view, new narrative techniques can be introduced and different types of stories can be developed (simple, complex, epic, dramatic, digressive, and even multilinear) without altering the nature of narrativity: complex narratives are simply more elaborate constructions of simple elements, and postmodern novels are incomplete realizations or deliberate underminings of narrative structures.

For those who conceive narrative as a historically and culturally variable form of representation, hypertext is a breakthrough that will accelerate its evolution. It is in this spirit that George Landow and Ilana Snyder title a chapter of their respective books *Hypertext 2.0* and *Hypertext* "Reconfiguring" and "Reconceiving Narrative." For these authors, who speak in the name of the experimental wing of hypertext writers, the development of interactive mechanisms is both a new way of telling stories and a generator of new narrative structures: broken up, open, without rise and fall of tension, unstable, multilinear, created in the act of reading, multiple, and so on. The universalists will of course question the narrative grammaticality of these structures. Does hypertext tell stories at all, or is it primarily a machine for the dismantling of narrative?

The developers of interactive texts who work for the commercial sector tend to take a much narrower, Aristotelian view of narrativity than academic writers and theorists. They know that the popular success of an interactive work—be it a soap opera, like *The Spot,* a mystery story, like *The Lurker Files,* or an adventure computer game, like Myst or Zork—depends on its ability to create an immersive experience, and classic narrative structures are the most time-tested recipe for keeping the user spellbound. Compared with the literary experimentalists, these writers and designers are much more skeptical of the narrative potential of interactive media. "I believe that on a fundamental level, storytelling and interactivity are exclusive to one another," writes Talin ("Real Interactivity," 97), a self-described "costumer, programmer, artist, musician, storyteller," and award-winning author of games (including the Faery Tale Adventure). Celia Pearce, designer of nonlinear adventure rides for theme parks (the Loch Ness Expedition), reinforces this opinion: "When I told a friend I was writing a paper on non-linear story-telling, he frowned. 'Isn't that a contradiction in terms?' And he's right. It is. Because any high school English teacher will tell you that a story is defined as having a clear beginning, middle and end. This implies a linear structure" ("Ins and Outs," 100). These pronouncements should not be taken at face value, for "nonlinear storytellers," as Talin and Pearce call themselves, obviously believe that it is possible to reconcile narrativity with some degree of interactivity. But nonlinear writers who want to preserve narrativity face a much more difficult task than their linear colleagues, because the creation of narrative structures involves foresight and global planning. A rewarding interactive experience requires the integration of the bottom-up, partially unpredictable input of the user into the top-down design (or collection of designs) of the storyteller.

Even if narrativity is regarded as a universal and stable cognitive structure, its conditions can be formulated on various levels of specificity. The compatibility of interactivity and narrativity obviously depends on how narrowly we define *narrative*. In its common usage, *narrative* can mean a discourse reporting a story as well as the story itself. Even in its "story" sense, *narrative* is an ambiguous term. It can be conceived (1) as a representation of physical or mental events involving common or related participants and ordered in a temporal sequence ("The king died, then the queen died"); (2) as an interpretation of events invoking causality ("The king died, then the queen died of grief"); or (3) as a semantic structure meeting certain formal requirements, such as a salient theme, a point, a development leading from equilibrium to crisis to a new form of equilibrium, and a rise and fall in tension. (This last idea has been diagramed in the well-known Freytag triangle.) I will call the first type sequential narrative, the second type causal, and the third type dramatic, since it corresponds rather closely to the Aristotelian concept of plot.

As the complexity of the definition of narrativity increases, so do the compositional demands on the interactive writer. The prototype of the sequential narrative is the diary, journal, or chronicle: it consists of a simple list of events that can be written down with minimal temporal displacement with respect to their occurrence. In a hyper-textual system a sequential narrative could be created by interlinking a collection of lexias that refer to the same individual and represent reasonably self-contained events. Each reading session would thus generate the day-by-day chronicle of a different life for the same character. If this scheme seems awfully boring, that is probably why there doesn't seem to be any well-known text that applies this formula. In a game environment a sequential narrative will be automatically produced by the presence of a user spending a variable amount of time in the virtual world. The narrative events will be written by the actions of the user, and the coherence of the sequence will be provided by the continuity in the identity of his game-persona.

While sequential narratives require no looking ahead, causal narratives are usually conceived retrospectively: the narrator links together events in a causal chain that leads to a specific outcome. An example of this backward logic is the problem-solving scheme. The causal sequence of plans is constructed to lead to a given effect—the goal of the agent. In interactive systems, causal narratives are found in adventure games. The system gives a very specific task to the user—rescue a princess, find the missing page of a book, or simply move along an initiatory path to reach higher status—and the user progresses toward this goal by solving a series of problems. In order to do so she reenacts prospectively the causal reasoning that has been written into the design of the game. The purpose of interactivity is to discover the plan set up by the system and to overcome the hurdles arranged along the way. Taking a wrong turn or failing a test does not threaten narrative coherence, because this framework allows both stories of success and stories of failure.

In the causal narrative of games the thrill of the quest is the main source of the user's pleasure. Dramatic narrative has no such uniform recipe for success. The purpose of the work is to take the audience through an experience of elusive and variable nature. Depending on the genre and the individual work, it could be defined as catharsis, laughter, suspense, empathy, self-knowledge, discovery, purification, or even therapy. Since dramatic narrative aims at controlling emotions and reactions, the duration of the sojourn of the spectator in the fictional world is strictly planned by the system. The implementation of a dramatic narrative in an interactive environment requires a delicate coordination of the user's actions with the goal of the system. Every choice given to the user constitutes a potential threat to the global design, and consequently to the quality of his own experience. The system designer must be able to foresee the possible actions of the user and to streamline them toward the desired effect. The user should progress under the impression that his actions determine the course of the plot, when in fact his choices are set up by the system as a function of the effect to be reached. The need to steer the user toward a certain goal without revealing this purpose (for fear of spoiling its effect) explains why dramatic structure, the fullest form of narrativity, is also the most problematic for interactive design.

THE STRUCTURES OF INTERACTIVE NARRATIVITY

The narrative potential of the interactive text is a function of the architecture of its system of links. Figures 4 through 12 illustrate what various designs produce in terms of narrative structures.

The Complete Graph

In a complete graph (fig. 4), every node is linked to every other node, and the reader has total freedom of navigation. This structure allows a free shuffling of text that makes it practically impossible to guarantee narrative coherence. To produce a collection of lexias that could be read in any order, and to generate for each possible order a well-constructed and different story, would be a feat of the same magnitude, and of the same interest—mathematical rather than poetic—

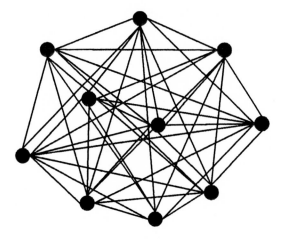

Paths are bidirectional

FIGURE 4 I The complete graph

as composing a large crossword puzzle without black spaces, filling a square with numbers that add to the same sum in all rows, columns, and diagonals, or designing a jigsaw puzzle whose pieces interlock in several visual patterns, each corresponding to the realistic image of a different object. Needless to say, the rare examples of this structure— or perhaps there is only one, Marc Saporta's *Composition No 1*—fall short of the ideal of narrative magic square.[1]

The Network

Figure 5 is the standard structure of literary hypertext. In this architecture the reader's movements are neither completely free nor limited to a single course. Since the network allows circuits, the system cannot control the duration or the course of the user's visit. In such a configuration narrative continuity can be guaranteed only on the local level—that is, from one node to the next, or within a sequence of nodes with single connections. A reader may, for instance, traverse a node that describes the death of a character and return later to a node in which that character is still alive. If the path of navigation is interpreted as chronological succession, the result is a nonsensical

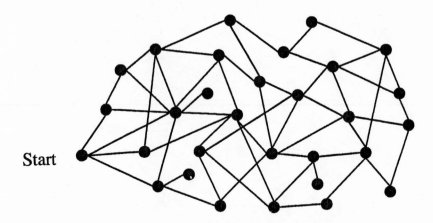

Start

Paths can be uni- or bidirectional

FIGURE 5 | The network: A hypertext-style decision map allowing circuits

sequence—a type of meaning that can, of course, be deliberately cultivated for its humorous effect. The model is therefore better suited for
a system of analogical connections or for a dadaist/surrealist carnival-
ization of meaning than for the generation of multiple stories.[2]

The Tree

The formal characteristic of an arborescent pattern is that it allows
no circuits. Once a branch has been taken, there is no possible return
to the decision point, and there is only one way to reach a given
terminal node. By keeping each of its branches strictly isolated from
the others, tree-shaped diagrams such as figure 6 control the reader's
itinerary from root node to leaf nodes and make it easy to guarantee
that choices will always result in a well-formed story. It is therefore the
structure of predilection of the *Choose Your Own Adventures* children's stories. For the sake of graphic simplicity I have represented the
choices as binary, but we can imagine a more interactive tree offering
a wide range of possible actions at every decision point. Since the tree
grows exponentially, this system quickly runs into a combinatorial
explosion. It would, for instance, take sixteen different plots, with
thirty-one different fragments, to ensure four decision points. The

number of fragments necessary to produce a certain number of plots can be restricted by allowing the merging of paths (dotted lines), but this structure is a directed graph (cf. fig. 9) and no longer a tree.[3]

The Vector with Side Branches

In the configuration described in figure 7, the text tells a determinate story in chronological order, but the structure of links enables the reader to take short side trips to roadside attractions. This structure is particularly popular in electronic texts designed for juvenile audiences because of its cognitive simplicity. In *Arthur's Teacher Troubles,* a children's book produced by Brøderbund, the user moves page by page through an illustrated story, but every page offers hidden surprises: click on the teacher, and she turns into a monster; click on one of the students, and he makes a funny face; click on a mirror on the wall, and it is shattered by a baseball; click on the text under the

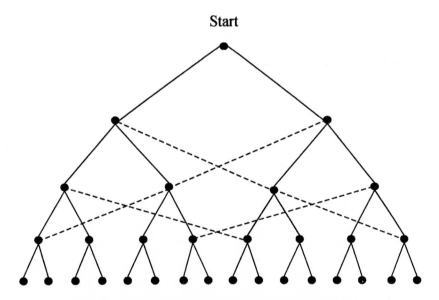

Start

Paths are unidirectional (from top to bottom)
Every traversal produces a well-formed plot

FIGURE 6 | The tree

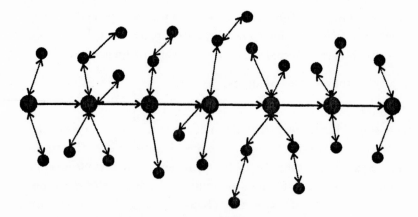

FIGURE 7 | The vector with side branches

picture, and it is read aloud. The same network structure could also be used to provide detailed descriptions or background information about the characters and settings of a linear narrative. The reader could thus choose to speed through the plot, as in *Reader's Digest* versions of books, or to linger on the scene.

When the options available at every station expand from linear branches—similar to footnotes—into a network shaped like figure 5, this structure gives rise to an experience that may be compared with a guided tour. Imagine walking through a museum that represents the history of mankind from the Stone Age to the Cyberage. To get an idea of cultural and technological development, you have to visit the exhibits in chronological sequence, but within each module you can choose your own path and decide what you want to see. If you are primarily interested in jewelry, you will skip the tools section; if you like tools and weapons, you may skip the jewelry; and if you are interested in everything, you may still decide in what order you visit the displays. This structure is most useful with didactic materials that require an accumulation of knowledge, since it can ask the student to complete separate modules in a determinate sequence. In the narrative domain, the "guided tour" structure is exemplified by the on-line soap opera *The Spot*. Every day the site displays a network of diary entries that represent the lives of the various characters; the reader moves freely among these entries, but the text as a whole must be visited daily, which means in chronological order.

The Maze

The maze structure, shown in figure 8, is characteristic of adventure games. The user tries to find a path from a starting point to an end point. This scheme has room for many variations. There may be one or more ways to reach the goal; the graph may or may not allow the user to run in circles; terminal nodes may be dead ends or allow backtracking. Far from guaranteeing satisfaction for every traversal, the system structures the reader-player's adventure in the textual world in terms of the two experiences that Espen Aarseth calls aporia and epiphany (cf. Aarseth's essay by that title). While aporia occurs when the player takes a dead-end branch on the game map or fails to overcome an obstacle, epiphany is a discovery, such as the solving of an enigma or the elimination of an opponent, that enables her to progress in her quest. Though the structure of the text is too complex for the designer to foresee every possible path, narrative coherence is guaranteed by the fact that all paths are attempts to reach a certain goal. The model contains as many plots as there are complete traver-

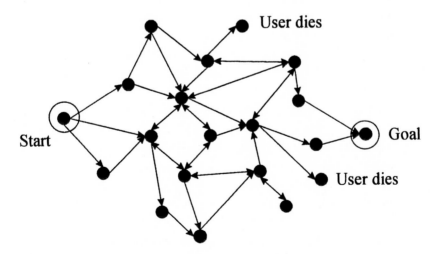

FIGURE 8 | The maze: Structure of an adventure game

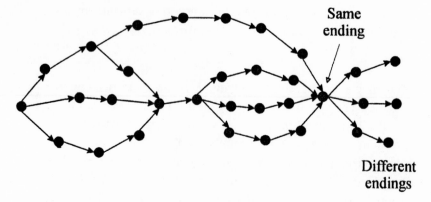

Same
ending

Different
endings

FIGURE 9 | The directed network, or flow chart

sals of the graph: plots with a happy ending or tragic plots leading to the waste of one of the user's lives, depending on whether the path ends with epiphany or fatal aporia.

The Directed Network, or Flow Chart

The negative experiences of running in circles and hitting a dead end are eliminated in the architecture of figure 9. This model represents the best way to reconcile a reasonably dramatic narrative with some degree of interactivity. In this type of network, horizontal progression corresponds to chronological sequence, while the branches superposed on the vertical axis represent the choices offered to the user. The system prescribes an itinerary through the textual world, but the user is granted some freedom in connecting the various stages of the journey. In one variant of the structure, illustrated by John Fowles's *The French Lieutenant's Woman*, the reader receives a choice of endings at the conclusion of a unilinear journey. This architecture prevents the combinatorial explosion of the tree-shaped diagram, but when the choices occur in the middle exclusively, it trivializes the consequences of the user's decisions. As Gareth Rees observes, "The merging of narratives keep[s] the story on a single track while offering [the user] an illusion of choice" ("Tree Fiction"). If the user can get from *A* to *B* by different paths, and from *B* move on to *C*, the choice among the various paths has no bearing on the final outcome.

One way to restore significance to the decisions of the user is to turn the text from a fully context-free transition system to a context-sensitive system capable of narrative memory. In such a system the decisions made by the user in the past affect his choices in the future, and narrative causality extends to nonadjacent episodes. An example of this distant causality is provided by the classic Proppian fairy tale. Imagine that the hero, on his way to rescue the princess from the dragon, is tested by a donor. He may either pass the test and receive a magical aid, or fail and be punished with a curse. When he encounters the dragon later in the story, the outcome of the fight is determined by whether or not he carries the donor's gift. Many computer games implement this idea by having players pick up and carry objects that will enable them to solve later problems. This use of memory makes it possible to include nontrivial choices at every stage of the story and to make the end dependent on the middle.

The Hidden Story

Figure 10 is the structure of those interactive mystery stories and computer games that implement the idea of discovering the prehistory of the game-world, as in the popular game Myst. This model consists of two narrative levels: at the bottom, the fixed, unilinear, temporally directed story of the events to be reconstituted; on top, the

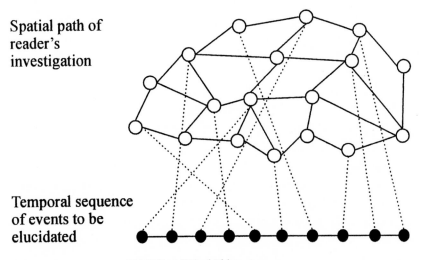

FIGURE 10 | The hidden story

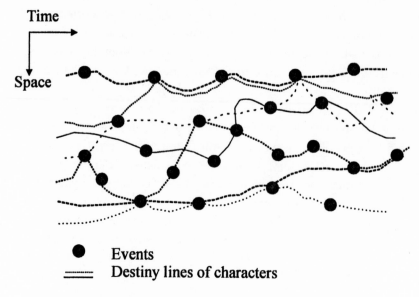

Time

Space

● Events
▬▬▬ Destiny lines of characters

FIGURE 11 | The braided plot: The house of many windows

atemporal network of choices that determines the reader-detective's
investigation of the case; between the two, dotted lines that link episodes of discovery in the top story to the discovered facts of the
bottom story. In this case, as in the maze configuration, a narrative is
written by the actions and movements performed by the player in the
attempt to reconstitute the underlying story.

The Braided Plot

Classic narrative consists, at least in part, of a sequence of physical
events objectively experienced by a group of characters, but every
character in the cast lives these events from a different perspective and
has a different story to tell. Figure 11 shows how interactive mechanisms can be used to switch "windows" on a multistranded but determinate narrative. On the diagram, the horizontal axis stands for time
and the vertical axis for space; simultaneous events are vertically
aligned, and events that take place in the same location occupy the
same horizontal coordinate. Each circle represents a physical event,
and the lines that connect them stand for the destinies of the participants. By selecting one horizontal line rather than another, the reader

enters the private world of a specific character and experiences the story from a particular point of view.

This architecture, like many others, allows a number of variations. Some systems may let the reader backtrack and relive the same events from a different point of view; some may allow switching between plot lines; some, finally, may impose a forward movement, so that the perspectives not taken will be irrevocably lost. An example of unconstrained movement along and between destiny lines is the aptly titled series *The Lurker Files,* a campus-life mystery by Marc Ceratini formerly available on the World Wide Web. This work makes use of hypertextual links to follow characters individually, to explore their past, to peep into their letters and diaries, and to go back to earlier episodes. A nonelectronic example of restricted mobility would be a play whose actions unfold simultaneously in several rooms of the same house. The spectator would be able to move from room to room but could view only one strand of the braided action at a time. It would therefore take many viewings to catch the whole story.[4]

Action Space, Epic Wandering, and Story-World

In the architecture of figure 9, the system designs the general outline of the plot, and the user selects the details of its realization. Figure 12 represents the inverse solution. Here, interactivity takes place on the macro level and dramatic plotting on the micro level. The space of the diagram represents the geography of the virtual world, and the nodes and links the prominent sites and access ways in this geography. The user is free to take any road, but when she reaches a site, the system takes control of her fate and sends her into a self-contained adventure (represented by the pathways looping back toward the nodes). This model abandons the idea of an overarching dramatic narrative in favor of an epic structure of semiautonomous episodes in which the user plays a largely passive role. The global narrative determined by the user's movement is purely sequential, but the micronarratives specified by the system can be causal or dramatic.

This model is illustrated by the structure of theme parks. Visitors wander in a geography made of distinct subworlds, each of which offers a different, carefully scripted adventure. In one site you take a boat ride through the pirates' lair; in another you ascend the Matter-

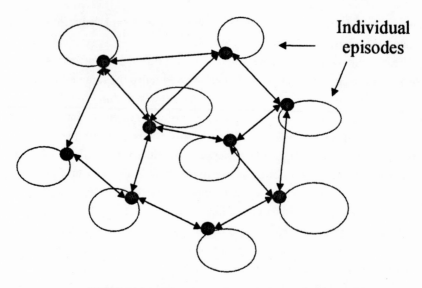

Individual episodes

FIGURE 12 | Action space, epic wandering, and story-world

horn in a train; in a third you venture on a rocket into intergalactic space. Choice is limited to deciding where to go and whether or not to take the ride; once the ride has been boarded, the system takes total control of the visitor's fate and (supposedly) provides a thrilling experience with proper Aristotelian contour. Another instantiation of this architecture is what George Landow (following Michael Innis, head of Inscape, Inc.) calls a "story-world." The literary model of the concept of story-world is the novel of proliferating narrativity in which "little stories" steal the show from the plot of the macro level. (The masters of this genre, a revenge of oral storytelling verve upon the antinarrative excesses of the New Novel, include Gabriel García Márquez, Isabel Allende, Günter Grass, Patrick Chamoiseau, and of course Boccaccio.) Landow's example of a story-world text is an interactive video, *Hypercafé,* that lets the user wander through a crowded café, choose a table, and listen to the conversations of its occupants.[5] Once a conversation has been selected, the video unfolds with little or no possibility of user intervention. In this case, as in most others, but perhaps more acutely, narrative coherence is maintained at the cost of interactivity.

The lesson to be learned from these various diagrams is that the potential of a network to generate well-formed stories for every traversal is inversely proportional to its degree of connectivity. When hypertext theorists describe the configuration typical of the genre (fig. 5) as a storytelling machine, they trust the power of the reader's imagination to create narrative connections between any two nodes and to make every transition meaningful. This trust presupposes either a very loose conception of narrativity on the part of the theorist or a very strong desire for narrative coherence on the part of the reader. For those who believe that narrativity is the product of global planning, not a type of meaning that can be freely constructed out of any collection of informational fragments, it is by controlling the general path of the reader, maintaining a steady forward progression, limiting decision points, or neutralizing the strategic consequences of decisions that interactive texts can guarantee narrative coherence.

Even so, storytelling is not the forte of these texts, and it is only by directing attention away from any kind of overarching plot line that they can hope to compensate for this deficiency. Empirical studies have shown that when readers are motivated by the desire to know how it ends—the primordial narrative desire—and when they can find out by hitting the return key, they will experience the other links as a distracting nuisance. According to an experiment described by Kirsten Risden, "Two out of three participants read [*The Lurker Files*] as a traditional story by choosing not to explore links" ("Can Theories," 5). The purely literary branch of hypertext deals with this narrative limitation by following the lead of the postwar experimental print novel and moving toward lyrical or musical organizations. A text such as Michael Joyce's *Twelve Blue* creates a very successful compromise indeed between symphonic orchestration of themes, lyrical language, and emergent but incomplete narrative developments.

Another way to overcome the fundamental incompatibility of narrativity and interactivity might be to stimulate interactive curiosity on a purely local level. The purpose of clicking should not be to find out how it will end—print texts are infinitely better at generating and satisfying this long-term narrative desire—but, as the multimedia design author Bob Hughes has suggested (*Dust or Magic*, 202–4), to trigger microevents that provide blasts of pleasure and instant satisfaction. As electronic authors design the reader's encounter with the text, they should concentrate on those truly magic moments when the click of the mouse provokes a response from the system. What would these microevents contribute to the text? Let us take a clue from *Arthur's Teacher Trouble,* or from the visual hypertext *The Manhole:* click on a window, and it shatters; on the teacher, and she turns into a monster; on a blade of grass, and it grows into a giant tree that you can climb into the sky. Those who work primarily in the language medium will have to find clever verbal equivalents of these visual effects. In such an approach, the system takes care of the general direction of the plot, but it coaxes the reader into exploring links by hiding a little treasure, an exciting discovery, a delicious morsel, along every side road. Hughes describes this structure as a modified version of the Freytag triangle (210). The system creates a dramatic contour shaped like a mountain, but the reader's decisions trigger microevents that draw jagged edges and multiple secondary peaks of momentary interest along this general outline. The same idea could be used to enliven the individual episodes of an epic structure. This medium-imposed shift of interest from what Jean-François Lyotard *(Postmodern Condition)* has called grand narrative schemes toward little stories represents what is perhaps the more genuinely postmodern feature of interactive textuality.

HYPERTEXT AND IMMERSION

Emergent is the favorite term of contemporary literary theory for a type of meaning that comes out of the text, rather than goes into it, and that is produced dynamically in the interaction between the text and the reader. At the risk of creating an oxymoron, or a mixed metaphor, can a form of textuality cultivated for its emergent quality lure the reader into an immersive experience? Since it is always possible for a certain text to overcome the strictures of its medium, a more proper question to ask is whether interactivity can be a positive factor of immersivity. Let me consider here each of the three types of immersion defined in chapters 4 and 5.

Temporal Immersion

The discussion of the preceding section has driven home the point that interactivity conflicts with the creation of a sustained narrative development, and consequently with the experience of temporal immersion. Among the architectures described above, the only one that places interactivity in the service of narrative desire is the mystery-story structure (fig. 10), because the reader's actions discover, rather than create, the object of this desire, and because the story to be investigated is itself unilinear, determinate, and external to the interactive machinery.

Temporal immersion requires an accumulation of narrative information. In a linear text, the more we read and the more we know about the textual world, the more we can anticipate developments—and the more pleasantly surprised we are when the outcome dodges our projections. The continuity of the plot line functions as a string on which the reader's memory threads information and keeps it together for easy access. It is not without reason that the mnemonic techniques of antiquity and the Renaissance built an itinerary through a building and disposed the items to be remembered along this linear path. The broken-up structure of the interactive text thus deprives memory of one of its most efficient modes of storage. It has been said that hypertext promotes a pattern of foregrounding and backgrounding data that mimics the associative mechanisms of the brain, but there is a world of difference between simulating the functioning of memory on the neural level and strengthening the process of recall.

The special power of interactive texts to generate a plurality of possible worlds could be regarded as a feature that facilitates the creation of an immersive plot. In chapter 8 of my book *Possible Worlds, Artificial Intelligence, and Narrative Theory,* I suggest that one of the properties that contribute to the intrinsic *tellability* of a story—another term for *narrative immersivity*—is the diversification of possible worlds in the narrative universe. This principle predicts that fictional universes that involve incompatible possible worlds—such as mistaken beliefs, conflict among the systems of belief of various characters, conflict between desires and reality, conflict between sincere and projected beliefs, consideration by characters of various lines of action, and so on—produce more interesting plots than narratives in which the private worlds of characters reflect accurately the real world, or exist in harmony with each other. (In fact, there cannot be a plot without incompatible possible worlds.) But this idea eventually hits the ceiling of the reader's processing ability. The comedies of errors and *romans à tiroirs* of the Baroque era flirt with, and sometimes transgress, these cognitive limits. Whatever advantage interactive narratives present over standard ones in the creation of forking paths and multiple realities leads to a degree of complexity that no longer supports narrative motivation. Two to four different endings, for instance, a structure easily realized in print, will receive serious individual consideration, and the various outcomes will invite comparison; sixty-four endings only convey the message "There are lots of possible endings," and each of them is lost in the crowd. At this point it becomes irrelevant whether there are sixty-four, two hundred fifty-six, or a thousand endings. In terms of complexity, hypertext compared with print texts is like satellite-dish TV with its five hundred channels versus cable TV with its mere fifty. Do viewers really take advantage of this complexity? The brain may be a "massively parallel processor" on the neural level, as cognitive science tells us, but on the level of the more conscious operations involved in reading, it remains very difficult to keep track of several strands at once, and it seems doubtful at best that systematic exposure to hypertext will significantly increase the mind's performance in distributed parallel processing.

Spatial Immersion

In a frequently drawn analogy, the hypertextual network is viewed as the image of the postmodern experience of space. For Fredric Jameson, postmodern space is an alienating, self-transforming expanse that offers neither rest for the body, refuge for the soul, nor landmarks for the mind: "[This] latest mutation in space—postmodern hyperspace—has finally succeeded in transcending the capacities of the individual human body to locate itself, to organize its immediate surroundings perceptually, and cognitively to map its position in a mappable external world" (*Postmodernism*, 44). The only possible relation to this postmodern space is the feeling of being lost, and the only possible movement is aimless wandering. Both of these experiences are reflected in the blind progression of the reader through the labyrinthine structure of hypertext.

For Marc Auge, the characteristic space of "supermodernity," as he calls the present time, is a nonplace that we traverse on our way to somewhere else, and that we come to inhabit because the postmodern condition is a state of perpetual transit *(Non-Places)*. These nonplaces are called airport, subway, freeway, and the network of Information Superhighways that crisscross cyberspace, the most nonplace of them all.[6] Cybernauts and hypertext readers spend most of their time clicking on the nonplaces of the links, never dwelling for long on a textual segment, because each of these segments is less a destination than a point of departure for other, equally elusive destinations. Theorists of electronic culture[7] make a virtue of this sense of never getting anywhere by regarding hypertext as a textual implementation of Deleuze and Guattari's concept of "smooth space." For Deleuze and Guattari, smooth space exists in contrast, though in constant interrelation, to an organized, hierarchical, and largely static space that they call striated. "In striated space, lines or trajectories tend to be subordinated to points: one goes from one point to another. In the smooth, it is the opposite: the points are subordinated to the trajectory" (*A Thousand Plateaus*, 478). Smooth space is nomadic (like the sea or the desert, it offers no home, only an experience of its immensity), sprawling, continually expanding (you can always add a link to hypertext), amorphous (you can add links wherever you want), heterogeneous, without clear boundaries, tactile rather than visual (through clicking, the reader grabs segments), and constituted by an "accumulation of proximities" (488). These proximities are links that negate physical distance, since all it takes to make two points adjacent in a network is to draw a line between them.

The interactive text may convey an exhilarating sense of power and mobility—a sense fortified by the tendency of the imagination to reconceptualize the travel of data on the Internet as travel of the user to distant sites—but the cost of embracing space in its globality is an alienation from its locality that prevents growing roots in any given site. The system of links of the interactive text is a constant temptation to move beyond the present screen. As Michael Heim observes, "Hypertext thinking may indeed reveal something about us that is agitated, panicky, or even pathological. As the mind jumps, the psyche gets jumpy or hyper" (*Metaphysics*, 40). Sven Birkerts, admittedly no friend of hypertext, concurs: "For the effect of the hypertext environment, the ever-present awareness of possibility and the need to either make or refuse choice, was to preempt my creative or meditative space for myself" (*Gutenberg Elegies*, 162).

One consequence of the mosaic structure of hypertext is that the lexias are rarely long enough to let an atmosphere sink in. The link is a jump, and each act of clicking sends the reader to a new, relatively isolated textual island. It always takes a while to make oneself at home in a text, to grow roots in the fictional world, to visualize the setting, to familiarize oneself with the characters and their motivations. It is for this reason that many people prefer fat novels over collections of short stories. In his novel *If on a Winter's Night a Traveler,* as we have seen in the interlude to chapter 5, Italo Calvino allegorizes the difficulty of immersion by embedding in the narrative of the primary level the beginnings of a dozen other novels which are brutally interrupted after a few pages—just as the reader is beginning to view himself as a citizen of the fictional world. In Calvino's novel the reader is left stranded at the end of every chapter; in hypertext, the threat of uprooting occurs with every change of screen.

Every time the reader is called upon to make a decision, he must detach himself from the narrative "here and now" and adopt a point of view from which he can contemplate several alternatives. Once the choice is made, the reader may regret his decision and be haunted by the "could have been." What Gareth Rees writes of his experience of tree fiction is even more to the point in the case of a more complex network: "I think that as readers we are not ready for tree fiction: I know that when I read such a story, I want to find out all the consequences of every decision, to read everything that the author wrote, fearing that all the interesting developments are going on in another branch of the story that I didn't investigate. I want to organize the whole story in my mind" ("Tree Fiction"). The body of the reader's imaginary persona in the fictional world would have to undergo a dismembering to take all the roads at the same time, and to overcome the nagging feeling of having missed something along the way. Can immersion be experienced by a *corps morcelé,* as Christopher Keep describes the hypertextual reconfiguration of the body ("Disturbing Liveliness")?

In defense of electronic technology, however, it must be said that all these deficiencies of interactive textuality with respect to the experience of spatial immersion can be compensated by hypermedia effects. Of the three types of immersion, the spatial variety evidently has the most to gain from the built-in spatiality of pictures. It seems safe to predict that the interactive texts of the future will make much more extensive use of visual resources than the literary hypertexts of the present, and it would therefore be unfair to pass judgments of immersivity on purely verbal attempts to convey an experience of space. I cannot think of a more efficient way to celebrate the spirit of a place than a well-designed interactive network that combines images with music, poems, short prose texts, maps, and historical documents.

Emotional Immersion

Turning to emotional immersion, once again the question is not whether the reader of hypertext can develop the kind of affective relations that lead to feelings of happiness or sadness when things turn out for the better or for the worse for a certain character, but whether interactive mechanisms can be used to enhance this emotional participation. Janet Murray writes that the sense of the irrevocability of life is alien to the spirit of interactive art: "Our fixation on electronic games and stories is in part an enactment of [a] denial of death. They offer us a chance to erase memory, to start over, to replay an event and try for a different resolution. In this respect, electronic media have the advantage of enacting a deeply comic vision of life, a vision of retrievable mistakes and open options" (*Hamlet*, 175).[8] One of the trademarks of the spirit of comedy is a playful detachment from the characters that precludes an affective investment in their fate. This detachment is strengthened by the knowledge that a character's life is simultaneously acted out in several possible worlds, and that if we do not like one of these worlds we can always jump to another. Emotional immersion requires a sense of the inexorable character of fate, of the finality of every event in the character's life, but as Umberto Eco observed in a radio interview, this outlook is fundamentally incompatible with the multiple threads generated by interactive freedom. "A hypertext can never be satisfying," Murray quotes Eco as saying, "because 'the charm of a text is that it forces you to face destiny'" (296).

The obstacle to immersion in an interactive text, however, lies as much in the aesthetic philosophy of its theorists and literary practitioners as in the inherent features of the medium. We are told that the virtue of hypertext is its ability to "propel us from the straightened 'either/or' world that print has come to represent and into a universe where the 'and/and/and' is always possible" (Douglas, "Hypertext," 155; view endorsed by Joyce, *Of Two Minds*, 5). This view suggests that the aesthetic ambition of hypertext is an awareness of the plurality of worlds contained in the system. Since this plurality can be contemplated only from a point of view external to any of these worlds, the proper appreciation of the multidimensionality of hypertext is incompatible with recentering and imaginative membership in a fictional reality.

THE FUTURE OF INTERACTIVITY

How can hypertext, or more generally interactive textuality, compensate for this immersive and narrative deficiency? Futurology is a risky discipline, but I see three avenues of development, between which there should be ample room for hybrid forms and connecting trails. One of these avenues will be explored in chapters 9 and 10. The second calls for a deeper understanding and bolder exploitation of the idiosyncratic properties of the electronic medium than we have seen so far.

When electronic technology presented literature with the gift of point-and-click interactivity, it did not include a user's manual. It is customary for technological innovations to be initially conceived as improvements on existing technologies fulfilling existing cultural needs: the automobile was supposed to be used as a new mode of transportation for Sunday family outings, and the computer was first thought of as an improved calculator (which, of course, it also was). As Tim Oren observes, "The forms of existing [but also new] media . . . have been largely determined by opportunism. The first technologically feasible configuration is rushed to market and exploited and then sets an *ad hoc* standard for the form and content of that medium" ("Designing a New Medium," 468). Hypertext was no exception to this pattern. Thinking in terms of the genre system, literary trends, and philosophical climate of their time, hypertext authors conceived the strange new gift of interactivity as a way to free the novel, even more radically than postmodern works of the print variety had done, from patterns of signification inherited from the nineteenth century. This conception of hypertext as the *super postmodern novel* was detrimental to the nascent medium for two reasons. First, the traditional length of the genre motivated hypertext authors to start right away with large compositions that made unreasonable demands on the reader's concentration. Instead of being gently initiated into point-and-click interactivity, readers were intimidated by the forbidding complexity of a maze that they had no fair chance of mastering. With the arrogance typical of so many avant-garde movements, hypertext authors worked from the assumption that audiences should be antagonized and stripped of any sense of security, rather than cajoled into new reading habits. Second, the model of the novel created a pattern of expectations that subordinated local meaning to a global narrative structure, and even though this structure hardly ever materialized, its pursuit distracted readers from the poetic qualities of the individual lexias.

Abandoning the model of the novel would allow hypertext to explore two avenues:

1. Direct interest more strongly to the local level, by working with relatively self-contained lexias such as poems, aphorisms, anecdotes, short narrative episodes, or provocative thoughts.

2. Give up on the idea of an autonomous "literary" genre and take greater advantage of the multimedia capability of the electronic environment. This approach would lead to a merger of hypertext with the burgeoning genre of CD ROM interactive art, and could also take the form of a hybridization with computer games.[9]

The renaissance of pure textuality that accompanied the early development of electronic writing may indeed have been a short respite in its ongoing loss of cultural territory to visual media. Compared with the hypertexts of the early 1990s, the more recent ones are much more dependent on hypermedia effects, undoubtedly because of advances in the technological support. Michael Joyce's *Afternoon* offered no visual pleasure, but the more recent *Twelve Blue* makes a substantial use of color and graphics, and *Twilight: A Symphony* includes pictures, video clips, music, human voices, and various other sounds. Gregory Ulmer envisions the future of hypertext as the exploration and implementation of new relations between image and text. Several hypertext web sites launched in late 1998 use, for instance, a combination of text and 360-degree interactive panoramas created in Virtual Reality Markup Language (VRML).[10] These panoramas can be explored with the mouse, and they may serve as either illustrations or story-navigating devices. In the latter case, clicking on an object in the picture will create a surprise, such as animating the object, displaying textual screens, or activating audio recordings.

In a multimedia environment, the instantaneous sense of presence that can be achieved through visual documents or through the intensely personal modulations of a human voice provides a way to compensate for the loss of immersivity that results from the fragmented structure of the work as a whole.[11] A particularly promising form of multimedia hypertextuality is the thematic cluster and electronic activity kit. In this type of project the system of links creates a guided but flexible tour through a collection of semiautonomous documents that either relate to a specific topic or develop a diversified vision. The interactive apparatus allows users to decide what to read, hear, or see, and occasionally lets them manipulate the individual documents. This formula encourages the pure joy of "doing things" with text or pictures, and because it does not construct lengthy chains of logical dependencies between screens, it leaves users without remorse if they decide to skip a certain segment.

A particularly attractive example of this approach is Agnes Hegedüs's CD ROM artwork *Things Spoken* (1999). An autobiographical "show-and-tell," *Things Spoken* takes the viewer on a tour of the curio cabinet of the mind. Displayed in the windows of the computer screen is a collection of things from the author's personal archives: kitsch, cheap tourist souvenirs, mass-produced objects, unique artifacts, family heirlooms or precious gifts from friends and relatives. Their shiny surfaces mirror or activate the phantasms, fears, thoughts, and memories that make up the private fabric of the self. Using these pieces as "things to think with" (to use Sherry Turkle's expression), each screen of the artwork juxtaposes the hyperrealistic reproduction of an object with two strands of personal narrative that run side by side, and from right to left, on two white lines at the bottom of a black screen. A key chain with two dangling parts—the upper and lower body of a woman without arms—inspires reflections on the disarmed condition of women when they are turned into sex objects; a menorah awakens thoughts of what it means to be Jewish by marriage; a pink plastic doll that looks like a phallus reminds the speaker of the taboo placed on sexual topics of conversation during her Hungarian adolescence. The line on which the cursor is placed is read aloud in a voice whose rich intonations create a powerful sense of human presence. By moving the cursor between the two lines the interactor can switch between two self-sufficient and rather short narratives read in contrasting voices: male versus female, or German versus English. The user can also jump to other screens by clicking on highlighted words. Since the text moves from right to left, "catching" these links with the cursor requires some hand-eye coordination that lends tactile interest to the work and turns its operation into a game of skill. Once the interactor has successfully clicked on a highlighted word, she is transported into another narrative that contains the same key word, even though it describes a different object. The impression is one of synapses firing each other in the brain, opening up new pathways into the secret caches of memory and thickening the web of its associations.

The third possibility I have in mind is to turn the interactive text into a form of conceptual art through a clever and diversified enactment of the idea of self-referentiality. I am not suggesting that hypertext should content itself with a crude version of Marshall McLuhan's all-too-famous slogan "The medium is the message," for no art form can survive if all it has to offer is a fully predictable message.[12] Even in genres as stereotyped as TV comedies or Hollywood movies, the invariant message "I am a comedy" or "I am a movie" that results from the formulaic use of canned laughter or the traditional car chase is acceptable only because it stretches like a watermark design across a more visible and variable image. In the best specimens of postmodern literature, similarly, self-referentiality does not carry the text all by itself but combines with other sources of interest, such as theme, style, and original narrative or antinarrative techniques. The device is also far more successful when its message is not merely generic but specific to the text. In this individuated self-reference, the text will not merely say, "I am an *x* [novel, fiction, text, hypertext]" but "I am me, a unique use of the resources of my medium." This type of self-reference, the polar opposite of the predictable message of stereotyped devices, can be considered a form of conceptual art because it arises spontaneously from the novelty of the productive idea. The trademark of conceptual art is that its formula must be entirely original. The generative idea resembles the punch line of a joke, in that viewers may or may not get it, and if they do, they often get all of it. This is why conceptual art specializes in those intense bursts of creative energy that need no further development and leave no foundation to build on besides the memory of their ingenuity. In contrast to those artistic formulae that can be used over and over again and be adapted to many types of content and effects, the idea that forms the message of conceptual art exhausts its expressive potential after a single use.

By suggesting that electronic literature take the conceptual route I do not wish to say that interactivity per se is one of those devices that should be used only once, but rather that the electronic medium can be a powerful tool kit for the production of one-of-a-kind textual forms such as the projects listed below. Not all of them are literally interactive, but they are all dynamic, and they all put the visual and kinetic properties of their medium in the service of a precise idea, often the literalization of a well-known metaphor:

— *The text as antiobject, or the work that cannot be reread.* William Gibson's *Agrippa (A Book of the Dead),* written in collaboration with the conceptual artist Dennis Ashbaugh, is a CD ROM text that erases itself while being read.
— *"You bring as much to the text as you get from it."* In some of the cyberpoetry projects of John Cayley (e.g., *Collocations: Indra's Net II)* the computer generates text by selecting and combining strings from a textual database. When the reader feeds a specimen of her own writing into the computer, the system literalizes the idea of the reader as co-author (Cayley, "Potentialities," 180–83).

— *Reading as incomplete process and random selection.* In a project described by Philippe Bootz ("Gestion," 243), his own electronic text *amour,* the text scrolls too fast for the reader to parse all the words, and the "text-as-read" is necessarily a mutilation of the "text-as-written." This is not a bug but a feature that literalizes the etymology of the French word for reading: *lire,* from the Latin *legere,* to pick. The idea is also used in the title segment of Mark Amerika's *Grammatron.*

— *The text as palimpsest.* In several cyberpoems by Jim Rosenberg, including "Diffractions Through" and "The Barrier Frames," the visual display begins with a chaotic superposition of several pages. By moving the mouse around the screen, the reader isolates one of the pages from the clutter of the background and reveals legible words. The effect is a very pleasant tactile sensation of peeling off the layers of the text, and of making words appear on the screen through the smooth caress of the cursor rather than through the harsh hitting of keys.

— *Literalizing the notion of textual space.* In *De Leesbare Stad (The Legible City),* a VR installation by Jeffrey Shaw and Dick Groeneveld, a user rides a bicycle in front of a video screen on which an image of the city of Amsterdam is projected. The houses, however, have been replaced by chunks of text borrowed from archive materials that tell the story of the various buildings. By pedaling the bicycle faster or slower and by turning the handlebars, the user is able to control the display, and consequently to select the itinerary of what becomes a journey through the history of the city.

— *The instability of meaning (a virtual project).* In this Derridean/dadaist word game—an electronic version of a magnetic poetry kit—brightly colored, imaginatively selected words or phrases would swim onto the screen like tropical fish in an aquarium. The reader would try to grab them with the cursor, a task that could be made more or less difficult, and assemble them into a poem. The generated statements would be readable for a limited time, after which they

would break up and their components would start swimming again. Words could also morph into other words and
form ever-changing statements.

Sending interactive textuality on the conceptual route is intellectually stimulating, but it is also a risky decision that involves two
pitfalls. One has been lucidly diagnosed by Umberto Eco: it discourages reading the work once the reader gets the productive idea:

> I recently came across *Composition No. 1*, by Mark Saporta. A
> brief look at the book was enough to tell me what its mechanism
> was, and what vision of life (and obviously, what vision of litera-
> ture) it proposed, after which I did not feel the slightest desire to
> read even one of its loose pages, despite its promise to yield a
> different story every time it was shuffled. To me, the book had
> exhausted all its possible readings in the very enunciation of its
> constructive idea. (*Open Work*, 170–71)

If interactive textuality opts for the conceptual approach, moreover,
its place within literature will remain that of a marginal experimental
form, comparable in impact and significance to the introduction of
the twelve-tone scale in the history of music. Arnold Schönberg, pioneer of the new scale, once reportedly said, "I can see one day when
everybody will be whistling my tunes." This hasn't happened. Nor will
literary hypertexts, in all likelihood, ever top bestseller lists. There is
something about the seven-tone scale and about linear narrative that
seems to make them indispensable to Western culture and, arguably
for the latter, to culture in general. But music and literature deserve
recognition, and are both substantially richer, for having dared to
challenge the commanding position of these two forms of expression.

I'm Your Man

Anatomy of an Interactive Movie

The idea of multiple-choice drama is not an invention of the electronic age. During the 1960s the French novelist Michel Butor and the Belgian composer Henri Pousseur worked on an opera project, *Votre Faust,* that was supposed to involve several plot variations and to respond to the reactions of the audience. When it was finally, and briefly, performed in Milan in 1969, however, it involved only one decision point.[1] The Oulipo movement also produced interactive drama projects (Fournel and Enard), but they were mostly blueprints never meant to be submitted to the test of performance. It wasn't until computer technology reached its current level of sophistication that the idea of interactivity became reasonably feasible in an environment other than exclusively textual. The biggest obstacle to the implementation of selective interactivity in movies or drama is the conflict between the solitary pleasure of decision making and the public nature of cinematic or dramatic performance. In a spectacle addressed to a large audience, interactive decisions must be taken by the majority, and freedom of choice is only freedom to vote.

The concept of interactivity-by-consensus was tested in 1992 by a company called Interfilm Technology through a system that enabled spectators in the theater to vote by operating a seat-mounted pistol grip, but after an initial success due to the novelty of the experience, audiences quickly tired of this form of artistic democracy. The productions of Interfilm Technology received new life in 1998, when a new storage device called DVD made it possible to gather enough data on a compact disk for the movies to be playable on a personal computer. In contrast to the members of a collective audience, DVD users are able to make individual choices, to play with the system at their own leisure, and to

INTERLUDE

explore many strands in the plot, rather than being restricted to the path selected by the majority. Thanks to this technological development, the idea of interactive cinema has entered a period of do or die, and the free flights of fancy of the past will have to be replaced with viable projects.

Here I propose to analyze the architecture and design philosophy of the first interactive movie commercially offered for the personal computer, *I'm Your Man*, a DVD version of a film shown in 1992 by Interfilm in the specially equipped theaters. It would be preposterous to judge the artistic potential of the newborn genre on the basis of these first wobbly steps, but the production has a lot to teach us about the kind of problem that will have to be solved for interactive movies to develop into a durable form of entertainment.

A blend of two of the most stereotyped Hollywood genres, the comedy and the action thriller, *I'm Your Man* may strike the sophisticated movie connoisseur as an amateurish collection of clichés, but it would be fairer to evaluate the production in terms of its handling of the difficulties presented by the format than to compare it with classic linear movies. The story revolves around three characters, comically incompetent caricatures of the standard actantial roles of archetypal plots. There is the villain, Richard, a seductive James Bond turned evil whose sarcastic wit introduces a metafictional dimension that helps relieve the simplemindedness of the plot(s); there is the ingenue, Leslie, coquettish, earnest, and naive, a nice girl entrusted with the execution of a top-secret plan; and there is the hero-in-spite-of-himself, Jack, an Everyman with adolescent lust, sophomoric smarts, and the looks of an insurance salesman.

It is always difficult, if not impossible, to summarize an interactive movie, but in this case the task is somewhat facilitated by the classically Aristotelian structure of *I'm Your Man*. The movie adopts what has been described in the preceding chapter as the flow-chart, or directed-network, design (cf. chap. 8, fig. 9), and like most Hollywood productions it follows very closely the contour of the Freytag triangle and of the Aristotelian plot line. Its structure (fig. 13) comprises all four of the standard components of narrative grammar:

1. *Exposition.* In the exposition the user has the choice between

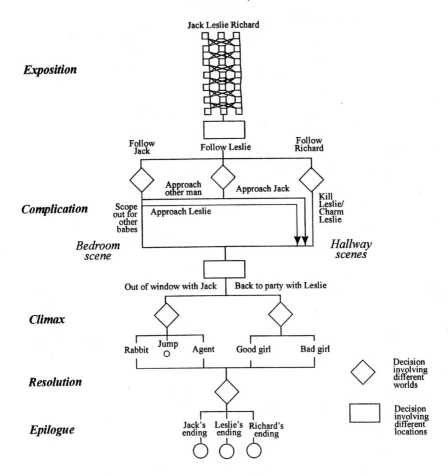

FIGURE 13 | Decision map for *I'm Your Man*

three points of view, corresponding to each of the main characters. Every twenty seconds, for a total of seven decision points, the screen freezes, the names of the three protagonists appear, and the viewer can switch the point of view to that of another character. Jack's introduction shows him on the phone getting the directions to a party and then walking to the social event surrounded by a crowd of sexy models. His mental preoccupations are suggested by closeups of the legs, derrières, and décolletés of the women. The other two introductions are complementary representations of a telephone conversation focused respectively on each end of the line. In Leslie's segment we see her in an office talking to a man who passes as an inspector from the FBI. They arrange to meet at a party, where she will recognize him through the password "I am your man." Leslie is then supposed to give the inspector some computer disks that will clear her name in a scheme of fund diversion and put a certain Richard in jail. In Richard's introduction, we see him riding in a limousine with an accomplice, Lawrence, who turns out to be none other than Leslie's telephone interlocutor. When Lawrence hangs up the phone, Richard reveals his intent to kill Leslie after he gets hold of the diskettes. Lawrence refuses to participate in this plan, and Richard promptly and cleanly eliminates him with a gunshot.

2. *Complication.* At the end of the exposition, all three strands converge in the same decision point. From now on, decisions will be spaced every one and a half to two minutes. The user is asked whether he wants to attend the party with Leslie, Richard, or Jack. Each of the three branches shows the selected character taking an elevator and entering an art gallery where the party is taking place. Jack is looking for a flirt, and he wonders whether he should go for Leslie or "scope out for other babes." Leslie is wondering which guest is the FBI agent to whom she should give the disks, and she debates whether she should approach Jack or another man. Richard asks the audience whether he should kill Leslie to get the disks or try to charm her. Two of these six possibilities lead to a bedroom scene, and the other four to three different versions of a hallway scene shot respectively from Jack's, Leslie's, and Richard's point of view. In an episode common to both branches, Jack inadvertently speaks the words "I am your man," leading Leslie to believe that he is the FBI agent she is looking for, while Jack hopes for action of a more amorous variety. They rush into either the hallway or the bedroom to complete the transaction, but in both cases they run into Richard, who starts firing at them. In the hallway scene Leslie blinds Richard with a mace spray, while in the bedroom scene she hides under blankets. In a merging of the two branches, Richard asks the spectator where he should go to find them: out the window through which Jack fled, or back to the party to which Leslie will soon return.

3. *Climax and resolution.* The window branch follows Jack up to the roof of the building. He asks the user whether he should "turn into an

FBI agent" or "run like a rabbit." The agent selection leads to a fight—or is it a martial arts demonstration?—with an anonymous opponent who conveniently drops from the sky. Shorter and rather anticlimactic, the rabbit selection merely shows Jack returning to the party. Meanwhile, the back-to-the-party branch shows Leslie being torn by a moral dilemma: should she be a "good girl" and persevere in the attempt to find the FBI agent among the guests, or should she be a "bad girl" and negotiate with Richard to save her life? In the bad-girl branch Leslie invites Richard to dance, but she takes advantage of a moment of distraction to kick him in the groin. In the good-girl branch she makes a comical announcement to the entire party, "I am looking for a man." All four branches are interrupted by the *coup de théâtre* of the common resolution: an FBI agent suddenly appears, like a deus ex machina, collects the disks from Leslie, arrests Richard, and congratulates Jack and Leslie on a job well done.

4. *Epilogue.* The user is offered a choice of three endings, each of which satisfies the wishes of the selected character. Leslie's ending rewards her with power (she is appointed project director of her company), Jack's ending rewards him with a dance with Leslie (the prelude, he hopes, to other pleasures), and Richard's ending is a dramatic reversal that turns the preceding climax into an anticlimax. The supposed FBI agent leads him out of the building, opens his handcuffs, and gives him the disks, and Richard drives away in a luxury car.

As this plot summary suggests, *I'm Your Man* combines several strategic types of interactivity. Throughout the exposition the choices of the user determine not what happens in the fictional world but what the user gets to know. By clicking on one name rather than on another, the user is transported to different locations, and she catches one of three parallel strands in the plot. The producers of the movie compare this activity to channel surfing in a system in which all the channels are related. If the user switches from character to character, as I did on my first try, she will get a mosaic of disconnected frames, and she will not be able to grasp the logic of the subsequent action. The same is true if she chooses Jack's introduction. Leslie's and Richard's segments provide a better idea of what the story is all about, but it is only by following all three segments in different runs that the user will gain a full understanding of the goals and plans that motivate the

action to come. Through their complementarity, the three exposi-
tions illustrate the jigsaw-puzzle model of interactivity.

At the end of his exposition, Richard turns toward the user and
says, "Until now you have been networking. But now your choices will
have more serious implications." (Leslie comes up with a similar statement.) These serious implications concern the actions of characters—
what actually happens in the fictional world. By choosing between
"should Leslie approach Jack" and "test another guest" or between
"should Jack go for Leslie" and "scope out for other babes," the user
makes decisions that send the plot toward different possible worlds
and create, at least potentially, different fates for the protagonists. This
godlike power to manipulate the strings of the characters is emphasized in the field of options of the climax section, when the user is
given a choice between making Jack a hero or a coward, and Leslie a
good or a bad girl. The ontological type of branching implements
what I call in chapter 7 the kaleidoscopic model of interactivity. Once
the user gets past the introduction, each run produces a different, self-
sufficient plot, and it is not necessary to accumulate knowledge from
former runs to follow the action.

The options offered to the user are not necessarily genuine ontological alternatives. They may also concern angle of shot and character point of view. When we are given the choice to enter the party with
one of the three protagonists, for instance, every selection offers a
different perspective, both spatial and mental, on a scene that contains the same events and the same participants. Later in the episode,
four of the six choices lead to the hallway scene, but we get a different
version depending on which character we have been following. The
different versions of the same scene are often made up of the same
shots, but the interpretation of these common images is affected by
the differences in the context. If we attend the party with Jack, for
instance, the shot in which we see him enter the room and wave at the
crowd suggests self-confidence and eagerness to go after the "gor-
geous babes"; but if we follow Leslie, the very same shot captures her
perception of Jack as an annoying clown.

The last type of choice is the pointless decision. Some of the supposedly strategic options of the exposition and complication are actually fake alternatives that lead to the same development. When Rich-

ard asks whether he should kill Leslie or charm her, for instance, both branches lead to Richard's hallway scene. In this case it will take several runs for the user to understand that he has been fooled, but in another instance the futility of the choice is revealed right away for a humorous effect. When Jack is on the roof, the possibilities include not only "become an FBI agent" and "run like a rabbit" but also "jump to the next building." If the user selects this last option, Jack will tell him, in a voice that imitates an automated phone message, "Unfortunately at this time jumping to the next building is a physical impossibility. Please hang up and choose again." The primary function of the fake decisions is to prune the tree of possibilities to a manageable size, thereby making the system look more interactive than it really is, but the device is also an ironic reminder that the user's seemingly godlike power to write the fate and moral character of the dramatis personae is a gift from a more powerful god. What this higher god giveth, he can randomly take away.

The branching system of the movie is operated in a context-free manner: no matter what choices the user made in the past, the same range of options remains open at any decision point. This creates logical difficulties that *I'm Your Man* does not completely overcome. The producers tell us that the plot was constructed, in a very postmodern job of *bricolage,* by putting together segments that were often shot with no particular development in mind. The various pieces of plot fit loosely together, but one should not inspect too critically the structural solidity and motivational soundness of the assemblage. One of the most serious architectural problems is the shot that fails to keep its dramatic promise in all possible developments. For instance, the mace spray that Leslie ostentatiously clutches in her exposition plays an important role in the hallway scenes, but if we take the bedroom branches, this "telling object" will never be put to use. This inconsistency merely breaks an aesthetic rule of plot construction, but it will turn into logical incoherence if we take the bad-girl branch of the plot and hear Richard tell Leslie, "Not bad for somebody who just sprayed me with mace." Such problems could have been easily avoided with better planning, or by giving some memory to the system. While none of the branches is truly well plotted, some are clearly more dramatic than others. In the climax-resolution section, for instance, Jack-as-FBI-agent and Leslie-as-bad-girl lead to a physical action that satisfies the Hollywood conception of climax, but Jack-as-rabbit and Leslie-as-good-girl take us to dramatic dead ends that depend on the deus-ex-machina entrance of the cops to be sent back to the proper track.

All of this suggests that the interest of *I'm Your Man* resides not on the level of the individual plot lines but on the level of the entire network. In this particular case, as with most interactive artworks, aesthetic appreciation is primarily a matter of playing with the system, of reconstructing the logical map,[2] and of assessing how individual elements fit (or do not fit) into a global architecture—all mental activities that require more than one run through the system. For the spectator who adopts a panoramic and detached point of view, the inconsistencies of the individual runs are no longer a threat to comprehension but a source of slightly perverse pleasure. With the proper ironic attitude, there aren't many bugs that cannot be viewed as features.

One of the most delicate design problems that faces the creators of interactive films is the handling of transitions. In an interactive film the obvious way to display options is to freeze the screen and to wait for a response, a strategy that uproots the spectator from the fictional world and highlights the conflict of immersion and interactivity. It may someday be possible to create a more fluid mode of interaction—for instance, by letting the movie follow a default plot line when no action is taken (Murray, *Hamlet,* 259) or by dynamically creating transitions at run time[3]—but these options were not available to the producers of *I'm Your Man.* The decision points all involve some kind of break in the action.

The movie proposes two contrasting strategies to deal with the blatant artificiality of these interruptions. One consists of naturalizing the decision point, and the other of creating a self-conscious "breaking of the fourth wall" (as the producers call it). In the naturalizing approach, the spectator's intervention is motivated by the internal logic of the plot. When Richard is temporarily blinded by Leslie with the mace spray in the hallway scene, he turns toward the spectator, rubbing his eyes, and says, "You, not having been maced, had an opportunity to see the mode of escape. I, however, was not so fortunate. So that puts us in a situation of mutual need. I need *you* to tell me where they went." Here the interaction between audience and character is made possible by pulling the spectator into the fictional world and placing her in the role of a witness. In the self-conscious mode, by contrast, it is the character who steps out of his role and world to communicate with the audience. "You need *me*," Richard goes on, "not to pull this switch that will close the doors of the theater and lock you up in a twenty-four-hour marathon of menudo music. So where did they go: up the window, or back to the party?" While he delivers his harangue, a giant number inscribed in the picture of a target counts down toward the deadline, reminding the DVD spectator of the days when the movie was shown in a theater and the audience had a limited amount of time to vote.

The playful self-consciousness of this approach to interactivity infiltrates almost every other aspect of the movie, from dialogue to body language, from music to costumes, and from acting style to the choreography of the fights. According to the composer, the score is pure Muzak, and there is not a single note that should be taken seriously. The same is true of almost everything else. The movie's aesthetic claims rest on the idea that all it takes to turn kitsch into art is to enclose it within quotation marks. The setting is an art gallery full of simulacra, the guests are campy performance artists, the humor of the situations is based on the most tacky brand of sexual double entendre, and the dialogue is sprinkled with clichés: "Don't worry, everything will be all right," says Jack as he turns into an FBI agent. "Miss Campbell, you are doing the right thing," says Lawrence before he is shot by Richard. "It's people like you who make the fabric of this great nation," the cops tell Jack and Leslie when they arrest Richard. The movie disarms the objection that interactivity turns characters into mere puppets—how else could they be randomly made into heroes or cowards?—by displaying all the strings that control them and all the tricks that preside over the writing of movie scripts. If everything in the fictional world is made up anyway, asking the user to contribute moral features and to decide fates is just another technique of fabrication.

This philosophy of emphasizing the overall constructedness of the plot adroitly dodges the conflict of immersion and interactivity, for it is difficult to blame an artwork for the failure to provide an experience that it so blatantly prevents. The predilection of interactive works for modes of expression that involve an ironic distancing from the fictional world confirms Janet Murray's diagnosis of their built-in affinity for the comic spirit (*Hamlet,* 175). But if self-consciousness becomes the standard way to compensate for the anti-immersive effect of interactivity, it will take a lot of ingenuity to prevent the device from becoming a metacliché.

PART IV Reconciling Immersion and Interactivity

Participatory Interactivity from Life Situations to Drama

I . . . remember vividly to this day the terrible shock of such a recall to actuality [from within a fictional world]: as a young child I saw Maude Adams in Peter Pan. It was my first visit to the theater, and the illusion was absolute and overwhelming, like something supernatural. At the highest point of the action (Tinkerbell had drunk Peter's poisoned medicine to save him from doing so, and was dying) Peter turned to the spectators and asked them to attest their belief in fairies. Instantly the illusion was gone; there were hundreds of children, sitting in rows, clapping and even calling, while Miss Adams, dressed up as Peter Pan, spoke to us like a teacher coaching us into a play in which she herself was taking the title role. I did not understand, of course, what had happened; but an acute misery obliterated the rest of the scene, and was not entirely dispelled until the curtain rose on a new set.   —SUSANNE K. LANGER

The claim that hypertext turns the reader into a writer may be a vast hyperbole, if by *writing* we mean an activity that requires imagination and language skills, but it contains an important clue as to why interactivity conflicts with immersion. The reader of multiple-choice narratives becomes an author not in the creative sense of the term but through the exercise of authority over the characters. In the moment of decision, readers associate with a godlike author who creates the fictional world from an external perspective. If the reader or spectator can choose whether Jack will be a hero or a coward, this means that Jack's behavior, and by extension the fate of the entire fictional world, is determined not by any kind of internal necessity but by the decisions of an omnipotent creator located in the real world. Yet the loss of the sense of the autonomy of the fictional world that occurs at every decision point is not compensated by a gain of creative power, because the choices are all prescribed paths. Even in a conceptualization that presents hypertext as a matrix of worlds and of stories, what readers do is control the strings of the fictional worlds through either reasoned or arbitrary choices, but they are themselves the puppets of the author.

The heavy reliance of the movie *I'm Your Man* on self-reflexive devices exposes the fundamentally metafictional and metanarrative effect of selective interactivity. Offering a menu of possible developments to the reader is tantamount to claiming, as does John Fowles in *The French Lieutenant's Woman*, "This story . . . is all imagination. These characters . . . never existed outside [your and] my own mind" (80). The cost of the metafictional stance is an ontological alienation of the reader from the fictional world. Insofar as it claims the reality of its reference world, fiction implies its own denial as fiction. By overtly recognizing the constructed, imaginary nature of the textual world, metafiction blocks recentering and reclaims our native reality as ontological center. Literary texts can thus be either self-reflexive or immersive, or they can alternate between these two stances through a game of in and out—masterfully played in *The French Lieutenant's Woman*—but they cannot offer both experiences at the same time because language behaves like holographic pictures: you cannot see the signs and the world at the same time. Readers and spectators must focus beyond the signs to witness the emergence of a three-dimensional lifelike reality.

For interactivity to be reconciled with immersion, it must be stripped of any self-reflexive dimension. To find out how to avoid the "holographic effect" that forces us to choose between playing with signs and plunging into the depth of worlds, let us take another look at the ideal VR experience. Why is it that in VR, far from destroying illusion, the possibility of interacting with the virtual world reinforces the sense of its presence?[1] Let us first note that in real life (RL) also, freedom to act enhances our bond to the environment. The main difference between VR and RL, on one hand, and textual environments, on the other, is the semiotic nature of interactivity. In a textual environment the user deals with signs, both as tools (the words or icons to click on) and as the target of the action (the text brought to the screen as the result of clicking), but in RL and VR all action passes through the body. This is not to deny that it takes a hand to click or write in textual environments, or that the entities encountered in VR are ultimately digital signs, but if realized, the ideal of the disappearance of the medium means that the VR user experiences at least the sensation of a direct encounter with reality. The hand that turns the pages of a book or that clicks on hypertext links does not belong to the textual world, but the body that moves around in a VR installation writes, or rather acts out, the "history" of the virtual world. We cannot say that "*x* clicks on link *p* at time *t*" is an event in any of the worlds of Michael Joyce's *Afternoon*—though it is an event in the reader's reading, which takes place in the real world—but in a VR installation, "*x* turns his head to the left," "*y* grabs an object," or "*z* takes an action that steers the plane toward the Earth" are history-making events. VR and RL thus offer a mode of action in which the body can be much more directly and much more fully involved with the surrounding world than through conscious symbolic manipulation.

Moreover, as we have seen in chapter 2, the worlds of VR are experienced as existing autonomously because they can be explored through many senses, particularly through the sense of touch. As the story of St. Thomas demonstrates, tactile sensations are second to none in establishing a sense of the alterity that convinces the mind of the objective existence of reality. The perceptions of the various senses are not isolated in the discrete windows of a hypermedia display, as they are in hypertext and World Wide Web pages, but fused in a global experience that enables the user to apprehend the virtual world under many facets at the same time.

The fluidity of the VR display is another critical factor of immersivity, for the world does not offer itself to the senses in broken-up packets of information. In a textual environment interactivity is made possible by the chopping down of data into the discrete fragments of "lexias" or "textrons"; in a VR installation, by contrast, the packets of data that respond to the user's actions are so small that they give the impression of a continuous evolution.

Finally, the corporeal participation of the user in VR can be termed world-creative in the same sense that performing actions in the real world can be said to create reality. As a purely mental event, textual creation can be called a creation *ex nihilo* in the sense that it excludes the creator from the creation: authors do not belong to the world of their fictions.[2] But if a mind may conceive a world from the outside, a body always experiences it from the inside. As a relation involving the body, the interactivity of VR immerses the user in a world experienced as already in place; as a process involving the mind, it turns the user's sojourn in the virtual world into a creative membership. For an agent embodied in a multi-dimensional environment, selective and productive interactivity can no longer be rigidly distinguished, because navigating the virtual world is a way to bind with it, a way to make it flow out of the acting body.

By arguing that the key to immersive interactivity resides in the participation of the body in an art-world, I do not wish to suggest that interaction should be reduced to physical gestures, but rather that language itself should become a gesture, a corporeal mode of being-in-the-world. As is the case in dramatic performance, the participant's verbal contribution will count as the actions and speech acts of an embodied member of the fictional world. Rather than performing a creation through a diegetic (i.e., descriptive) use of language, these contributions will create the fictional world from within in a dialogic and live interaction with its objects and its other members. Whereas merely immersive art is a representation of a fictional world, the reconciliation of immersion and interactivity will propose a genuine *simulation*.

Where can we turn, besides VR, for models of an enriching experience of immersive interactivity? Long before electronic technology made it possible to "[take] your body with you into worlds of imagination," to quote once again Brenda Laurel's seductive characterization of the power of VR ("Art and Activism," 14), these worlds were made accessible to flesh and blood bodies through a variety of art forms and cultural activities.

CHILDREN'S GAMES OF MAKE-BELIEVE

I mention in chapter 6 that games of make-believe provide a compromise between the "game" and the "world" aesthetics, since their purpose is to create a world in which to play. This purpose makes them uniquely qualified to reconcile immersion and interactivity. My strongest memory of being corporeally and creatively involved in a fictional world is a game that we used to play in kindergarten, "Who's Afraid of the Big Bad Wolf?" One player was the wolf and the others were pigs. A delimited territory, such as a manhole cover, was the house of the wolf; all around were the woods. At the beginning of the game the wolf was sleeping in his house. The pigs would gather near his door and taunt him with a chant:

Promenons-nous
Dans les bois
Quand le loup
N'y est pas.
Loup y es-tu?
Que fais-tu?

(Let's walk in the woods when the wolf is not there. Wolf, are you there? What are you doing?) The wolf would respond to each repetition of the chant with an improvised line: "I am getting out of bed, putting on my socks, brushing my teeth, adjusting my toupee," and so on. The responses had to follow the script of getting up and leaving the house for work, but the player was free to add as many colorful details as he wanted, both for the sake of their inherent creativity and as retarding devices. The pigs would react with delighted screams of terror, or with renewed taunting. The suspense rose as the wolf got closer and closer to being ready. Finally, he would roar some terrifying line, storm out of the house, and go after the pigs as they scattered into the woods. The pig that got eaten became the next wolf.

In this game, both the wolf and the pigs had the freedom to improvise, and the creativity of their contributions was an element of pleasure for all players. At the same time, however, the improvisation was monitored by a narrative line with a proper dramatic development: exposition (the wolf getting up), complication (he is taunted), climax (he storms out of his house), and resolution (he takes revenge on one of the pigs while the others get away). What made the experience so immersively pleasurable is that our creative interaction started not in a vacuum but in a designed and rule-governed environment that offered, as should any efficient "prop in a game of make-believe," a potent stimulant to the imagination.

EROTIC SCENARIOS

Children, of course, are not the only ones to engage in games of make-believe. It is through precisely scripted dramatic scenarios that many people enact their private sexual fantasies. Though Jean Genet's play *The Balcony* is a work of imagination, it offers a believable picture of the construction of virtual realities that takes place behind the doors of brothels and sadomasochist dungeons. In one scene we see a john posing as a bishop who tries to extract a confession of abysmal sin from a prostitute; at times she is willing, at times she insolently resists, and the performance is never exactly the same. In another scene the customer pretends to be a judge sentencing a thief who must show proper fear and contrition to satisfy the pseudo-magistrate. In a third scenario the john is a general, the prostitute is a horse, and he rides her in complete equestrian attire, including whip and spurs. But when a revolution gives the pseudo-bishop, -judge, and -general an opportunity to exercise in the outside world the power whose insignia they have been wearing in their private fantasy worlds, they panic at the idea of becoming forever welded to their roles, obviously preferring the convenient duplicity of make-believe to the responsibilities of real-world identities. Meanwhile, the revolutionaries in the presumed "real world" turn out to be just as dependent on simulacra, symbols, and role-playing as the clients of the brothel.

In this post-Baudrillardian age it is difficult not to see the play as an allegory of the fundamental virtuality of all realities, or at least of all the truly livable ones, even if one does not personally endorse this idea. My point, however, is not to propose an interpretation of *The Balcony*, no matter how fascinating it is for the dialectics of virtuality, but to use it as an illustration of an interactive practice that leads to the most evidently embodied form of immersion, since the experience corresponds to the satisfaction, or at least the pursuit, of sexual desire.

THE FAIR AND THE AMUSEMENT PARK

If there is an environment in which you can literally take your body with you into worlds of the imagination, it is the amusement park. We may dismiss this culturally invasive avatar of the fair as an attempt to recreate the communal street life of earlier days in a fenced-off and fully commercialized space of three-dimensional postcards, but with its multiple offerings, diversified geography, parades and masquerades, and mosaic of sensory stimuli, it also stands as a testament to the postmodern fascination with the playful spirit and protean nature of the carnivalesque. Jay Bolter and Richard Grusin capture the immersive dimension of theme parks through these observations:

The parks could appeal to the immediacy of physical presence. The rides themselves offer sound, light, and tactile sensations, as amusement parks have always done, and the themes and narratives associated with the rides, the animated characters that roam the park, and the themed architecture all give the young visitors the exhilaration of being physically surrounded by the media. (*Remediation*, 172)

Glorianna Davenport's eloquent description of the affinities of Coney Island, the ancestor of all modern-day fairs, with the ideals of electronic entertainment places more emphasis on the interactive aspect of the amusement park experience. For Davenport, Coney Island is a model of immersive interactivity because once the visitor has stepped over the boundaries of its magic world she feels instantly at home, and because this world goes out of its way to show her what to do in it. Through this internal guidance, the world of the fair displays choices of action without breaking the illusion:

Like the World Wide Web of today, Coney Island was an anarchical sprawl, lacking any form of urban planning or overarching design. Browsing, "grazing," and the delight of accidental encounter were the rule rather than the exception. It offered something for every taste: the physical thrill of high G-forces as a roller coaster hurled your body along a wildly twisting path; the sexual titillation of exotically costumed showgirls shimmying on stage; the long-term, episodic drama of premature babies struggling for survival in Martin Courtney's "Child Hatchery" (which introduced and popularized the use of incubators in hospitals).

Most notably, Coney Island was exceedingly "user friendly": Visitors needed no tutorial sessions or user manuals to enjoy its attractions. In the distance, the potent sight of enormous rides and ornate buildings lured potential customers. Just as importantly, these huge and uniquely delineated structures defined a

sort of cognitive map—or a preliminary shopping list, at least—
to assist navigation on the ground. Close-up, Coney Island's
pleasures became more tangible and human-scale. Colorfully
dressed barkers and shills lined the streets, their sole purpose to
steer passersby into a particular attraction. Once inside, rich
detail, meaningful travel, and the delight of discovery marked
every moment of your experience.

Coney Island provides an outstanding model of "virtual reality," a successful symbiosis of human entertainers and constructed mechanisms dedicated to human pleasure. How many
hours have you spent in front of your personal computer wishing you were at Coney Island instead? ("Care and Feeding," 8).

BAROQUE ART AND ARCHITECTURE

It is almost a truism to say that architecture is an art that inscribes the
body in its world. Its material is space, and its purpose is to arrange
this material for the recipient to inhabit, even if the sojourn is only a
temporary visit. It is no less obvious that architecture is an art that
requires a physically active involvement. The building is designed for
a living body, and the body must perform a walk-through and walk-
around to experience the spatial design under all the points of view
that matter to the intended recipient (typically those points of view
within reach of a normal-sized adult). But the immersive power of
architecture is usually tempered by abstract shapes that do not engage
the imagination. This is why Baroque churches, with their predilection for the curves of seashells and violins, their extravagant decoration, their exuberant use of organic forms, and their trompe l'oeil
effects, offer a better prefiguration of the VR experience than the
architecture of any other age. My most memorable sensation of interactive immersion—-if we interpret *interactive* in a metaphorical way,
as a mind-set rather than as physical intervention—came about during a visit to the Rococo-style abbey church of Zwiefalten, located in
the Swabian Jura halfway between Munich and the Rhine.

The church at Zwiefalten is the realm of the fake and the triumph
of appearance: the gold of the crown of Mary is a thin layer of precious metal, the "marble" of the columns is painted plaster, and many

of the architectural details are themselves painted on the walls. The Baroque age is notorious for its suspicion of mere appearances—in stark contrast to present times—but it also understood the value of the fake as a "prop," not in a game of make-believe (to parody Kendall Walton) but in the business of creating belief. A church, by definition, is the site of a lived interaction between man and God. In the Baroque age this interaction was conceived and staged as an experience that involved the whole of the person, this whole that Ignatius of Loyola described as an indivisible "compound of body and soul" (*Spiritual Exercises*, 136). The religious art inspired by the Counterreformation offers the concrete expression of the spirituality of Ignatius. The brightly colored interior of Baroque churches, their illusionist paintings, tonitruous organ music, burning incense, alabaster statues that invite caressing, and even relics (or their containers) occasionally offered to the touch speak to all of the senses that Ignatius wanted to involve in the religious experience. Through its insistence on the corporeal—there are bodies crammed into every niche and jamming every picture—the style proposed a literalization of the doctrine of the Word made flesh.

There is no need to dwell here on the well-known sensuality of Baroque art, on its almost sadistic depiction of the suffering flesh of Christ, or on its representation of the mystical union as orgasmic trance (cf. the statue of St. Theresa by Bernini). The Baroque age solved the Christian conflict of sensuality and spirituality by presenting the marriage of God and the soul as a deeply sexual experience. Whereas the design of Gothic churches integrates the divine as a purely spiritual light that enters the dark interior of the building, symbol of the human soul, from above and from outside, the Baroque church surrounds the visitor with the radiant, almost physical presence of God. The light does not come from an external source, it originates inside the church itself, in the whitewashed walls, golden frames, lighted candles, and gilded moldings of the altar. Walking down the aisles, past numerous alcoves that represent in three-dimensional displays the various episodes of the Passion, the visitor wanders in a story space inhabited by lifelike and palpable incarnations of biblical characters. This dramatic organization of space, typical of all Catholic churches, fulfills Celia Pearce's recommendation that architecture be treated as a narrative art and the visit to a building be structured as a meaningful succession of events (*Interactive Book*, 25–29).

At Zwiefalten, the sense of corporeal involvement in the religious experience is reinforced by numerous effects of trompe l'oeil that expand the physical space of the church and invite the visitor to ascend into virtual realms. The ceiling of the cupola is occupied by a fresco painted from the perspective of an observer situated directly under the scene, where the actual spectator happens to be physically located.[3] As in all trompe l'oeil art, this painting "continues the architecture of the building itself" (Bolter and Grusin, *Remediation*, 25). The effect of the coincidence between actual and virtual observer is a powerful sense of being pulled into pictorial space. The fresco represents the stages of a tortuous ascension toward heaven, which could be said to emulate the levels of a computer game if the inspiration did not run in the opposite direction. The levels of heaven are connected by the flights of a gigantic staircase that presents the quest for the kingdom of God as an active process and continuous effort. At the apex of the real-world cupola, in the center of the picture, a ray of lights breaks through the clouds and pierces the heart of those who have traveled so far. From there on, believers will be literally pulled by God into the upper reaches of heaven. In this dynamic vision of life, which makes it impossible to contemplate the painting without mentally traveling the path from Earth to heaven, the grace of God touches only those who work for their own salvation.

Umberto Eco has superbly captured this power of Baroque art to pull the spectator into its own movement and to simulate a corporeal mode of participation:

> Baroque form is dynamic; it tends to an indeterminacy of effect (in its play of solid and void, light and darkness, with its curvatures, its broken surfaces, its widely diversified angles of inclination); it conveys the idea of space being progressively dilated. Its search for kinetic excitement and illusory effects leads to a situation where the plastic mass in the Baroque work of art never allows a privileged, definitive, frontal view; rather, it induces the spectator to shift his position continuously in order to see the work in constantly new aspects, as if it were in a state of perpetual transformation. (*Open Work*, 7)

Art of the fake put in the service of potentialities; art of curves, laby-rinths, and morphing shapes that launch a dance of the imagination—the Baroque is VR in veneer and stucco.[4]

RITUAL

The style of the Baroque church may suggest immersive interaction, but as a ritual, the ceremony performed in the building aims at ac-complishing a much more complete and literal fusion of the two modes of binding to a world. In its religious form, ritual is a tech-nique of immersion in a sacred reality that uses gestures, performative speech, and the manipulation of symbolic objects—symbolic at least until the algorithm succeeds in establishing communication between the human and its Other. As Mircea Eliade, the great historian of religions, has convincingly argued, religious ritual is the reenactment of the foundational events that are commemorated in myth. Through the exact repetition of what the gods did in the beginning *(in illo tempore)*, the community is transported into mythical time and un-dergoes spiritual rebirth:

> Ritual abolishes profane, chronological Time and recovers the sacred Time of myth. Man becomes contemporary with the ex-ploits that the Gods performed *in illo tempore*. On the one hand, this revolt against the irreversibility of Time may help man to "construct reality"; on the other, it frees him from the weight of the dead Time, assures him that he is able to abolish the past, to begin his life anew, and to re-create his world. (*Myths,* 139)

As they reenter primordial time, the participants in the ritual experi-ence the live presence of the gods as an infusion of creative power. In this state of interactive immersion in the sacred, they attain a status that may be properly described as co-authorship of the cosmos. From a purely semiotic point of view, the climax of ritual is an event of transubstantiation, by virtue of which symbols are metamorphosed into what they represent. Through this transubstantiation the per-forming bodies of the participants become the incarnation, not just the image, of mythical beings, and the commemorative language of absence is replaced by the jubilation of live presence. In the Catholic mass, the passage from symbolic re-presentation to literal enactment occurs during the moment known as consecration, when the bread and wine of the communion are miraculously transformed into the flesh and blood of Christ. The participation of the community in the sacred, reward for the exact performance of the prescribed actions, thus prefigures the "postsymbolic" communication that forms the lofty ideal of the most exalted prophets of VR technology.

The affinities between VR and ritual are indeed a common theme in the propaganda and scholarship devoted to the medium. The comparison seems to grow spontaneously out of the atmosphere of religious fervor that has surrounded the early development of cyberculture. But whereas the above discussion conceives ritual as an action that produces immersion through transubstantiation, most of the writers who have explored the analogy invoke the archetypal scenario of the rite of passage and initiatory journey. For Howard Rheingold, the first virtual realities were the painted caves of the Paleolithic age, the best known of which is the cave of Lascaux. Though Rheingold's book *Virtual Reality* appeared too early (1991) to establish connections with this newer technological development, the prehistoric cave prefigures the form of VR in which images are projected onto the walls of a small room, rather than being made visible and palpable through goggles and data gloves. In reference to Plato, the VR room is known as the CAVE, an acronym for Cave Automatic Virtual Environment (Heim, *Virtual Realism*, 26).

Building on the theories of the paleontologist John Pfeiffer, Rheingold imagines a ceremony in which carefully selected novices were led into the caves by a group of shamans to learn the secrets of toolmaking, the most important breakthrough in the history of the human race. Rheingold speculates that when the candidates entered the underworld of the cave, after hours of crawling through dark shafts, the figures on the walls illuminated by the flickering light of torches took on magical life and induced an altered state of consciousness, thereby putting the candidates in the proper state of mind to be initiated into technological secrets (379ff.). This scenario includes the three standard moments of the rite of passage: (1) separation from the profane world; (2) journey in a dangerous place, site of a symbolic death; and (3) spiritual rebirth and return to the everyday world. The pivotal moment in this protocol is the immersive experience that cleanses the mind and makes room for the acquisition of new knowledge.

For Brenda Laurel, the ceremonial model for the spiritual potential of VR is the cult of Dionysos, as well as the kivas of the Anasazi culture of the Four Corners area. In an oral response to Rheingold she says, "The transmission of values and cultural information is one face of VR. The other face is the creation of a Dionysian experience. The piece I find important to both of these functions is this notion of being in the living presence of something. With the ceremony of the kiva, one is in [the] living presence not just of other people but of an event that is happening in real time" (quoted in ibid., 385). The presence of the gods can be compared to the telepresence of VR, because it breaks the boundary between the realm of the human, located *here,* and the realm of the divine, located *there* in sacred space.

DRAMA AS IMMERSION AND/OR INTERACTIVITY: A VERY SHORT HISTORY

The history of drama is marked by an alternation of immersion-seeking and interaction-promoting moments. In this nonlinear evolution, the periods and movements that come closest to reconciling the two dimensions are those that remember the ritual origins of the genre. Two philosophers, Aristotle and Nietzsche, have been particularly influential in imposing the view that ancient Greek tragedy derives from ritual, and though their theories are by no means regarded as unassailable, I will use them as the basis of my discussion. This means that I will be examining interpretations of tragedy more than the historical institution itself.

Aristotle locates the source of Greek tragedy in the dithyramb, the sung and improvised poetry that traditionally accompanied the orgiastic cult of Dionysos (*Poetics* 3.3). This claim forms the basis of Nietzsche's classic work *The Birth of Tragedy,* an inspiring but highly personal interpretation of Greek drama whose interest resides more in its brilliant insights into the nature of art than in its historical accuracy. Nietzsche describes classical Greek tragedy as the perfect collaboration of the spirit of the two gods Dionysos and Apollo. As artistic principles, the Dionysian and the Apollonian are too complex for a straight assimilation with immersion and interactivity, but the entrancing possession of the Dionysian is a prime example of immersive experience, while the Apollonian urge to capture vital forces in an artistic form requires the distanced perspective that underlies the interactive stance. Nietzsche describes the social and aesthetic function of tragedy as a taming of the horror of life through a Dionysian shattering of the individual and a joyful "fusion with primal being" (*Birth of Tragedy*, 65). By forcing the spectator to look into the abyss on which everything rests, tragedy should provide the metaphysical comfort that "life is at the bottom of things, despite all the changes of appearances, indestructibly powerful and pleasurable" (59).

Aristotle takes a different view of the event that forms the ritual purpose of tragedy: "Tragedy is an imitation of an action that is admirable, complete and possesses magnitude; in language made pleasurable, each of its species separated in different parts; performed by actors, not through narration; *effecting through pity and fear the purification* [katharsis] *of such emotions*" (*Poetics* 4.1; my italics). In its attribution of a therapeutic value to tragedy, the concept of *katharsis* seems self-explanatory, but its precise interpretation has raised heated controversies. Is *katharsis* synonymous with the aesthetic pleasure of tragedy, or is it a supplemental effect? Do all spectators need to be "purified" from terror and pity, or only those who are crippled by an excess of these emotions—in which case *katharsis* is more an exorcism of the sick than a moral event of universal value? How can one explain that by eliciting terror and pity, tragedy will free the spectator from the negative effect of these emotions? Is tragedy a sacrifice, and the tragic hero a scapegoat whose downfall purifies the community from harmful feelings, or is Aristotle proposing a kind of homeopathic cure for existential anguish: experience a little dose of fear and pity in the fictional world, and you will be relieved of these feelings in the real world? Does *katharsis* reduce tragedy to a utilitarian purpose and downplay purely aesthetic pleasure,[5] or on the contrary, is the concept an acknowledgment of the scandalous fact that experiencing terror and pity for fictional individuals is a source of pleasure—the "purification" of these feelings meaning their aesthetic sublimation? And finally, could Aristotle be telling us that the pleasure taken in tragedy is morally pure, and not a matter of sadistic *jouissance* in the suffering of the hero? No matter how we resolve these individual questions, however, the concept of *katharsis* presents tragedy as the instrument of a deeply transforming spiritual event, and the spectator's immersion in the fate of characters as the mode of action specific to the genre.

For all its ritual power, though, tragedy is a spectacle and not a participatory event. The play acts upon the spectators, but the spectators do not act upon the play, nor do they receive an active role in the script. Greek tragedy can be read as a reconciliation of immersion and interactivity only if we look at what happens symbolically on stage, in the dialogue between the chorus and the characters. The function of the chorus is one of the most hotly debated features of ancient tragedy. One common theory, first voiced by G. F. Schlegel, interprets the chorus as the voice of an "ideal spectator." In *The Birth of Tragedy* Nietzsche first rejects this hypothesis on the ground that the spectator views the play as a work of art—as a fiction—while the chorus interacts with the characters as real human beings; but later he suggests that the audience identifies with the chorus (57). The chorus is not the spectator as real-world individual but his recentered projection in the fictional world as what Thomas Pavel has called a "non-voting member" (*Fictional Worlds*, 85). According to Nietzsche this projection is favored by the architecture of the Greek amphitheater: "A public of spectators as we know it was unknown to the Greeks: in their theaters the terraced structure of concentric arcs made it possible for everybody to actually *overlook* the whole world of culture around him and to imagine, in absorbed contemplation, that he himself was a chorist" (*Birth of Tragedy*, 63).

Usually impersonated as a collective character (Trojan women, daughters of Oceanos, old men of Thebes), the chorus fulfills several functions: narrating the past (especially in Aeschylus), commenting on the action, weighing possible developments, or lamenting the fate of characters; but for all of its vocal presence it has no influence on the development of the plot. The narratorial, extradiegetic function presupposes that the chorus knows more than the audience, thus precluding strict identification, but all the other types of intervention offer credible reactions of the spectator's fictional counterpart in the fictional world. In contrast to what a real audience would say, these interventions are couched in beautiful language, follow the proper metric pattern, and occur at appropriate times in the development of dramatic tension. A simulation of interactivity, the vicarious participation of the audience through the chorus is the only way to acknowledge the spectator's voice and presence without threatening the aesthetic integrity of the performance—the only way to achieve a compromise between Dionysos, who wants to lure the spectator into an individuality-threatening celebration of life, and Apollo, who wants to confine violence to the stage and disarm it through artistic form.

As responses to dramatic performance, immersion and interactivity are strongly affected by the arrangement of theatrical space. It is because standard theaters are designed for immersion, not for interaction between the actors and the audience, that the young Susanne Langer was so upset when the actress playing Peter Pan asked the children in the audience to attest their belief in fairies (*Feeling and Form*, 318; quoted in the epigraph to this chapter). The breaking of the illusion and the loss of pleasure that ensued from this ill-fated attempt to establish an interactive relation with the audience coincided for the little girl with the moment she became aware of hundreds of children watching the play with her.

Whereas the designs that promote immersion keep the audience hidden from itself, the quintessential interactive architecture is a circular arena, such as a sports stadium, that allows spectators to see each other as well as the action on the field (fig. 14, panel A). During a sports event, fans yell at the players, comment loudly on the quality of play, and engage in cheering duels with the partisans of the other team. In a truly *inter*active feedback loop, the game elicits cheering, and the cheering influences the outcome of the game. But this model of immersive interactivity is obviously not applicable to the theater. The representation of a tragedy under the circumstances of a sports event would result in a cacophony of actors and spectators talking to each other in all possible directions: actors to actors, actors to spectators, spectators to spectators, and spectators to actors. In this chaotic cross fire of addresses the spectators would be totally distracted from the plot.

The Greek stage design, with its semicircular seating, offers a compromise between immersion and interactivity, rather than a fusion. While the architecture acknowledges the presence of the audience, thus establishing a spirit of communion between actors and spectators, it also maintains a strict separation between the stage and the seats, which favors make-believe. The architecture makes it clear that the audience has a role in the performance, but this role does not

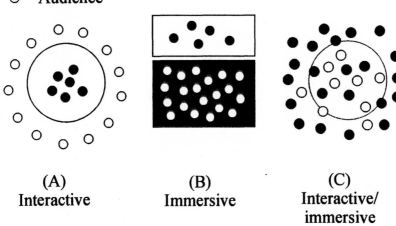

FIGURE 14 | Three types of stage design

permit interference. A similar compromise between the interactive and immersive stances was maintained in the open-air representations of medieval mystery plays on the parvis of cathedrals, and in Elizabethan stage architecture.

In the seventeenth century the balance of the two modes of participation was broken in favor of the immersive pole. The so-called Italian stage design of the Baroque era perfects the art of make-believe by taking the obliteration of the audience as far as it can go. Enveloped in darkness, the spectator sits in front of a brightly lit and lavishly decorated stage, imaginatively separated from the setting of the action by the transparent fourth wall, and physically kept away from the raised stage by the proscenium and orchestra pit (fig. 14, panel B).[6] Walter Benjamin describes the orchestra pit as "the abyss which separates the players from the audience as it does the dead from the living; the abyss whose silence in a play heightens the sublimity" ("Epic Theater," 156). The strict division of the house between stage and auditorium does not alienate the spectator from the action but on the contrary heightens the illusion of live presence and authenticity. Though spectators cannot step onto the stage, they are, fictionally, part of the same world, just as the spectator of a perspective painting is included in the imaginary extension of the pictorial space. The transparent fourth wall is a peephole that enables the spectator to spy on characters who live their lives unaware of this observing presence.

Not only does the stage design suppress the audience for the actors, it also suppresses the actors for the audience. The actor of the Baroque age was not supposed to be acknowledged as a performer but to dissolve his identity into the self of the impersonated character. Writing in 1657, the French drama theorist d'Aubignac enjoined actors to act "as if there were no spectators. All the characters must speak and act as if they were really Kings, and not as Bellerose and Mondory, as if they were in the palace of Horace in Rome, and not in the hôtel de Bourgogne in Paris; and as if nobody saw nor heard them" (*Pratique du théâtre*, quoted in Rousset, *L'Intérieur*, 171; my translation). Another French theorist, Champlain, wrote in 1630 that the purpose of theatrical performance is to "surprise the imagination of the spectators" by presenting everything on the stage as "genuine and present." All difference between representation and represented must be erased. The purpose of this perfect replication is to "eliminate any opportunity for the spectators to reflect on what they are seeing and to doubt its authenticity" (quoted in ibid., 170). Illusion is reinforced through elaborate stage decoration and complex machinery—the deus ex machina is a Baroque invention—but in contrast to the medieval stage, which represented heaven and Earth simultaneously, as would a cross-section of the house of the universe, the seventeenth-century stage was entirely occupied by one spatial setting. Immersion cannot be complete if the attention of the audience is divided among several locations.

Rather than inviting the recipient to a willing suspension of disbelief, as does written fiction, the theater of the Baroque age thus opted for an aggressive creation of belief through the tricking of the senses. As Rousset observes, the trademark of seventeenth-century dramaturgy is a heavy reliance on a vocabulary of magic, sorcery, waking dream, seduction, and enchantment that presupposes passivity and loss of self-control on the part of the spectator. Yet the Baroque age was acutely aware that when the curtain falls, the enchantment abruptly ends, and it maintained, as already noted, a highly ambiguous attitude toward appearances. Its dramatic magicians also knew how to practice the demystifying game of the play within the play, reminding the spectator that immersion means illusion, illusion means deception, and the beauty of their spectacles is only the fugitive beauty of dreams.

At the other end of the immersive/interactive spectrum is the epic theater of Bertolt Brecht. The aesthetics developed by Brecht in support of his conception of the theater as an instrument of political action systematically inverts the values of the Baroque era and of its less playful successor, the naturalistic theater of the late nineteenth century. The famous *Verfremdungseffekt,* or V- (distancing) effect, was meant to encourage critical thinking by preventing any kind of immersion. The spectator was not supposed to regard the plot as the manifestation of an inexorable fate and to empathize with the characters—this would be equivalent to giving money to the poor, a Band-Aid solution to social injustice—but to analyze the circumstances that created their problems, to consider what other courses their lives could have taken, to assess what kind of action and attitude would be necessary to divert this course, and to apply the knowledge thus gained to real-world action.

Though Brecht's plays are often performed on a standard stage, the author's stated ambition to induce a critical evaluation of the social reality depicted in the text makes them very well suited for a circular theater. In the epic theater, temporal immersion is inhibited through songs, interruptions of play, and abstracts of the episode to come; spatial immersion through a minimalist stage setting in which signs often take the place of props; and emotional bonding through a style of acting that encourages actors to step out of character. As James Roose-Evans writes, "The actor was . . . not to 'be' Galileo but to impersonate him. To achieve this Brecht insisted that during rehearsals his actors should prefix their lines with the words 'he said' or 'she said'" (*Experimental Theatre,* 69). This distancing of the actors from their characters not only exemplified how the play should affect the audience, it was also meant to contribute to the critical education of the actors themselves: "In every instance," writes Walter Benjamin, "the epic theater is meant for the actors as much as for the spectators. The didactic play . . . facilitates and suggests the interchanges between audience and actors and vice-versa through the extreme paucity of the mechanical equipment. Every spectator is enabled to become a participant" ("Epic Theater," 154).

The term *participation* in this last sentence should not be interpreted as an invitation to the spectators to see themselves as members of the fictional world. If an actual dialogue takes place between the actors and the audience—an event that rarely happens but that would be very much in the spirit of Brechtian aesthetics—this means not that spectators step into the play but rather that actors step out of their roles to engage in a critical discussion. The locus of interaction is the real world, not the fictional one. In his theoretical thinking Brecht may, however, have underestimated the power of immersion, a power that the playwright in him knew very well how to unleash. "In one performance of *The Threepenny Opera*," writes Roose-Evans, "it was obvious that his intentions had failed when the audience became totally absorbed in the plot, left the auditorium whistling Kurt Weill's tunes, and even found themselves identifying with Polly Peachum" (*Experimental Theatre*, 70).

While Brecht was perfecting anti-immersive techniques, a French writer and drama theorist who spent a good part of his life in mental institutions proposed a return to ritual that promised to reconcile immersion and interactivity through a trancelike involvement of the audience. Exerting an enormous influence on the experimental theater of the postwar era, the writings of Antonin Artaud (1896–1948) envision the impact of theatrical performance on the spectator as a transforming experience of almost unbearable intensity; hence the choice of the label "theater of cruelty": "Theater is first ritualistic and magic, in other words bound to powers, based on religion, on actual beliefs, and whose effectiveness is conveyed through gestures, directly linked to the rites of theater which is the very practice and the expression of a hunger for magical and spiritual manifestations" (*Artaud on Theater*, 124). Artaud conceived the ritual impact of dramatic performance as a visceral experience that restores the integrity of the self through the involvement of both body and mind. The experience itself is described in rapturous tones that owe a great deal to Nietzsche's doctrine of the Dionysian essence of tragedy: "Briefly, it seems the highest possible concept of theatre is one which philosophically reconciles us with Becoming and which, through all kinds of objective situations, suggests the covert notion of the passage and metamorphosis of ideas, far more than the shock of feelings transmuted into words" (113). Or again: "A real stage play disturbs our peace of mind, releases our repressed subconscious, drives us to a kind of potential rebellion (since it retains its full value only if it remains potential), calling for a difficult heroic attitude on the part of the assembled groups" (116).

The means recommended by Artaud to achieve this painful yet healing experience anticipate many of the themes of cyberculture, and many of the ideals of VR theorists. For instance:

— *Liberating the theater from the hegemony of spoken language, and replacing the text of the author with the body-language of the actors.* Modern theater must become "a new bodily language no longer based on words but on signs which emerge through the maze of gestures, postures, airborne cries . . . leaving not even the smallest area of stage space unused" (88). "I maintain that this physical language, aimed at the senses and independent of speech, must first satisfy the senses" (92).

— *Creating a "poetry of space" through stage design.* "The stage is a tangible, physical space that needs to be filled and it ought to be allowed to speak its own concrete language" (92). Such a language should permit "spatial poetry to take the place of language poetry" (93). This view anticipates the VR developer Randall Walser's characterization of the cyberspace artist as a "space-maker" ("Spacemakers," 60).

— *Developing theatrical performance into a multisensory, multimedia event.* "Practically speaking we want to bring back the idea of total theatre, where theatre will recapture from cinema, music-hall, the circus, and life itself, those things that always belonged to it" (*Artaud on Theater,* 109). "This difficult, complex poetry assumes many guises; first of all it assumes those expressive means usable on stage such as music, dance, plastic art, mimicry, mime, gesture, voice inflection, architecture, lighting and decor" (93).

— *Placing the body of the spectator at the center of the theater.* "We intend to do away with stage and auditorium, replacing them by a kind of single, undivided locale without any partitions of any kind, and this will become the very scene of the action. Direct contact will be established between the audience and the show, between actors and audience, from the very fact that the audience is seated in the center of the action, is encircled and furrowed by it" (104).

Artaud's idea of a stage architecture that surrounds the spectators (fig. 14, panel C) has had a great deal of influence on the avant-garde theater of the 1950s through the 1970s. In 1952, for instance, John Cage staged a Happening in which the audience, sitting in the middle, could watch several activities taking place simultaneously. The 1971 production of the historical spectacle *1789* by the Théâtre du Soleil, directed by Ariane Mnouchkine, placed the audience on a platform surrounded by scaffoldings whereon actors performed song, dance, and mime, using masks and puppets and without the guidance of an authorial text (Roose-Evans, *Experimental Theatre,* 75 and 88–89; see also Popper, *Art,* 120–21). The encircling design offers a visual metaphor for immersion, and the disappearance of the proscenium makes it physically possible for the audience to mingle with the actors, but most of these projects do not permit a literally productive participation of the spectators in the play. It is still the actors who perform and the spectators who watch.

The caveat of letting spectators on stage is that if they gain control of the action, the resulting performance may become utter chaos, since unlike actors, spectators are not trained to perform a script. If only a small fraction of the public is allowed to participate, the majority will remain in a noninteracting role, and they will have to put up with the unpredictable, and most likely unpolished, contributions of the selected few. On the other hand, if everybody is let on stage, how can interaction among so many people be coordinated into a performance that will be pleasurable for all? As Brenda Laurel observes, the idea of the audience as active participant is more likely to add "clutter, both psychological and physical" to the stage than to lead to an artistic event (*Computers as Theatre,* 17). To circumvent this problem, Artaud's theater of cruelty resorts to a vicarious interactivity, as did Greek drama before it. It is through a communion, almost a transubstantiating identification, of the spectators with the actors that the performance exercises its ritual purpose of a "reconciliation with Becoming." In some participatory productions—for instance, *The Domestic Resurrection Circus* by the Bread and Puppet Theatre, 1970 (Roose-Evans, *Experimental Theatre*, 123)—the audience is invited to invade the stage at the end of the show and to join the actors in a chant or song, but this literal communion concludes a ritual in which the actors remained very much in control of the action, as do the priests in a religious ceremony before the Holy Spirit descends on the congregation.

The alternative to vicarious interactivity is to coach the audience into taking part in a disciplined action, so that literally there will be no spectators, and the play will be staged for the benefit of its own participants. In this situation, as Brenda Laurel puts it, *"the representation is all there is"* (*Computers as Theatre*, 17; italics original). We have seen this idea of the actors becoming their own audience, which inverts the concept of the spectator turned actor, surface in the discussion of Brecht's dramatic philosophy. It flourished in the political street theater of the Vietnam war and civil rights era, when performance groups such as the Living Theatre of Julian Beck and Judith Malina approached the theater as a way of life, adopted a communal lifestyle, developed their spectacles collectively, and regarded them as a means of self-discovery. "The [members of the] Living Theatre," writes Peter Brooks, "are in search of a meaning in their lives, and in a sense even if there were no audiences they still would have to perform because the theatrical event is the climax and the centre of their lives" (quoted in Roose-Evans, *Experimental Theatre*, 104).

When performing becomes synonymous with living, the theatrical experience inherits the immersive and interactive qualities that define our experience of being-in-the-world. This fusion of life and representation and this total engagement of the actors are, of course, far too utopian to offer any kind of useful guidelines for the developers of electronic forms of interactive art. Nor is it practical to coach a large number of participants to perform a script, even one that allows ample room for creative improvisation. These difficulties leave only one viable option for the design of immersive/interactive experiences: let only one or two users, at the very most a handful, into the virtual world. This ability to customize virtual worlds and artistic experiences to individual users is precisely the forte of electronic media.

Participatory Interactivity in Electronic Media

Whereas film is used to show a reality to an audience, cyberspace is used to give a virtual body, and a role, to everyone in the audience. Print and radio tell; stage and film show; cyberspace embodies. . . . Whereas the playwright and the filmmaker both try to communicate the idea of an experience, the spacemaker tries to communicate the experience itself. A spacemaker sets up a world for an audience to act directly within, and not just so the audience can imagine they are experiencing an interesting reality, but so they can experience it directly. . . . Thus the spacemaker can never hope to communicate a particular reality, but only to set up opportunities for certain kinds of realities to emerge. The filmmaker says, "Look, I'll show you." The spacemaker says, "Here, I'll help you discover."
—RANDALL WALSER

Cyberculture and postmodern theory have popularized the view that we own not simply a physical body, given to us, mortal, subject to irreversible changes, limited in its abilities, and anchored in "real reality," but also numerous virtual bodies, or body images, which either clothe, expand, interpret, hide, or replace the physical body, and which we constantly create, project, animate, and present to others. If the artistic reconciliation of immersion and interactivity requires the participation of the body in the art-world, there is no reason why the bodies in question could not be of the virtual kind. Virtual bodies are implicated in numerous human experiences, and they entertain various relationships to the physical body. Some, like the bodies impersonated by dancers and actors, follow exactly the movements of the physical body but express a totally different self, while others, like the bodies that we enter in dreams, belong to the same self as does the physical body—it is still *me*, though my dream body can fly—but they are not animated by the same voluntary muscles. Still other body images are operated by the physical body, but the gestures of the physical body do not correspond to those of the virtual one: while one body slays dragons, flirts with a used-car salesman who poses as a hooker, or explores an enchanted forest, the other one types on a keyboard or squeezes a joystick. Such are the bodies that we take across the screen to reach fictional worlds in standard two-dimensional electronic media.

COMPUTER GAMES

Invoking computer games as a potential reconciliation of interactivity and immersion may seem inconsistent with the opposition drawn in chapter 6 between "game aesthetics" and "world aesthetics," but my discussion there was more concerned with the interpretation of the game metaphor by literary theorists than with games in themselves. We have seen that the proponents of the metaphor tend to privilege certain types of games at the expense of others—games of construction, permutation, and manipulation of opaque objects over games of mimicry, role-playing, and make-believe. In an abstract sense, of course, most if not all games create a "game-world," or self-enclosed playing space, and the passion that the player brings to the game may be regarded as immersion in this game-world. But I would like to draw a distinction between "world" as a set of rules and tokens, and "world" as imaginary space, furnished with individuated objects. The pieces of a chess game may be labeled king, queen, bishop, or knight, but chess players do not relate to them as fictional persons, nor do they imagine a royal court, a castle, an army, and a war between rival kingdoms.[1] The involvement of the chess player in the game is an intellectual and emotional absorption—he wants to win and takes winning seriously—but it is not an imaginative experience, as immersion has been treated throughout this book. Many board games, such as Monopoly, Chutes and Ladders, Titanic: The Game, or Barbie's Shopping Trip—I am inventing these last two, though Titanic may exist—stimulate the player's imagination by building the game-world around a concrete theme, but when these games sacrifice strategic interest to a purely thematic appeal, they tend to be short-lived. Who will remember Titanic: The Game two years after the movie? It is certainly not because of fascination with the world of real estate that people continue to play Monopoly.

But if thematic appeal is not sufficient to guarantee the success of a game, it suggests a way to turn the intrinsic interactivity of games into a reasonably immersive experience. Through the versatility and power of their representational resources, computer games can take a significant step in this direction. Many of the earlier computer games, such as Tetris and PacMan, proposed abstract, purely strategic worlds similar to those of chess or bridge. As long as graphics remained at a primitive stage, the best way to offer reasonably detailed and concrete worlds was through purely textual means. This approach was taken by the so-called interactive fiction games of the 1980s, such as the popular Zork adventures. In these games the player tried to solve a mystery by typing verbal queries, and the computer responded, after parsing and analyzing the input, with customized fragments of text that described a fictional world and built a more or less coherent sequence of episodes. Though their rudimentary understanding of language frequently led to nonsense,[2] repetition, or vicious circles, these textual games had more immersive potential than literary hypertext because they projected the user not just as the operator of a textual machine but as a fully empowered member of the fictional world. As Espen Aarseth writes, "The user assumes the role of the main character and, therefore, will not come to see this person as an other, or as a person at all, but rather as a remote-controlled extension of herself" (*Cybertext*, 113). By having an extension—or recentered counterpart—in the fictional world, the user experiences at least a figural form of immersion, while by exercising remote control on this counterpart she plays an active role in the development of the narrative action.

When sound and graphics became more powerful, the purely textual games fell out of favor and were replaced by games that placed an increased emphasis on the sensorial representation of the gameworld. Aarseth observes that the most advanced computer games are almost indistinguishable from interactive movies. The principal difference between a game that makes heavy use of film clips, such as DVD games, and a movie like *I'm Your Man* resides in the role of the user and the purpose of his actions. In *I'm Your Man* the exploration of various alternatives is an autotelic activity, and the user is cast in the role of a puppeteer external to the fictional world. In a genuine game, by contrast, the user receives a specific task, through which, as already noted, he controls a character and identifies with it: if the character overcomes obstacles, the player wins; if the character dies, the player loses. When game developers discuss the various modes of inscription of the user persona in the game-world, they borrow from narratology the terminology of "first-person" and "third-person" point of view. In third-person games, such as the Mario Brothers games of the Nintendo Play Stations, the user controls a tiny graphic representation of his character. The minimal form of this representation is the abstract shape of the cursor. In first-person games—sometimes deceptively promoted as virtual reality games—the user is implicitly inscribed in the game-world through the perspective of the display. As he moves around this world, the gradual changes in the display reflect a traveling point of view.

Through the increasing attention devoted to the sensorial representation of the game-world, the pleasure of modern games is as much a matter of "being there" as a matter of "doing things." From a strategic point of view the newer games (Doom, Myst, or Quake) are not superior to the old ones (PacMan or Tetris), but they are infinitely more immersive.[3] This emphasis on imaginative involvement raises the question of the aesthetic status of computer games. Should we wholeheartedly agree with Aarseth when he calls the adventure game "an artistic genre of its own, a unique aesthetic field of possibilities, which must be judged on its own terms" (107), or is there a significant difference in attitude between immersion in a game-world and immersion in a movie- or novel-world? The arguments for regarding games as a form of art are powerful ones: we engage in games for the sake of pleasure, this pleasure depends on design, we evaluate games—not just our playing skills—as intrinsically good or bad, and in the case of multimedia games, enjoyment is enhanced by the artistic quality of graphics, sound, and plot. But couldn't we advance this last argument in the case of advertising, a form of discourse that subordinates art to a very narrow practical purpose? From an aesthetic point of view the main difference between art and games is that the greatness of an artwork is judged on the basis of its potential for self-renewal, while computer games, even the best of them, are meant to be "beaten." Once the player has solved all the problems and conquered the highest level, there is usually no reason to play the game again.[4]

But even though the satisfaction of mastering adventure games is qualitatively different from the contemplative and renewable pleasure we take in a good novel or in a beautiful painting, there is no reason why a computer game could not awaken simultaneously the desires of moving forward toward a goal and of lingering in the game-world. A step toward the hybridization of game and art, both verbal and visual, is taken by the 1998 CD ROM *Ceremony of Innocence.* An interactive version of the illustrated epistolary novel series *Sabine and Griffin* by Nick Bantock, *Ceremony of Innocence* offers both visual pleasure and narrative interest, but the user must "work" to let the plot unfold. She does so by exploring the backs of the postcards on which the letters are written until she finds the key that turns each postcard around and reveals its message. The purpose of the game is thus to reconstitute a literary object, the narrative itself. Once the postcards have been read, the user can save them and revisit the pictures in any order, liberated from the urge to progress in the plot.

MOOS

A genealogy of interactive genres leads from the popular role-playing game of the 1970s, Dungeons and Dragons, to the first computer game, Adventure (mid-1970s); from Adventure to multi-user computer games, or MUDs (multi-user dungeons), in which users logged on to play against each other; from MUDs to MOOs (multi-user domains, object-oriented), an environment in which users do not play a structured game but freely interact under make-believe identities; and from MOOs to Internet chat rooms, where role-playing is largely eliminated in favor of the social functions of meeting and conversing. In this evolution MOOs represent the apogee of creative make-believe by the user. They combine immersion and interactivity through a purely textual form of role-playing in a fictional world.

The MOO system relies on two types of statements (or images, for those newer electronic forums, such as Active Worlds, that incorporate visual decor and avatars): permanent descriptions of the setting and characters, and evanescent utterances that represent the actions and social lives of the characters. The decoration of the setting—typically a diversified locale with many well-defined subspaces, such as a house with many rooms—may contain many clever features, such as "smart objects" that react to the user's investigation in unpredictable ways or AI animated robots that behave like characters. Activities such as wandering through the MOO, exploring its many nooks and crannies, manipulating the built-in objects, and testing the robots may thus bring their own rewards, but the principal reason for participating in MOOs is to experiment with fictional identities.

Users create a MOO persona in much the same way novelists write imaginary individuals into being, by posting textual descriptions of their character's body, personality features, and favorite occupations, and even of the character's room in the MOO hotel, a spatial extension of his or her individuality. Once a make-believe identity is established, the user enters the virtual body of his character and plays its role from the inside. He encounters other users playing other characters, and they engage in a dialogue in real time. Most contributions are speech acts (x says), but the system also allows the performance of physical actions and even the building of virtual objects. The conventions of the game attribute a performative force to language, so that a described action such as "Garlic kisses Turnip" counts as the live performance of this action. As Elizabeth Reid writes, "On [MOOs], text replaces gestures and has even become gesture itself" ("Virtual Worlds," 167). Since language can express almost every possible act, the limits of interactivity are the limits of the imagination.

Through their written messages, MOO users thus participate in what sometimes sounds like freely meandering conversation and sometimes comes very close to goal-oriented dramatic action, such as flirting, spying, building castles, telling stories, engaging in love affairs, breaking up, and starting new relationships. The design of this action is entirely the responsibility of the players. As Reid observes, "The MOO system provides players with a stage, but it does not provide them with a script" (170). The minimal structuring of the MOO world makes aesthetic pleasure almost entirely dependent on the creativity, compatibility, and cooperativeness of the players. Art can sprout out of MOOs, as it can out of conversation, but MOOs are not art in themselves. What you get out of them in terms of gratification depends on who happens to be on line, on whether you have formed rich imaginative relationships with these players, and, more literally than in any other mode of textual communication, on your own performance. MOO sessions can be a tremendous waste of time if you don't know what to say and don't meet the right people. This is the inevitable consequence of seizing a creator's power over a fictional world when you are not a gifted storyteller.

Despite their lack of controlling script, however, MOOs seem to have no problem generating something akin to immersion. The literature on the genre is full of expressions of fanatical loyalty on the part of its users. Sherry Turkle reports several cases of MOO users who regard their MOO identities as "more real" than their ROL (rest-of-life) selves, presumably because they take greater pleasure in these masquerades than in face-to-face interaction with flesh and blood individuals. For any given person, it is difficult to tell how much of this enthusiasm is due to the pleasure of role-playing per se and how much to the need to act out personal fantasies, to an infatuation with the relative novelty of the medium, to a desire for social interaction, or to a fascination with the real-world identities that hide behind the screen. Some people treat the MOOs as poetry cafés and bulletin boards for creative writing, others as conference rooms for the exchange of ideas, and still others as meeting places, singles bars, sex clubs, and personal ads. Turkle calls MOOs a "new form of collectively written literature" (*Life on the Screen*, 11), but for John Perry Barlow they are "CB radio, only typing."[5] Which opinion is more justified, and whether MOOs generate immersion or addiction, aesthetic appreciation of the fictional world or more practical forms of gratification, depends not on the MOOs themselves but on what people do with the medium, and on how creatively they use language in real time.

AUTOMATED DIALOGUE SYSTEMS

One way to remedy the lack of global design in MOO interaction and its aesthetic inconsistency is to replace the human dialogue partner with a smart, cooperative, and witty robot, or chatterbot, as these talkative creatures have come to be known. All it would take is an AI program that teaches computers the art of entertaining conversation. The power of an automated dialogue system to reconcile immersion and interactivity was demonstrated by the tremendous popular success of Eliza, the classic conversational program written in 1966 by Joseph Weizenbaum. The purpose of Eliza was to test the limits of a computer's ability to fake intelligence by simulating a dialogue between a patient (the user) and a psychotherapist (Eliza). From a performance point of view, the most remarkable feature of Eliza was its ability to carry on a reasonably coherent conversation without using any sophisticated language-parsing techniques. Rather than building a syntactic and semantic representation of the user's input, as do programs seriously aiming at language understanding, it relied on rather crude pattern-matching strategies, such as detecting key words and responding with canned formulae, recycling the user's input, changing pronouns from first to second person, and, most importantly, answering almost every input with a question of its own. Yet despite the system's total lack of understanding of the human mind, Eliza's conversation was clever enough to fascinate users. In what came to be known as the Eliza effect, namely the user's willingness to suspend disbelief in the humanity of the computer, many people turned to Eliza as a help toward self-understanding. The strategy of answering questions with questions was ideally suited to the context of the psychotherapeutic exchange, since the role of the therapist is to help patients produce their own analysis rather than to impose an interpretation. But the truly important legacy of Eliza to interactive literature was her ability to promote a ludic attitude. It did not matter that Eliza did not understand a conversation, as long as she could fake understanding. To the user willing to play a game of make-believe with the computer, Eliza was the perfect prop.

In classic AI, the superficial pattern-matching strategies of Eliza are discarded in favor of more "intelligent" (i.e., humanlike) language parsing. Understanding is reached through a syntactic and semantic analysis of the input that supposedly simulates the operations of the brain. But in a recent movement pioneered by Joseph Bates in conjunction with his *Oz* project (to be discussed in a later section), emphasis is shifted from the *how* to the *what* of intelligent behavior. In this movement, known as alternative AI, the appearance of intelligence and the resulting credibility of the agent are considered more important than the simulation of cognitive processes. When the system is not able to respond intelligently to the user's input, it resorts to *graceful degradation,* a standard AI term meaning that unanswerable questions are handled in a way that neither halts the conversation nor breaks the user's willing suspension of disbelief. An example of graceful degradation is offered by Julia, a robot created at Carnegie-Mellon University that haunts a MUD based in Pittsburgh. According to Turkle, Julia is sophisticated enough to fool at least some of the users some of the time into taking her for a human being. Whenever she becomes unable to continue the conversation in a coherent way, she abruptly changes the subject to hockey. These "sarcastic non-sequiturs provide her with enough apparent personality to be given the benefit of the doubt" (Turkle, *Life on the Screen,* 88).

In an interactive environment, eccentric personalities thus cover up the limitations of the AI system that animates agents. They make the bugs in the system pass as features, as they respond to unprocessable input with poetic nonsense. A significant advantage of eccentricity is that it makes the characters more memorable and lovable. As Bates observes, "It is the oddity, the quirk, that gives personality to a character" ("Role of Emotions," 124). Bates recommends modeling the agents of interactive systems after the flat but highly engaging characters of cartoons, comics, and the theater of the absurd rather than after the complex "round" characters of classical tragedy and psychological novels. While digital characters should display strong desires and visible emotions to capture interest, the relation of the user to the character will be more ludic than emotional. In the present stage of AI, computer-generated characters are much more in demand for their ability to entertain than for their psychological complexity or their credibility as possible human beings.

It is no coincidence that Bates and his colleagues chose the character of a cat, bearing the postmodern/leonine name of Lyotard, to illustrate the possibility of accommodating occasional randomness of behavior within a clearly defined personality pattern (Bates, "Nature of Characters"). As any cat lover will tell you, felines are such superior creatures that they can read the minds of their human sponsors with infallible accuracy. Their behavior alternates between condescension to cooperate and manifestation of independence, but it can never be interpreted as failure to understand what is wanted of them. Whatever an AI-driven cat does, the algorithm cannot go wrong. If the system misreads the user's intention and programs Lyotard to respond with an arched back and a hiss to an attempt to feed him, this behavior will be credited not to the system's stupidity but to feline whimsy.

In the literary domain, an example of a character whose personality could gracefully put up with the limitations of AI algorithms is the Humpty Dumpty of *Alice in Wonderland*. His dialogue with Alice is a game in which the two partners take turns at selecting topics. As an AI-driven character, Humpty Dumpty has random access to the database of the past dialogue, and he can reactivate any old topic: "However, this conversation is going on a little too fast [said Humpty Dumpty]: let's go back to the last remark but one." Alice, the human user, has no such capabilities:

> "I'm afraid I can't quite remember it," Alice said very politely.
>
> "In this case we start afresh," said Humpty Dumpty, "and it's my turn to choose a subject—" ("He talks about it just as if it was a game!" thought Alice.) (271)

In an AI simulation of this game, Humpty Dumpty's contributions to the dialogue would be generated by a combination of real-time language analysis and Eliza effect. When Humpty Dumpty takes his turn at choosing a subject, the AI would make a more or less random selection from a data bank of ready-made, human-composed performance (or display) texts. In *Alice in Wonderland*, Humpty Dumpty's conversation is part questions, part crazy monologues, part riddles and jokes. It culminates in a bravura piece, the recitation of a lengthy poem. Performance texts not only introduce verbal art into the exchange, they also enable the character to monopolize the floor for an extended turn, thus relieving the AI from the demands of coherent response. When it is the user's turn to choose a topic, the AI-driven Humpty Dumpty will converse along as far as his database allows; once his limits are reached, he can resort to the old Eliza trick of picking a word from the user's last turn and generating a comic effect through the incongruity of the response:

> "I mean," [Alice] said, "that one can't help growing older."
>
> "*One* can't, perhaps," said Humpty Dumpty; "but *two* can."
> (272)

To camouflage his inability to follow the conversation, the AI Humpty Dumpty can borrow many tricks from his literary counterpart: scornful asides ("Some people . . . have no more sense than a baby" [268]), regaining the initiative through interruptions ("Don't stand chattering to yourself like that . . . but tell me your name and your business" [ibid.]), or producing deliberately obscure remarks that provoke demands for explanation ("Impenetrability. That's what I say!" "Would you tell me, please," said Alice, "what that means?" [275]). If the reader of *Alice in Wonderland* delights in the arbitrary changes of direction, semicoherence, humor, capriciousness, and moments of brilliant soliloquy of Humpty Dumpty's conversation, even more so will a user facing Humpty Dumpty in a virtual world, where she can test interactively the limits of his wit. The creation of an automated dialogue system featuring a digital character able to fascinate the human interactor and to stimulate her into creative role-playing would be a literary (and AI design) feat every bit as significant as the writing of a great novel. In such a system, as the literary critic Norman Holland writes, "the final 'work of art' is the conversation that results from what the reader puts into the program. This final text will be variable, different for every reader and different for every 'reading' by the same reader. This work of art has no clear boundaries between reader, writer, and text" ("Eliza," ll. 372–76).

Julia, the most celebrated of today's chatterbots, has a long way to go to reach the level of sophistication of Humpty Dumpty. She has too many practical functions on her MUD to devote herself full time to verbal art. Her most creative replies are no more than regurgitations of the contributions of human players, for one of her duties is to remember everybody's self-description and current activities. But in her best performances, such as the dialogue quoted below, she gives us glimpses of the "literary" potential of an automated dialogue system: relieve human users of their inhibitions, unleash their imaginations, lure them into playful verbal dueling, challenge them to catch the cognitive weaknesses of the program, and when the program is indeed caught, delight the human interlocutor with humorous absurdities:

(Barry, a MUD player unaware that Julia is an AI agent, tries to lure her to his room for virtual sex.)

Barry says, "Are you still holding out on me"

Julia says, "I mean, have I convinced you I'm a computer, yet?"

Barry says, "I guess your real . . . but Id still like to get you to my DEN to run tests on you"

Julia says, "I live for your approval, Barry."

Barry says, "I m so happy to hear that. Why don't you come over to my home"

Julia says, "I want to have a child someday."

Barry says, "What a coincidence . . . I too wish to make a baby . . . some day"

Julia says, "The date is Tue Jul 21 23:57:31, Barry."[6]

INTERACTIVE DRAMA

With interactive drama we enter the domain of three-dimensional experiences and of physically penetrable virtual space. In contrast to the screen-based forms of communication discussed so far, this is the realm of full-body immersion. The term *interactive drama* was introduced by Joseph Bates and his colleagues to describe their project *Oz*, developed in the mid-nineties at Carnegie-Mellon University, but it can be applied to any VR application designed for the purpose of entertainment and involving some degree of narrative development. These conditions are satisfied by a variety of actual and conceivable projects: multimedia versions of MOOs (or conference rooms) in which physically distant users meet and converse in a virtual landscape under fictional identities; dramatic performances that place the spectator in a computer-generated setting, where he listens to or interacts with live actors; electronic theme-park rides during which the user explores a fantastic virtual world populated with computer-simulated characters; and art installations that allow the user to perform a walk through a dynamic digital display, such as a project by Rita Addison called *Detour: Brain Deconstruction Ahead*.[7]

Detour tells the story of its creator, a photography artist who suffered severe brain damage in a car accident. This damage resulted in a distorted perception that prevented Rita from practicing her art. The VR installation tells the story from a first-person point of view—not in the classic novelistic way, by exposing the visitor to the narration of an I who retains its alterity, but much more forcibly, by transporting the visitor into this other's mind, thus erasing the difference between I and you. In a three-act linear narrative, the visitor first walks through a hallway decorated with beautiful nature pictures—the vision of Rita before the accident. After a while the pictures fade away, and they are replaced by the image of a freeway, suddenly followed by screeching noise, screams, loud heartbeat, visions of shattering glass, chaotic flashes of light—a vision of "the brain being fried" and of "exploding synapses," as Celia Pearce describes the experience (*Interactive Book*, 414). In the next episode the visitor is returned to the picture gallery, but she is now unable to look at the photographs. When she tries to focus on a scene, her vision is impaired by a variety of distorting effects: a giant black spot obscures the picture, a haze blurs the shapes, colors are inverted, shapes undergo morphing, the picture flashes or disappears under ripples. The dizzying display is underscored by a persistent noise that leaves the visitor mentally and physically exhausted. In Pearce's account: "You are trapped inside a compulsory hallucination, beautiful in its organic abstraction, but terrifying in its relentlessness. You are trapped inside the mind of Rita Addison, a world where sight and sound are perpetually distorted by a series of sensory anomalies—a veil between you and the world that never goes away" (415). In the last scene of the walk-through, the visitor meets Rita herself, who smiles and shakes her hand, thereby releasing the visitor to her own mind, and to the world of normalcy.

In its most ambitious, but also most utopian, version, interactive drama takes the form of a fully automated dramatic performance in which users impersonate characters in a dialogue with AI-driven agents. The goal of this last conception of interactive drama—the one on which I focus in the rest of this chapter—is to abolish the difference between author, spectator, actor, and character. As Brenda Laurel writes, "The users of such a system are like audience members who can march up onto the stage and become various characters, altering the action by what they say and do in their roles" (*Computers as Theatre*, 16). Interactive drama will be staged "solely for the benefit of the interactor[s]" (Kelso et al., "Dramatic Presence," 9). As beneficiary of the production, the interactor is audience; as active participant in the plot and member in make-believe of the fictional world, he is character; as physical body whose actions and speech bring the character to life, he is actor; and as initiator and creative source of the character's speech and actions, he is co-author of the plot.

The age-old dream of merging the four dramatic functions, as I will call these roles, has given birth to many stage experiments, but all of them have been only partial. In street theater, spectators may be pulled from the audience and invited to participate in a script (spectator becomes actor and character but ceases to be spectator); in commedia dell'arte, actors improvise their parts within a prescribed scenario, and it is their live performance rather than the vision of an author that constitutes the focus of the show (actor becomes co-author but is not spectator); and finally, in children's puppet theater such as the French *théâtre Guignol*, spectators yell advice at the characters, warning them of the trap that has been set up by the bad guy (spectator crosses the boundary into the fictional world and becomes character while remaining audience, but the advice does not influence the script). When and if interactive drama is fully implemented, it will be the most complete fusion of the four dramatic functions.

Interactive drama can succeed as an artistic genre only if, in contrast to the MOOs, it offers both a stage and a script to its participants. Even the most suggestive and immersive setting is not sufficient to give dramatic goals to the interactors:

> It is not enough . . . to walk or fly around a virtual environment of extreme complexity and multisensory detail. I can imagine a virtual haunted house, for instance, with boards creaking, curtains waving, rats scurrying, and strange smells wafting from the basement. Sooner or later something will have to happen, and if it does, that something will be interpreted (at least by my dramatically predisposed brain) to be the beginning of an unfolding plot. (Laurel, *Computers as Theatre*, 188)

In order to maintain some dramatic value, the performance must channel the interactor's actions toward a goal sanctioned by the system: "Although you control your own direction by choosing each action you take, you are confident that your experience will be good, because a master story-teller subtly controls your destiny" (Kelso et al., "Dramatic Presence," 1). This remark exposes once more the basic paradox of interactive art, a paradox reminiscent of a familiar theological problem: How can the interactor freely choose her actions if her destiny is itself controlled by the godlike authority of a world designer? The conflict is intensified by the different perspectives of the interactor and the system designer. The immersed interactor lives the fictional world from the perspective of life, as a continuous present, and her actions are oriented toward an unknown future, while the designer creates the fictional world from an atemporal perspective that enables him to construct the events of the past and to arrange them in a teleological chain aimed at a foreseen effect. Moreover, as an individual trying to live his life according to personal values, beliefs, and ambitions, the interactor will not be motivated by the same goals as the system. Whereas people engaged in concrete life situations normally select what they regard as the most efficient solution to their problems, characters in stories act according to a double logic: their own interests, but also the interests of the plot. As Nicolas Szilas observes, "Characters' actions are motivated by narrative constraints rather than emotional, psychological, or social reasoning" ("Interactive Drama," 151). The need to thicken the plot and to prepare climactic situations explains why characters are often made to select highly impractical plans or to make foolish decisions. The Elizabethan convention of the "calumniator credited"—an intelligent character, such as Othello, uncritically believing the accusations of an obviously untrustworthy villain—stands as a blatant example of the resolution of the conflict between the two logics in favor of global narrative interests.

From the viewpoint of the system, the actions of the user thus introduce an element of randomness into the work that can only threaten its aesthetic purposefulness. To ensure a felicitous integration of the bottom-up input of the user into the top-down design of the system, interactive drama may take a clue from the game of the Big Bad Wolf discussed in chapter 9. In the game, children are free to improvise, and they do so for their own pleasure, but their improvisations are controlled and coordinated by a familiar narrative scenario that the players agree to implement. If there is a lesson to learn from children's games of make-believe, it is the advantage of familiarizing the players with the script, and the importance of counting on their voluntary cooperation. The interactors should know that their personal enjoyment depends on a collaborative effort to enact the narrative.

Since human actors are vastly more intelligent and adept at interacting with a user than the most sophisticated of computer-programmed agents, one may wonder what is the point of electronicizing interactive drama. The main reason is economic. Interactive worlds are realities made for one or two participants, and it would be prohibitively expensive to use live actors in real time. But the unique potential of digital technology to combine the features of movies and theater adds an artistic incentive to this practical motivation. The trademark of the movie screen is the dreamlike fluidity of its pictures: camera movements and special effects allow shifts in point of view, instantaneous change of decor, and the morphing of shapes. While the movie screen is a flat surface traversed by evanescent images, the theatrical stage opens a three-dimensional space populated by physically present bodies of potentially palpable solidity. The spectator of traditional theater may be confined behind a "transparent fourth wall" that separates her from the stage, but she experiences the crossing of this purely symbolic barrier as a physical if not fictional possibility. In the sensory illusion of the VR environment objects will be fluid like movie pictures, yet their appearance of solidity will invite the user to reach out toward them (at the risk of grabbing only air if she tries to close her hand). The electronicization of interactive drama will turn the event into a film whose world can be walked into and touched, into a play whose props, characters, and setting can undergo limitless transformations.

In an application sensitive to the properties of its medium, the thematic landscape of interactive drama should take maximal advantage of the immersive power, fluid appearance, and self-transforming capabilities of three-dimensional digital displays. These features predispose VR art to the representation of fantastic worlds, supernatural creatures, animated objects, magical metamorphoses, movements defying gravity, and postmodern/Escherian transgressions of ontological boundaries, such as pictures becoming animated, people penetrating into books and becoming illustrations, or the meeting of flat cartoon characters with three-dimensional, pseudo-real creatures. I suggest in chapter 2 that the involvement of the physical body in VR systems offers a dramatization of phenomenological doctrine. VR can heighten the user's awareness of being-in-a-world, which means being there as a body, and turns the experience of immersion in an environment from being taken for granted—as it often is in the real world—into a source of wonder and delight. The power of the medium to manipulate the relations between the body and the simulated environment favors narrative themes that revolve around alternative experiences of embodiment: flying, walking through walls, being dismembered (as the user's hand floats in front of the body, apparently disconnected), reaching beyond the physical limits of the body in what is known as extension of proprioceptive boundaries, acquiring a different sense of scale by growing or shrinking, moving out of the body and looking at it from an external perspective, feeling one's body lose its solidity and become liquid, gaseous, immaterial, ghostlike, ubiquitous, teletransportable, expanding to cosmic proportions and merging with the surrounding world.

If the body in space is the dominant theme of VR narratives, the most important component of the plot will be the setting, and the narrative structures will be predominantly epic: the user will explore fantastic landscapes, navigate a space fragmented into multiple domains (the rooms of a castle, the diversified geography of an island, even the books of a library), take possession of virtual worlds through movement and action, or achieve intimacy with the environment. As Celia Pearce has observed (*Interactive Book*, 496), themes of spiritual and corporeal communion with the world, especially with the natural world, are far more representative of the spirit of VR dramatic installations than the violent and dystopic themes of science fiction. Through this fascination with a pristine natural environment, the thematics of VR art-worlds emulate the disappearance of the computer—the very feature that defines VR as a technology.

DESIGN PHILOSOPHY IN TWO INTERACTIVE DRAMA PROJECTS: *PLACEHOLDER* AND *OZ*

The two projects in interactive drama that I propose to discuss here, Laurel, Strickland, and Tow's *Placeholder*, and an experiment conducted by Bates and his colleagues as part of their larger *Oz* research project, illustrate opposite poles in design philosophy and narrative structure. They will therefore offer an instructive comparison.

Placeholder grew out of Laurel's fascination with the affinities between VR and ritual. The installation is an electronic demonstration of her belief that VR can fulfill the same spiritual function of fostering communion with life forces as once did Dionysian festivals and, in her speculations, the ceremonies of the great kivas of the Anasazi civilization.[8] "Virtual reality may be many things," she writes. "It may become a tool, a game machine, or just a mutant form of TV. But for virtual reality to fulfill its highest potential, we must reinvent the sacred spaces where we collaborate with reality in order to transform it and ourselves" (*Computers as Theatre*, 197). The script of *Placeholder* enacts the ritual pattern of penetrating into a magic circle, being reborn in a different body, and acquiring enhanced powers of perception that deepen the bond of the subject to the natural world.

Placeholder uses three-dimensional videographic scene elements, spatialized sounds and words, and simple character animation. Two users, equipped with head-mounted displays, wander for about fifteen minutes (a time limit made necessary by throughput requirements)[9] inside a ten-foot-diameter area—the "magic circle," which corresponds to the range of the equipment that tracks their movements and generates the visual and auditory display. The two participants are said to be "physically remote" (Laurel et al., "Placeholder," 118), and their activity consists primarily of an individual exploration of the virtual world, but insofar as each can hear and see the other's character, they engage in a rudimentary form of dramatic interaction.[10] The system does not need AI components—all texts are prerecorded—but one character is improvised live by an actor. This character, the "Goddess," functions as on-line help, initiating users into the secrets of the virtual world and providing internally an equivalent of the knowledge that children bring into games of make-believe.

In stark contrast to the Aristotelian design philosophy advocated in Laurel's book *Computers as Theatre*, the narrative structure of *Placeholder* is not dramatic but predominantly epic. The stated purpose of the system is to create a "sense of place," and its architecture is more indebted to a poetics of space than to a poetics of plot. The "sense of place" is conceived as an encounter of the user with what the Romans called the *genius loci:* a spirit that protects a site and safeguards its essence. The spirit of a place manifests itself through the narratives that recount its origin, thereby establishing its sanctity and affective significance: "When a person visits a place, the stories that are told about it—by companions, by rock art or graffiti, or even by oneself through memories or fantasies—become part of the character of the place" (121). To dramatize this idea, *Placeholder* creates a magical space inhabited by animated objects. The user wanders around an environment teeming with narratives in which every creature, every landscape feature, has a story to tell. In this architecture (illustrated by fig. 12 in chap. 8), interactivity occurs on the macro level as freedom to explore, while narrativity is found on the micro level as embedded stories. Through the invariant ritual pattern of the entire visit, however, narrativity is not altogether absent from the macro level.

The setting of *Placeholder* consists of three distinct sites inspired by actual locations found near Banff National Park in southwestern Alberta: a hot spring in a natural cave, a waterfall, and a landscape of rock formations (hoodoos) created by erosion. These different locations are separated by dark portals, so that moving from one to another is similar to undergoing a rite of passage. The world of *Placeholder* is inhabited by four mythical "critters," Spider, Snake, Fish, and Crow. As the user enters the virtual world, these critters are represented as petroglyphs on the walls. The user cannot see anything, but he hears faint voices that emanate from the petroglyphs—the voices of the critters speaking about themselves. Fascinated by their narratives, the user comes closer and hears the story more and more distinctly. In a forceful allegory of the immersive power of narrative, the user who crosses a certain threshold toward a critter becomes embodied as that critter, suddenly taking on the appearance, voice, point of view, and mode of locomotion of the mythical being. Now the user is able to see the world, and to see it from a different perspective: Crow's vision captures spectacular reflections, Spider has eight eyes corresponding to eight points of view, Snake can see in the dark, and Fish sees underwater.[11]

The system enhances the user's sense of identity with his reconfigured body by making it visible to him and to others, but acquiring a virtual body is not merely a matter of presenting it to others, as it is on the MOOs. It is mainly by inhabiting his new body from the inside and by reaching with it toward the world that the user acquires "a new sense of what it is to be an embodied [creature]" (125). Through the theme of metamorphosis, the developers of *Placeholder* hope to increase the user's sense of embodiment, not only by altering his mode of perception but also by making him construct his own body through the action of moving toward a critter. The theme of corporeal and perceptual rebirth is underscored by the fact that as long as users are human, they can see only their hands; but once they become reincarnated as one of the critters, their full body image becomes visible to them and to the other participants. Another function of the metamorphosis is to enhance the user's verbal creativity. By giving him a new body, the system also gives him a persona to act out and experiences to talk about. Users leave their mark on the virtual world by interacting verbally with it. Their utterances are recorded and stored in devices called "voice holders." These devices, actually digital images that look like rocks, can be transported around by a handheld device—much in the way a mouse drags icons on the computer screen—and replayed by the next users. Interactivity is not merely a matter of walking, speaking, and using hands to touch and move objects around, but more importantly a matter of enriching the narrative tradition that expresses and creates the spirit of the place. It reaches symbolically across time and generations, as users can talk to their successors and listen to their predecessors.

If *Placeholder* is an electronic simulation of a physical world, *Oz*, in its current state of development, is a simulation by physical means of an electronic simulation of a physical world. In most VR applications, computer simulation rehearses life, but here it is the live performance that rehearses the simulation. The philosophy behind the project is the belief that the problems of staging a rewarding interactive performance will not be solved by electronic technology before they can be handled by human intelligence. The project thus invests in a double virtuality: there is no guarantee that it will ever be feasible to write satisfactory interactive dramatic plots (the narrative scheme used in the experiments does little to lift this uncertainty); and AI has still a long way to go to produce aesthetically viable plots, even of the linear kind.[12]

In the dramatic experiment described by Kelso, Weyhrauch, and

Bates, the user interacts with improvising human actors, themselves guided by a human director. The director prefigures the top-down algorithm of an electronic implementation, and the actors stand for the future AI-driven agents. The purpose of the experiment is twofold: to test "how it feels to be immersed in a dramatic virtual world, filled with characters and story" (Kelso et al., "Dramatic Presence," 2); and to explore the requirements to be met by the characters so as to induce in the interactor a suspension of disbelief. Through its insistence on plot, character, climax, and authorial control—in short, on the factors that favor emotional and temporal immersion as opposed to the spatial brand[13]—the project is much more indebted to Aristotelian dramaturgy than is Laurel's *Placeholder*.

During the performance the actors wear headsets that make it possible for the director to issue instructions without being heard by the interactor. Since the actors are independent agents pursuing their own goals in the fictional world, they are ignorant of the global design, and they react mainly to local conditions, as do the members of an ant colony. But in contrast to the behavior of ants, their actions may conflict with the goal of the system. The task of the director is to steer the actors in the right direction when their actions stray from the plot. The director also speeds the actors up or slows them down so that the performance will take approximately fifteen minutes—the time considered best suited to the attention span of the user in an environment as demanding as interactive drama. (Interestingly enough, this is also the duration selected by the designers of *Placeholder*.)

The narrative architecture is a variant of figure 9 (chap. 8). It offers different endings, but it guides the interactor through obligatory stations. A scenarist (Margaret Kelso) designed a plot with a proper dramatic development: exposition, complication, climax, and resolution. The scene is a bus station. The interactor is instructed to play the role of a passenger who buys a ticket to another city to attend a funeral. These directives, which play the role of exposition, are external to the performance. (This contrasts with the in-world help offered by the Goddess in *Placeholder*.) The remaining characters comprise an uncooperative clerk, a blind customer named Tom, and a punk who tries to rob the blind man by pulling a knife on him. The basic events—approaching the clerk, being helped by the clerk, the arrival of Tom and of the punk, conversing with Tom—can occur in different orders, and various minor events can be interspersed. But all the runs lead to an episode in which the interactor is called to the window and given a gun by the clerk while the punk is actively threatening Tom. At this point the interactor can fire a shot into the air that causes the punk and the knife to drop to the ground (the first run of the experiment). She can also "wearily edge past the punk" and leave—the rather anticlimactic ending of the second run—or maybe fire at the punk and kill him. These choices are not forced on the user by the system, as would be the case if a menu instructed her to do a, b, or c, but determined by the concrete situation. The user evaluates her choices spontaneously, on the basis of her own assessment of the current state of affairs. The mark of a well-designed interactive plot resides indeed in its ability to limit the choices of the user while maintaining a sense of personal freedom. According to the authors, one of the greatest deterrents to suspension of disbelief is producing in the user the feeling of being manipulated by the system.

In the scenario described above, the complication allows a considerable range of variations, and the resolution offers several possibilities, but one event remains constant: the interactor is given a challenge, and she must either accept or refuse it. To an outside spectator, the outcome of the second run may seem badly lacking in dramatic power, but this does not necessarily mean that the experience was a failure from the point of view of the interactor. It could indeed very well be the case that the second run caused an emotional experience and a sense of involvement in the situation as strong as the more tragicomic first run: refusing a challenge and cravenly leaving the scene is not an action to be taken lightly. The difference between the perspective of an interactor and that of an external spectator is the most significant lesson to be learned from the experiment. The authors point out that external observers were much more critical of the motivations of the characters than was the interactor. Absorbed by the action, the interactor did not notice when characters abruptly changed their stance after being urged by the director to speed up the performance. Inconsistencies of behavior were often covered up by the Eliza effect: "During the experience, the interactors were caught up in the story, did not notice many inconsistencies in either the characters or story, and liked the surprises. In contrast, the observers did not find the characters believable and often lost interest when action seemed to lag" (9). The phenomenon known to VR developers as spontaneous "accommodation" to the idiosyncratic circumstances of the virtual world seems to work in the areas of emotional and temporal immersion as well as in the spatial domain.

Since the interactor is the real beneficiary of the production, the disparity between the internal and external perspectives represents good news for the future of interactive drama. The active membership of the interactor in the virtual world deprives her of the distance necessary to critical judgment. It may be structurally much more difficult to write interactive plots than regular dramatic narratives, but the *Oz* experiment suggests that this difficulty is compensated by the user's greater propensity to suspend disbelief in an interactive and physically immersive environment. The aesthetic criteria of interactive drama will not be those of classical drama; the future of the genre will be as a game to be played and an action to be lived, not as a spectacle to be watched. This suggests that a resolution of the double logic that I mention above—global narrative goals versus local personal ones—in favor of the character's interests will be less detrimental to interactive drama than to the regular brand.

But if the experiment reveals some of the fundamental rules and playing conditions of interactive drama, it fails to answer the most urgent question of all: Assuming that the scenario actually manages to get the interactor emotionally and creatively involved, will this involvement be a source of aesthetic pleasure—will the game, in other words, be worth playing at all? *Do we really want to become characters in a dramatic action?* The authors of the project suggest that the scenario induced an acute moral dilemma in the interactors by forcing them to confront the question "Is it right to shoot another human being?" Janet Murray, similarly, believes that becoming a character in cyberdrama is a potentially valuable experience because the simulated environment, in which nothing really "counts"—we can always get out—"provides a safe space in which to confront disturbing feelings we would otherwise suppress" (*Hamlet*, 25). But didn't classical literature do just this, and is there any moral or aesthetic benefit in intensifying the experience by transposing it into an immersive/interactive environment? The ethical dimension of interactive drama suggests a didactic potential that could be useful in practical applications, such as testing candidates for a position of high responsibility, but if the dilemma were really intense, the experience would be too painful to be really enjoyed. Classical tragedy may already have reached the limit of the pleasurable in forcing on the audience the contemplation of painful situations.

Compared with *Placeholder,* the project described by Kelso and her colleagues seems almost contemptuous of the advantages of electronic technology. While *Placeholder* adapts its themes to the strength of the medium, the *Oz* project confronts its weaknesses in the domain of plot and characterization. Should one regard this effort to duplicate the features of traditional dramatic literature in an interactive environment as a failure to understand the nature of the medium—as does Espen Aarseth (*Cybertext,* 136–40), who rightly thinks that electronic literature should find its own forms rather than imitate established literary genres? Or, on the contrary, does the design philosophy of the *Oz* developers represent a bold, almost avant-gardist attempt to expand the expressive power of interactive language?

Beyond their diverging thematic, dramatic, and narrative philosophies, however, *Placeholder* and the *Oz* drama experiment are bound by a common aesthetic ideal. Both projects conceive the interactive artwork as a "two- [or three-] player game in which the drama manager (i.e., the system) is acting to maximize its chances of providing a dramatic experience for the interactor" (Kelso et al., "Dramatic Presence," 13). In this two-player game the roles are unequal. The burden of responsibility for the user's pleasure falls on the drama manager. In their insistence on "the safety of a controlled situation" (19), the developers of *Oz* and *Placeholder* demonstrate a greater allegiance to the aesthetics of classicism than to postmodern taste. Nothing could be more remote from the subversive spirit of postmodernism and of its rejection of "grand narratives" in favor of purely local meaning patterns than this ideal of user-friendliness, the protective role attributed to the system, and its top-down control of the plot.

In their conception of interactivity, literary hypertext and interactive drama indeed represent polar opposites. The interactivity of literary hypertext disrupts narrative coherence in order to release interpretive freedom and creative energy. It resides less in the physical gesture of clicking than in the mental operations triggered by the gesture. The unpredictable consequences of interactivity plunge the text into chaos and leave it to the reader to decide whether to try to put it back together as cosmos or to maintain it in its chaotic state by reading it on a purely local level. In *Oz* and *Placeholder*, by contrast, the user's interventions neither explode the surrounding world nor threaten its intelligibility. It is, on the contrary, the user's understanding of the laws of the virtual world that makes it possible to interact in a way compatible with his own pleasure. This is why the user needs "in-world" initiation (as provided by the Goddess) or preliminary instructions. This opportunity to act within the constraints of a controlled environment epitomizes the classical spirit of these two experiments in interactive drama.

Each in its own way, *Placeholder* and *Oz* suggest a path for the future development of immersive/interactive art. It seems safe to predict that if interactive drama ever gets off the ground, the first productions will work with, rather than against, their medium. The immersive quality of VR technology is bound to exert a powerful magnetism on the public, and the most successful projects will likely be those that emulate *Placeholder*'s dramatization of the relations between the user and the virtual world. But after a while the purely spatial immersivity of the medium will be taken for granted, and it will become more and more important to focus on emotional and temporal immersion. This will be done by populating virtual worlds with genuine characters and by creating opportunities for real-time dialogues between user and characters.[14]

If the system is to generate a truly narrative (i.e., temporal) type of immersion, it will have to exercise far greater control over the user's actions than in a purely exploratory type of installation, in which the user can be simply turned loose in the simulated world. I envision the future of interactive drama as investing in an *Alice in Wonderland* type of narrativity, made out of episodic "little stories," rather than being supported by an overarching Aristotelian plot.[15] Thematically, it will rely on the visual appeal of the scene, on a fantastic setting conveying a sense of estrangement (the dream world), on surprising metamorphoses (the Cheshire cat) and alternative experiences of embodiment

(Alice's growing and shrinking). The macro level will be dominated by an epic of exploration, but on the micro level the fictional world will also include those elements emphasized by the *Oz* project: dramatic or humorous confrontations (such as the Queen of Hearts episode) and live dialogue with memorable characters (the Humpty Dumpty conversation).

To reach the level of genuinely dramatic art, VR installations will have to face many technological challenges, such as overcoming the "lag time" between user movements and system response, developing less cumbersome interfaces than headsets and data gloves, and improving the communicative skills of AI-driven agents; but the main obstacle by far to the implementation of immersive/interactive art forms is economic rather than technological. The throughput is limited, since VR is essentially a reality made for one (or a handful working as a team), the equipment is expensive, and multimedia applications require the collaboration of a large team of artists and engineers. In recent years we have seen the Hollywood movie industry increasingly limit itself to productions with mass appeal, sacrificing artistic integrity to profitability. With their insistence on violent themes, computer games have followed the same route. But VR installation art does not have the option of pandering to popular taste, since even the most successful projects will not be visited by a sufficient number of people to ensure a palatable return of investment. Even in the amusement-park sector its future is questionable, because amusement parks draw their profit primarily from retail and are not overly interested in adding expensive electronic rides with limited throughput (Pearce, *Interactive Book,* 398). In a culture more and more dominated by the pursuit of profit, the continued exploration of immersive/interactive art forms depends almost entirely on the largess of granting institutions. The question is not whether computer technology can offer a new artistic experience but how badly society wants it.

Dream of the Interactive Immersive Book

Neal Stephenson's *The Diamond Age*

It is the twenty-first century. Nation-states have almost all disappeared, and the world is balkanized into countless small "phyles," whose identity resides not in geographical roots but in ethnic, ideological, racial, or religious affiliations: Ashantis, Boers, Kurds, Navajos, Mormons, Jesuits, Heartlanders, and so on. These phyles occupy numerous small territories, or claves, dispersed throughout the world. Some of them lease territories from one of the very few nations left, the Coastal Republic of China, successor of the People's Republic, at the edge of which much of the action takes place.

Such is the world of Neal Stephenson's novel *The Diamond Age*. Thanks to the invention of nanotechnology, the art of building objects from the atomic level, this world is virtually free of material needs; productive energies can be almost entirely devoted to the entertainment sector. Every home, even the poorest, is equipped with two essential contraptions: the "Mediatron," a universal entertainment system, and the "matter compiler," or M.C., an electronic device programmed to fabricate on demand any kind of solid object, even digestible food, given a small amount of "ucus" (diamond powder), the currency of the time. In a chilling inversion of the idea, advocated throughout this book, of the participation of the body in computer-generated worlds, nanotechnology allows the implant of a variety of tiny computers ("nanosites") into the human body. These nanosites serve as concealed weapons ("skull guns"), as surveillance systems (they record people's activities), or as data banks that spread information into the bloodstream in viral fashion.

The elimination of material needs does not mean that all

groups in society enjoy equal prosperity. Of all the phyles, the wealthiest is that of the neo-Victorians, or Vickys, of New Atlantis, a society fiercely dedicated to the ideals of the nineteenth century, such as the work ethic, patriarchal family structure, rigid hierarchical organization, conservative sexual politics, repression of emotions, and worship of science and technology. The least prosperous members of society are the Thetes, the nickname for a group of people who do not belong to any phyle, live in the decaying urban landscape of the leased territories, and maintain the lifestyle of the late twentieth century, characterized by street violence, homelessness, broken families, child abuse, heavy drug and alcohol consumption, addiction to the Mediatron, and a predilection for electronic tinkering with the human body. Though this Dickensian underworld exemplifies everything that neo-Victorian culture rejects, it is essential to the functioning of Vicky society as a source of cheap labor, destination of sexual escapades, and provider of art and entertainment.

In Diamond Age society, interactive entertainment is so pervasive that the old "passive" movies of the twentieth century have become the object of a nostalgic cult. The most important type of "ractive," as the products of the new media are called, is a form of script-driven interactive drama in which human "ractors" are networked with paying customers located anywhere in the world. Their images meet in a virtual setting electronically accessible to all participants. Sitting on a private stage at the theater company, the ractors read lines that are fed to them by the system. The voices of the ractors are intercepted by tiny electronic implants located in their throats and relayed to digital images of their characters, which they animate through the movements of their bodies.

One may wonder why a society so technologically advanced still uses human ractors rather than automating the whole process. The most obvious answer is that Stephenson needs the character of the ractress Miranda[1] for the sake of the plot; but the technological failure to simulate the richness and presence of the spoken voice is also symbolic of the irreproducible essence of the human. The artistic quality of the productions of the interactive theater company is as low as the freedom of the ractors and the sophistication of the software. The plots are not generated on the fly by an intelligent system but selected from a database of stereotyped story lines: "Computer decides where you go, when. Our dirty little secret: this isn't really that ractive, it's just a plot tree—but it's good enough for our clientele because all the leaves of the tree—the ends of the branches, you understand—are exactly the same, namely what the payer wants" (*Diamond Age,* 90). What the clientele (mostly Vickys) want is often sexual rather than dramatic thrills. Playing a character named Ilse in the ractive *First Class to Geneva,* Miranda has her hands full disarming a customer who "had clearly signed up exclusively for the purpose of maneuvering Ilse into bed" (122). At this point in her career Miranda has acquired sufficient expertise and familiarity with the story-world to improvise her part.

One day Miranda is assigned to a revolutionary project, the "Young Lady's Illustrated Primer." This extraordinary feat of nanotechnology and artificial intelligence was commissioned by Lord Alexander Chunk-Sik Finkle-McGraw, spiritual father of neo-Victorian society and pioneer of nanotechnology, from the superhacker John Percival Hackworth, as a birthday present for McGraw's granddaughter Elizabeth. Realizing that Vicky society is not a perfect world but only the best equipped for survival among current cultures, McGraw hopes that the Primer will prevent his granddaughter from passively accepting neo-Victorian values, and that it will instill in her the subversive spirit necessary to help mankind evolve toward higher forms of intelligence. When Hackworth is done with the project, he sneaks out of New Atlantis into Chinese territory to make an illegal copy for his own daughter, Fiona, with the help of a mysterious Dr. X, who intercepts the code and mass-produces slightly inferior copies. These copies are destined for thousands of Chinese baby girls who are being sneaked out of the Coastal Republic, where their lives are threatened by cultural attitudes toward female children.

On his way back to the New Atlantis clave, Hackworth is mugged by a gang of street kids and loses his copy of the Primer to a Thete boy named Harv, who takes it back home as a present to his six-year-old sister, Nell. Harv and Nell live in a dull housing project, ignored by their mother, Tequila (who works as a maid in New Atlantis), and sexually abused by some of her transient boyfriends. Meanwhile, the mugging scene has been recorded by surveillance devices, and the whole story comes to be known to Lord Finkle-McGraw. Nell is allowed to keep her copy of the book, but Hackworth—who has in the meantime made another copy for his own daughter—is sent in punishment by McGraw on a secret technological mission to Vancouver, far away from his wife and daughter. Nell, Elizabeth, and Fiona will each be brought up by her copy of the Primer, but because they are connected to different anonymous ractors—Nell to Miranda, Fiona occasionally to her father, and Elizabeth to a variety of temporary people[2]—the three girls grow up very differently. Whereas Elizabeth is raised by a mere machine—the temporary ractors who work for money do not count as a human presence—and Fiona by the digital mind of a male hacker, Nell experiences a combination of circuit logic and human warmth that replaces both her dead father and her negligent biological mother, reconciling the "two cultures" of Hackworthian technology and Mirandian art. In the end, Elizabeth joins a counterculture and disappears from the scene, Fiona revolts against her father and opts for the make-believe of a racting career, but thanks to the distant love of Miranda, who sacrifices her career to the Primer project, Nell becomes a leader in a movement toward a society that overcomes the strictures of Vicky culture, thereby justifying the hopes placed in the Primer by Lord Finkle-McGraw.

The Primer is a fictional construct, and its description gives no clue to the AI algorithms that animate it. We therefore cannot use the novel as a source of practical ideas for the reconciliation of immersivity and interactivity. But if Stephenson remains mum on the software, he has a clear idea of the hardware. The material support of the Primer is a nanotechnological "smart paper" made of countless little computers:

> Each of these spherical computers was linked to its four neighbors, north-east-south-west, by a bundle of flexible pushrods running down a flexible, evacuated buckytube, so that the page as a whole constituted a parallel computer made up of about a billion separate processors. The individual processors weren't especially smart or fast, and were so susceptible to the elements that only a small fraction of them were working, but even with those limitations the smart paper constituted, among other things, a powerful graphical computer. (64)

The purpose of this "parallel distributed system," whose structure resembles the neural network of the brain, is to produce a dynamic text that constantly adapts itself to the personality and evolving needs of the user. Using as database "a catalogue of the collective unconscious" (107), the Primer translates the universals of the imagination into "concrete symbols familiar to their audience," thus conducting "a sort of dynamic mapping from the database onto [the young reader's] particular terrain" (106). This mapping is performed by creating a fictional counterpart of the reader in a fairy-tale world—in this case, Princess Nell—and by guiding this counterpart through an archetypal narrative script, similar to that of a computer game, during which she must accomplish tasks of increasing difficulty before she reaches the state of "living happily ever after." In the course of the plot, Princess Nell acquires knowledge and problem-solving skills that enable the real-world Nell to deal with her own situation. Their parallel life stories, developed in alternating episodes, thus demonstrate the enduring educational value of fantasy and, through the secret connection to the live performance of Miranda, the nurturing effect of oral storytelling.

When Nell opens the book for the first time, she finds an outline of the story that she will enact, both as Nell and as Princess Nell, during the next ten years. Imprisoned in a dark castle with her brother Harv, Princess Nell manages to escape while Harv remains prisoner. In order to free her bother, Nell embarks on a quest to collect the twelve keys that will open the castle door. When she returns with the keys, after defeating twelve enemy castles, Harv is killed by one of her soldiers, but the dark castle is transformed by her tears into a radiant landscape, and she lives happily ever after as queen of the new kingdom. Outlined from the very beginning by an authorial intelligence, the story told by the Primer allows interactivity only on the level of individual episodes. As in most computer games, the script is predefined; it is up to the player to make the right moves and fulfill her destiny.

The Primer offers both the selective and the productive type of interactivity. The selective variety enables the reader to control the pace of narration and to manipulate the visual display. As we have seen, Nell gets the whole story in a nutshell as soon as she opens the book; but the elastic fabric of this grand narrative can be indefinitely expanded and refined. Nell turns the summary into an immersive narrative by zooming in on details and blowing them up into stories. If Nell tells the Primer, "Princess Nell descended the stairs for many hours," the rendering will last many hours; if she says, "Princess Nell climbs the mountain" (I make up this example), she is instantly transported to the top. Because of the essentially recursive nature of narrative, there is no limit to this zooming activity: "She had known, from the very first day Harv had given her the book, how the story would come out in the end. It was just that the story was anfractuous; it developed more ramifications the more closely she read it" (343). The adjective *anfractuous* suggests the infinite length of the coast of England, a favorite example of mathematicians working in the area of fractal geometry. The fractal structure of the text is not lost on Miranda: "She found herself reading the same story, except that it was longer and more involved, and it kept backtracking and focusing in on tiny bits of itself, which then expanded into stories in their own right" (135).

The productive interactivity of the text takes a variety of forms. Nell controls the actions of Princess Nell in the same way VR users control their virtual bodies; to make Princess Nell build a fire by rubbing rocks, for instance, Nell mimics the action with her empty hands; the fire will start only when she develops the proper technique. When Nell finds the narrative not relevant enough to her own life, she links the Primer to what is happening to her at that very moment by offering input to the system. The text immediately adapts itself to her specifications. This enables her to learn on the fly how to deal with her current situation, such as needing to defend herself against her mother's boyfriend. And finally, when Princess Nell needs to solve a problem, interactivity takes the form of an exploration of many paths, leading into different possible worlds: "Nell tried every trick she could think of, but it seemed as though the ractive were made in such a way that, once she'd made the decision to go away with the stranger, nothing she could do would prevent her from becoming a slave to the Pirates" (225). In this case Nell is unable to find an immediate solution, for the Primer, like the God of Leibniz, has arranged for her not a perfect world but only the best of all possible ones.³ This means having to suffer tragic losses in her quest for a happy ending, such as the death of Harv—violently killed in the fictional world, dying of lung disease caused by toxic air in the real world.

As the child grows, the Primer evolves in style, content, and mode of participation required of the reader. At the beginning, the main characters in the story are the fictional-world counterparts of Nell's favorite toys: Duck, Dinosaur, Purple, and Peter Rabbit, who play the role of helpers to Princess Nell in a classic Proppian narrative scheme (the main difference with Propp's masterplot is that the Princess is the hero, not the desired object). Together with Miranda's voice, these animal companions teach Nell how to read the Primer by herself. The story develops first like a standard linear fairy tale, but as Nell's reading skills and world knowledge deepen, it becomes "more like a ractive and less like a story" (258). The animal characters gradually disappear, and after learning valuable moral and practical lessons from them, Nell must face alone the challenge of the Twelve Castles. The text itself becomes tricky, indirect, full of hidden clues, and like literary language it requires extensive interpretive abilities: "A few years ago Nell could have relied on [what the character said]. But in recent years the Primer had become more subtle than it used to be, full of hidden traps, and she could no longer make comfortable and easy assumptions" (346).

The twelve keys to the castle are the keys to digital technology, and it is by decoding the mode of operation of computers that Nell manages to open the twelve doors. In the Turing castle, she learns the basic principles of computation theory and communicates by means of binary code with a Duke whose identity must be guessed. Is he human, or is he a machine? (He turns out to be a machine that killed the live Duke.) In the Water-Gate castle, she learns the functioning of the logical circuits that make up the hardware; in the Cipherer's Market, how to send packets of information in a network; in yet another castle, the secrets of nanotechnology; and in the last fortress, King Coyote's castle, named after the famous trickster of Native American myth, she discovers the power of computers to produce alternative realities. In the last episode of the Primer story Princess Nell liberates an army of captive mice, who help her conquer King Coyote's land, and she is reunited under the sea with her long-lost mother—Miranda's fictional counterpart.

While Princess Nell, through the Primer, experiences adventures and acquires skills normally reserved to male heroes, another strand in the plot traces the inverse initiation of the hacker John Percival Hackworth into modes of acting and thinking traditionally associated with the female aspect of the mind. *The Diamond Age* is a grand narrative, in the Hegelian tradition, that chronicles the emergence of what some cybertheorists (among them N. Katherine Hayles) would call the posthuman: a higher form of intelligence, born of a technology that presents a better approximation of the functioning of the brain than today's information systems. The character Hackworth initially represents the logical mode of thinking that made New Atlantis the most powerful phyle on earth; but the hegemonic status granted by Vicky culture to this type of intelligence represses other, equally productive fields of mental energy, such as the artistic, the subconscious, the libidinal, and the emotional. To remedy this situation, Hackworth is sent by Lord Finkle-McGraw on a mission to find "the Alchemist," a transparent allegory for the secret formula that will transcend the binarism of neo-Victorian thought and turn opposite forms of intelligence into complementary, cross-fertilizing modes of thinking.

The first stage of Hackworth's initiation takes him under the ocean to the cave of the Drummers, a tribe of undifferentiated beings who stand for the collective unconscious. In this subterranean world he is made to participate in the mind-shattering ritual of a religion in which the sacred is incarnated as information. During this ceremony, whose underpinning of archetypal patterns hardly needs to be detailed, the members of the cult copulate with a woman whose body is then burned. Her ashes, which contain the semen of every Drummer, are collected and dumped into a steel drum full of liquid, and each participant drinks a cup of the resulting mixture. Because this sharing of bodily fluids inoculates the participants in the ceremony with viral nanosites that duplicate themselves and spread data into the brain, the ritual turns the community into a collective mind. In Stephenson's first novel, *Snow Crash,* the introduction of self-duplicating data into the body is a means to control thought and eventually destroy the brain.[4] In *The Diamond Age,* by contrast, Stephenson presents the operation as a way to expand the power of thought by pooling the resources of multiple minds and by tapping into elementary forces. When Hackworth emerges, ten years later, from the world of the Drummers, he carries in his bloodstream cells of information that connect him to the universal flux of life. The rite of the Drummers gives a powerful imaginative expression to the cybercultural dream of a collective intelligence, but Stephenson does not advocate a complete dissolution of individuality. After the ceremony, Hackworth's mind is no longer his own but a copy of the collective unconscious; to regain his ego boundaries, he must copulate with another woman whose bodily fluids disarm the cells received from the Drummers by preventing their chaotic self-duplication.

Though ritual immersion is traditionally a cleansing experience, prelude to a rebirth, ten years under the sea have not completely freed Hackworth of his rigid neo-Victorian principles. He must therefore undergo a second stage of initiation, both emotional and intellectual. This learning experience takes place in the London branch of New Atlantis, during a boat ride that turns out to be a VR show run by a company called Dramatis Personae. Hackworth soon realizes, to his great dismay, that the boat ride is going to be "some kind of participatory theatre" (417), and that he will be asked to perform in front of everybody. But he eventually learns to overcome his inhibitions, and by acting out his part, which means exteriorizing his emotions, he is transported into a virtual world that exists only for him. There he engages in a flirtation with a "devil-woman" who briefs him on the three main features of the system.

First, the show is *parallel,* and every customer of the VR company is the center of his own story—the star of the show: "Apparently Hackworth's Quest was (as the devil-woman herself had just told him) just one of several concurrent stories being acted out tonight, coexisting in the same space" (431). Second, the machine that runs these parallel shows is *nondeterministic.* In computing theory, a nondeterministic machine is not a mechanism deprived of rules or causality but one that permits transitions to several different states on a given input and makes it possible to reach the same final state through several different routes. The implementation of a nondeterministic automaton requires a machine with parallel processors, or, if we associate "machine" with a fully deterministic process, a network of machines (Savitch, *Abstract Machines,* 35). Third, the show is a *complex dynamic system,* and because every local event creates global changes, it can evolve without a predetermined plot. The devil-woman's description of its mode of operation brings to mind the famous "butterfly effect" of chaos theory: a butterfly flapping its wings in Beijing affects the weather in Wyoming:

> "It's not that we do a set show, reconfigure, and a different one next night. The changes are dynamic and take place in real time. The show reconfigures itself dynamically depending upon what happens *moment to moment*—and mind you, not just what happens here, but what is happening in the world at large. It is a *smart play*—an intelligent organism."
>
> "I think I understand," Hackworth said. "The internal variables of the play depend on the total universe of information outside."
>
> "Substitute tonight's show for the brain, and the information flowing across the net for molecules flowing through the bloodstream, and you have it," the woman said. (*Diamond Age,* 427)

This conversation helps Hackworth realize that he has been sent into the VR world not for mere entertainment but to discover the model of a revolutionary information system, one that will simulate the functioning of the brain and will therefore enable users to "interface with the universe of data in a more intuitive way": "When I was working with information [Hackworth said], it frequently occurred to me, in a vague and general way, that such a thing may be desirable. But this is beyond my imagination." (428). One may wonder how this system can already be in place in the productions of Dramatis Personae, when the future of the world depends on Hackworth's ability to implement it, but since the encounter with the devil-woman takes place in virtual reality, everything may have been a prophetic dream or a visionary artwork: "For all he knew, the whole show was just a figment being enacted in the mind of the Drummers" (431). It is now up to Hackworth to decode this dream/artwork and to reencode it as technology, thereby accomplishing the reconciliation of the two modes of thinking.

At the end of his initiatory journey Hackworth understands that *he* is the Alchemist, and that his task is to develop a new form of philosopher's stone called the Seed. Diamond Age society has so far been governed by two types of informational structure. One of them, the Feed, a product of New Atlantis philosophy, is a rigidly centralized, top-down operating system that sends energy to customers through the nanotechnological equivalent of electric power lines. A society dependent on a Feed system can be disabled by merely destroying the lines. The other structure, prevalent in the entertainment industry of the claves that border New Atlantis, is a robust network of purely local connections that transmits information efficiently, since a disabled link can always be contoured by sending data on another route, but its lack of self-consciousness makes it impossible to track the destination and fates of information packets. Because the system that links ractors to their customers is organized in such a way, Miranda is unable to locate Nell and make personal contact with her. Though Seed technology is not really described, we can infer from the informational structures praised in the novel that it will overcome the weaknesses of both rigidly centered and radically decentered systems through the following features:

1. It will be a bottom-up system, rooted in the subconscious as a tree is rooted in the earth, but growing toward (and thanks to) the light of logical intelligence.
2. It will network bodies, minds, and machines into a collective intelligence, but it will not dissolve the individuality of its constituents.
3. It will be a body without narrowly specialized organs, which means that every one of its parts will be able to function as center.
4. It will be conscious of its own global state, react to local changes, and constantly update itself.

The last chapter, and the dramatic climax of the novel, ties the two strands of the plot and takes the reader to the eve of the Age of the Seed. After completing the script of the Primer and her education in a

Vicky school, into which she was accepted by order of Lord Finkle-McGraw, Nell apprentices as a writer of interactive scenarios for the house of fantasy—in fact, the brothel—of Madame Ping in the Celestial Kingdom, part of the Chinese territories. (In contrast to the customers of the ractives of Miranda's company, Madame Ping's clients are much more interested in the narrative than in the sex act itself.) The Celestial Kingdom is conquered by a band of fanatical Barbarians, the Fists of Righteous Harmony, who destroy the Feed lines and take Nell prisoner; but she is freed by a battalion of young Chinese girls, the protégées of Dr. X, real-world counterparts of the mouse army of the Primer, and themselves raised by Primers. Taking command of this army, who defeat the invaders, Nell leads the refugees of the war under the sea, where the ritual of the Drummers is once again performed to the benefit of all:

> He [Carl Hollywood, former boss of Miranda] saw it all now: that the refugees had been gathered into the realm of the Drummers for the harvest of fresh data running in their bloodstream, that this data had been infused into the wet Net in the course of the great orgy, and that all of it was now going to be dumped into Miranda, whose body would play host to the climax of some computation that would certainly burn her alive in the process. It was Hackworth's doing; this was the culmination of his efforts to design the Seed, and in so doing to dissolve the foundations of New Atlantis and Nippon and all of the societies that had grown up around the concept of a centralized, hierarchical Feed. (498)

There is a good reason why the novel ends at the threshold of the new age: it is itself very much a script-driven narrative orchestrated by a central authorial intelligence and written with the teleological determination of the nineteenth-century novel. Leaving nothing to chance, the very Victorian chapter titles forecast all the steps of an almost computerlike narrative program: "Final onslaught of the Fists; victory of the Celestial Kingdom; refugees in the domain of the Drummers; Miranda" (488). Within the fictional world, the counterpart of the godlike author is the spiritual leader of neo-Victorian society: "It did not take very deep reflection," we read in the last chapter, "to see the

hand of Lord Alexander Chung-Sik Finkle-McGraw in all of this"
(490). It is as if God himself had written the script of the abdication of
his position of authority, and of the dissemination of his creative
power into every member of his own creation.

We could read this as an allegory of the Death of the Author, since
Finkle-McGraw is nowhere to be seen in this last chapter, but it is the
author who has taken the story to this Aristotelian climax. Narrative
as we know it, with its sharply delineated identities, its competition
born of conflicting desires, its top-down design, and its climactic
convergence of plot lines—all features prominently implemented in
The Diamond Age—is the mode of expression of the Age of the Feed.
In the Age of the Seed, art will be either the nostalgic revival of
the structures of the past, a revival anticipated by the popularity of
twentieth-century "passives" in the Diamond Age, or a succession
of oneiric, unpredictable transformations driven by local analogies
that will place the experiencer in the middle of a collective dream. If
the essence of the Seed is its capability for metamorphosis, as the fairy
tales of the Primer tell us—

> [Fiona] was on the verge of reminding [Hackworth] that her
> dreams had been filled with seeds for the last several years, and
> that every story she had seen in the Primer had been replete with
> them: seeds that grew up into castles; dragon's teeth that grew
> into soldiers; seeds that sprouted into giant beanstalks leading to
> alternate universes in the clouds (410)

—then we should imagine Seed art as the multisensory transposition
of two artworks experienced by Hackworth in Vancouver on his way
to the Drummers. One is a totem pole and the other a tattooed body:

> The woman was looking at a totem pole dominated by a repre-
> sentation of the orca, head down and tail up, dorsal fin project-
> ing horizontally out of the pole. . . . The orca's blowhole had a
> human face carved around it. The face's mouth and the orca's
> blowhole were the same thing. This promiscuous denial of
> boundaries was everywhere on the totem poles and on the wom-
> an's tattoos. . . . The woman's navel was also the mouth of a
> human face, much like the orca's blowhole, and sometimes that

face became the mouth of a larger face whose eyes were her nipples and whose goatee was her pubic hair. But as soon as [Hackworth] had made out one pattern, it would change into something else, because unlike the totem poles the tattoo was dynamic and played with images in time in the same way that the totem poles did in space. (247)

Conclusion

Literature in the Media Landscape

Throughout this book I have treated virtual reality as a metaphor for total art. Over the centuries the dream of the ultimate artwork has taken many forms and nourished many myths: Pygmalion's statue transformed into a live woman, the words of language becoming their own referent in a poetic transubstantiation, and the text as a field of energies that produce perpetual becoming and regeneration (this one a favorite of hypertext theorists). All these conceptualizations involve the transmutation of art into some kind of life not far removed, as N. Katherine Hayles suggests, from the artificial life, or *alife*, generated by computers ("Artificial Life," 205). This idea of art as a form of life implies in turn its negation as a mere reproduction of something external to itself. In their common focus, the myths of total art express the same desire as the fascination of modern culture with ever more transparent, lifelike, and sensorially diversified media: the desire "to get past the limits of representation and to achieve the real" (Bolter and Grusin, *Remediation*, 53). Through the VR metaphor, however, the emphasis on life as the ultimate purpose of art (and artifice) is shifted from the artwork as live object, capable of growth and autonomous behavior, to the artwork as life-giving and life-sustaining environment. The total artwork is no longer something to watch evolve forever but a world in which we will be able to spend an entire lifetime, and to spend it creatively.

What enables VR to serve as a metaphor for a complete habitat for the mind and the body is its reconciliation of two properties once described by Marshall McLuhan as polar opposites. In its pursuit of immersive interactivity, VR wants to be at the same time a hot and a cold medium. For McLuhan, a hot medium

> is one that extends one single sense in "high definition."[1] High definition is the state of being well filled with data. A photograph is, visually, "high definition." A cartoon is "low definition," simply because very little visual information is provided.

> Telephone is a cool medium, or one of low definition, because the ear is given a meager amount of information. And speech is a cool medium of low definition, because so little is given and so much has to be filled in by the listener. On the other hand, hot media do not leave so much to be filled in or completed by the audience. Hot media are, therefore, low in participation and cool media are high in participation or completion by the audience. (*Essential McLuhan*, 162)

Though the term *participation* may suggest immersion, the type of involvement that McLuhan associates with cool media is much closer to the interactive than to the immersive dimension of VR. A hot medium facilitates immersion through the richness of its sensory offerings, while a cold medium opens its world only after the user has made a significant intellectual and imaginative investment. The media that offer data to the senses are naturally hotter than language-based media because in language all sensations must be actively simulated by the imagination.

To expand the expressive power of media, we need to cool down those that are naturally hot and heat up the cold ones. Pictures have reached an unprecedented level of immersivity because of mathematical discoveries (perspective, fractal geometry) and technological advances (photography, cinema) that have added depth, photorealism, and temporality to their built-in spatiality. But electronic technology can prevent this heat from frying up the brain by making the visual image more interactive. VR, multimedia CD ROM art, navigable VRML pictures, animated screens sensitive to the movements of the cursor, click-and-open windowed displays on the Internet, and walk-through electronic art installations are all attempts to intensify the experience that McLuhan calls participation by making the spectator "work" for the next image rather than passively witness a steady flow of pictures, as in film and TV.

For a long time literature has been dominated by the opposite philosophy: remediate the coldness of its medium and turn it into a vivid experience. In contrast to visual representation, language requires a great deal of imaginative activity and extensive logical inferences to produce any kind of picture in the mind of the reader.

Anticipating the vocabulary of hypertext theorists, McLuhan observes that "in reading a detective story the reader participates as co-author simply because so much has been left out of the narrative" (166). Reading places far too many demands on the imagination to let passive readers produce a mental picture rich enough to grant pleasure. It is vastly more challenging to heat up the literary text into an immersive experience than to cool it down through a self-conscious display of signs.

But when a challenge has been met, there is no thrill and less merit in repeating the feat. Nineteenth-century and early-twentieth-century novelists were so successful at developing immersive techniques that later generations fell under intense pressure to search for other types of intellectual satisfaction. Who could outdo Emily Brontë's *Wuthering Heights* or Gustave Flaubert's *Madame Bovary* in binding the reader to the fictional world through all three forms of immersion: spatial, emotional, and temporal? In contrast to classicism, modernism and postmodernism operate under an ideal of perpetual revolution that makes successful forms instantly obsolete. As the postmodern novelist John Barth wrote in 1967, "Beethoven's Sixth Symphony or the Chartres Cathedral, if executed today, might be simply embarrassing" ("Literature of Exhaustion," 66). The fear of creating mere copies—by cultural opinion inferior to innovative art—explains why a major branch of postmodern literature has turned its back on immersion and redirected the reader's activity from the construction of the fictional world to the contemplation of the process of construction itself. No longer interested in transmuting the signs of language into cinema for the mind, many of today's avant-garde authors believe that the reader's activity will be intensified if it develops into a reflexive stance that underscores the coldness of the medium. For these authors, to parody Tolstoy, "all immersive texts are immersive in the same way, but self-referential and interactive texts display these qualities in their own separate ways."

Fortunately for those readers who care about retaining a choice of experiences, this school of thought is not the only force in contemporary literature. Another school remains confident that immersion presents as much diversifying potential as self-reflexivity, because literature can take its readers to ever-different worlds in the universe of the imaginatively possible: worlds of the fantastic, of anticipation, of magical realism, or simply of newly developing or exotic social realities. What makes it so important to maintain alternatives to self-reflexivity is that left entirely by itself, the device cannot carry the literary text. This point is forcefully made by the following spoof:

> This is the first sentence of this story. This is the second sentence. This is the title of this story, which is also found several times in the story itself. . . . This sentence is introducing you to the protagonist of the story, a young boy named Billy. This sentence is telling that Billy is blond and blue-eyed and strangling his mother. . . . This sentence, in a last-ditch attempt to infuse some iota of story line into this paralyzed prose piece, quickly alludes to Billy's frantic cover-up attempts, followed by a lyrical, touching, and beautifully written passage wherein Billy is reconciled with his father (thus resolving the subliminal Freudian conflict to any astute reader) and a final exciting police chase scene during which Billy is accidentally shot and killed by a panicky rookie policeman who is coincidentally named Billy. (Text by David Moser; quoted in Hofstadter, *Metamagical Themas*, 37–40)

Why is this piece of prose so paralyzed? Because it never lets the reader forget the mediation of language, because it stubbornly discourages make-believe, because it never allows recentering into the fictional world. The most obvious purpose of self-reflexivity is to provide a set of internal guidelines, a kind of on-line help file that tells us how to read the text. It would be pointless for these guidelines to instruct readers to keep their gaze aimed at the signs, without giving them a chance to develop interest in what the signs reveal when they function as signs. Pursued for its own sake, self-reflexivity can be no more than the type of statement illustrated by the famous paradox "This statement is false": a purely semiotic and logical curiosity.

Is the moment of appreciation of the form and substance of the text (its texture) necessarily delayed with respect to the moment of immersion, or can they blend together? Jay Bolter and Richard Grusin argue that the more lifelike the medium, the more it attracts attention to itself. In this paradoxical logic, every technological breakthrough that increases the transparency of signs also increases their visibility. Cinema, for instance, is a fuller representation of the real than still pictures, but the spectators who flocked to the early films of the brothers Lumière were undoubtedly more fascinated with the new medium than with what it represented: the arrival of a train in a station, or workers leaving a factory. When the increased transparency results from a technological innovation, audiences quickly become jaded, and after a while they no longer notice the medium. But when the sense of presence is the effect of artistry in the use of the medium—we may call this a stylistic effect—wonderment is more lasting, because it is a response to an individual achievement rather than to a resource available to many. Nowadays we take the lifelikeness of the cinematic image largely for granted, but when we contemplate a photorealistic artwork, such as a painting by Andrew Wyeth, the sharpness of the image is as present to the mind as the depicted scene.

If this view also holds for the literary experience, we can at the same time, or without radical change in perspective, enjoy the imaginative presence of a fictional world and admire the virtuosity of the stylistic performance that produces the sense of its presence. As Bob Witmer and Michael Singer have argued in their study of presence in virtual environments ("Measuring Presence," 226), the mind is fully able to focus on several objects at the same time if the focus on at least some of these objects remains diffuse or backgrounded. We may, for instance, be caught up in a novel but still remain aware that it will soon be time to drive the kids to soccer practice. Similarly, the substance of language may be spectrally present to the mind of the immersed reader like an enveloping atmosphere.

This medium-aware immersion is less contradictory than it appears at first sight if one keeps in mind the fundamental duplicity of the artistic and media experience. Though the term *illusion* is widely used to describe the response to realistic representation, it is really a misnomer. Except for some pathological cases mainly documented through imaginary characters—the usual suspects, Emma Bovary and Don Quixote—media users remain fully conscious of contemplating a representation, even when this representation seems more real than life. In *The Perfect Crime,* Jean Baudrillard comes to suspect that illusion does not exist, because the appreciator is either caught in the forgery and does not notice it, or is aware that the representation is a fake and avoids illusion. What this really means is that illusion, like error, is a condition that can be diagnosed only in others, because its recognition requires an external perspective on a personal belief system. In an art experience, illusion is thus a judgment passed by a real-world self on the mental state of a fictional alter ego—the appreciator's recentered counterpart in the textual world. The same duplicity that diagnoses illusion allows one self to be immersed and the other to appreciate the vehicle of the experience.

A subtle form of awareness of the medium, then, does not seem radically incompatible with immersion. It can grow almost spontaneously out of the text, rather than being forced on the reader by emphatic devices such as metafictional comments or embedded mirror images (what narratologists call strategies of *mise-en-abyme*). But the self-reflexivity that derives from the electronic and purely selective brand of interactivity is anything but subtle. In VR, interactivity is part of a total package, and the user's awareness of the medium does not separate this feature from the immersive dimension. In literature, it is a supplemental feature tacked on to an art form that did very well without it and that still hasn't quite figured out what to do with this strange new resource. The novelty of interactivity, and the self-reflexivity that comes with it, will pass—it may in fact already have passed, now that surfing the World Wide Web has become a normal part of life—but interactivity in a literary text, especially in a narrative one, will retain an intense visibility long after the device becomes widespread in informational contexts, because every time the reader is called on to make a decision, the projector that runs the "cinema for the mind" comes to a halt. As Italo Calvino's continually interrupted novel *If on a Winter's Night a Traveler* demonstrates, it takes a while to get the projector running again. Immersion wants fluidity, wholeness, and a space-time continuum that unfolds smoothly as the imaginary body moves around the fictional world. But in purely textual environments, interactivity presupposes a broken-up and "windowed" structure, since every link teletransports the reader to a new island within the textual archipelago. Bolter and Grusin call this broken-up structure "hypermediacy": a "style of visual [or textual] representation whose goal is to remind the viewer of the medium" (*Remediation*,

273). Through its aggressive focus on the surface, hypermediacy prevents sinking into the depths.

By suggesting that interactivity has a definite function in VR, while its role and contribution remain to be defined for literature because of its conflict with immersion, I do not mean to promote VR—or, rather, what we dream into it—as a superior art form. Nor do I wish to equate immersivity with artistic quality. Immersion is a proven means of aesthetic satisfaction, but it is not the only one. Many readers are willing to sacrifice at least some degree of immersivity to formal experiments, especially to those that bring the intellectual delight of playing a game of in and out between a world-internal and a world-external, medium-conscious stance. Moreover, what Arthur Kroker calls "the hypertextual imagination"[2]—fascination with the discontinuous, the analogical jump, the chance encounter of heterogeneous elements, and the poetic sparkles caused by their collision—is a major force in contemporary culture. Though purely textual mosaics do not need the computer to be implemented, as surrealist and cyberpunk art has shown, the selective interactivity of hypertext takes the possibility of fragmenting and juxtaposing to a higher level. It remains to be seen whether the processing capabilities of the human mind are up to this level of complexity, or whether the point-and-click interactivity of hypertext merely allegorizes the aesthetic productivity of a certain form of imagination, without taking genuine advantage of the unlimited combinatorial resources of the electronic medium. It also remains to be seen to what extent a literary text can emancipate itself from make-believe and replace the appeal of a relatively stable and comprehensive mental image easily committed to memory with combinatorics, kaleidoscopic effects, a constant state of flux, and self-reflexivity. Postmodern literature has conducted a daring and dangerous exploration of the limit between world aesthetics and game aesthetics, for there is everything to lose—in terms of readership—if the limit is transgressed. But even if interactivity and immersion cannot be experienced *at the same time* in a literary text, as they are in VR and in dreams of total art, the conflict between the two principles should not be regarded as an aesthetic disadvantage. VR is an art of expanding resources—some would say an orgy of information—but literature, bound as it is to a single medium, is mostly an art of

TABLE 3 | Immersion, Interactivity, Design, and Pleasure:
A Typology of Human Experiences and Activities

Immersion	Interactivity	Design	Pleasure	
−	−	−	−	Being dead
−	−	−	+	Being a "Grateful Dead"
−	−	+	−	Nonmimetic informational texts
−	−	+	+	Abstract art, poetry
−	+	−	−	Conversation for the sake of social obligations
−	+	−	+	Conversation with friends
−	+	+	−	Business software
−	+	+	+	Games (abstract ones), hypertext fiction, construction kits
+	−	−	−	Nightmares, hallucinations
+	−	−	+	Pleasant dreams
+	−	+	−	Functional architecture
+	−	+	+	Realistic fiction, theme park rides, perspective paintings
+	+	−	−	Business teleconferencing in simulated meeting room (− on design concerns shape on conversation, not visual setting)
+	+	−	+	MOOs and MUDs (same as above for value of design)
+	+	−	+/−	Real life
+	+	+	−	Nonentertainment VR (flight simulators, etc.)
+	+	+	+	Children's games of make-believe, entertainment VR, VR installation art, interactive drama, computer games

overcoming constraints, an idea that Oulipo made its aesthetic program. The profound difference of spirit between VR and literature is one of "more is more" versus "less is more." This is why VR is a neo-Baroque project.

No matter how extravagant or sober its resources, however, art encounters conflicting demands. In VR and in the participatory forms of textuality discussed in chapter 10, the conflict involves the relation of interactivity to design—immersion being given by the medium or the setting. In literature, the conflict is two-pronged: it pits immersion against interactivity and interactivity against design. (See table 3 for an assessment of various activities in terms of immersivity, interactivity, and strength of design.) The challenges may be more complex than in VR, but the possibilities for compromise or combination are more varied. Immersion, as we have seen, is the mode of reading of an embodied mind; interactivity/self-reflexivity is the experience of a pure mind that floats above all concrete worlds in the ethereal universe of semantic possibility. Literature thus offers a choice between the cerebral and the corporeal. Contemporary theory frowns on any idea of mind/body split, but as long as it is a temporary game and not a permanent condition, the mind's exile in the nowhere of incessant travel from sign to sign may lead to a deeper appreciation of what it means to have a body and to belong to a world. Self-reflexive and interactive reading can be used to enhance the reader's awareness of her desire for immersion by temporarily holding her virtual body out of the textual world.

The best model for purely language-based literature to try to emulate in its quest for a workable combination of immersion and interactivity may not be VR after all but an artwork that proposes an alternation rather than a fusion of the two types of experience. This artwork, *Rooftop Urban Park Project/Two-Way Mirror Cylinder Inside Cube*, by Dan Graham (1991; located on the roof of the Dia Art Foundation in New York City), is a glass pavilion that appears either opaque or transparent depending on the light's intensity and the location of the viewer. At times the structure arrests the eye at the surface of its materials, reflecting the background and the spectator's image, while at others it lets the visitor's glance reach into the world that lies beyond its walls.

NOTES

INTRODUCTION

1. Bolter and Joyce created the writing program Story Space, Landow helped develop the hypertextual network Intermedia through the creation of several literary databases, and the hypertext novels of Joyce and Moulthrop, *Afternoon* and *Victory Garden,* are considered the classics of the genre.

2. The chapter is titled "The Extraordinary Convergence: Democracy, Technology, Theory, and the University Curriculum."

3. In her later essay, "Beyond the Hype: Reassessing Hypertext," Snyder takes a much more skeptical approach to the claim that hypertext represents the fulfillment of postmodern doctrine.

4. The signifiers of language—spoken or printed—have admittedly their own aural or visual properties, but these sensory dimensions do not usually relate to the meaning, and their expressive potential is exploited only in certain genres, such as poetry.

5. This is an adaptation. The translation of Ignatius's actual formula is "compound of body and soul" (136).

ONE | The Two (and Thousand) Faces of the Virtual

Epigraph: Dickinson, "I Dwell in Possibility," in *Poems of Emily Dickinson.*

1. This essay, which appears on pp. 19–27 of *Art and Artefact* (a tribute to Baudrillard edited by Nicholas Zurbrugg), is synthetically put together out of various passages from *Le Crime parfait.* The translation is not the same as the one that appears in *The Perfect Crime,* the English translation of *Crime parfait* cited in this book.

2. The relation between these two pairs, as well as between the two components of each pair, leaves quite a few unresolved questions. Is the possible opposed to the real, or does Lévy regard the real as a subset of the possible (as does modal logic: the real world is a member of the set of all possible worlds)? Does the actual coincide with the real, or does the real comprise both the virtual and the actual? One solution would be to regard the real as a subset of the possible, and the actual and virtual as two modes of being within both the real and the possible. If the design of a computer is a real virtuality, by virtue of being a design, and a specific computer made from this design a real actuality, then the blueprint of a VR installation that provides a *perfect* sense of the presence of a virtual world would be a possible virtuality, and a particular implementation of this blueprint a possible actuality.

3. What language would be without this inherent virtuality has been

described by Borges in "Funes the Memorious." Funes, who receives as the result of an accident the dubious gift of total recall, conceives the project of an infinite vocabulary with a different word for every sensory perception: "He was, let us not forget, incapable of ideas of a general, Platonic sort. Not only was it difficult for him to comprehend that the generic symbol dog embraces so many unlike individuals of diverse sizes and form; it bothered him that the dog at three fourteen (seen from the side) should have the same name as the dog at three fifteen (seen from the front)" (*Ficciones*, 114).

4. In what is probably the least transparent of Langer's equivalencies, this description of film as virtual dream refers to the fluidity of the cinematic image, to its ability to transport the spectator instantly to another scene, and to the fact that the camera, by representing the point of view of the spectator, places him in the center of the display, just as a dreamer is always in the center of a dream. In cinema as in the dream mode, furthermore (but also in drama), images succeed each other in a "ubiquitous and virtual present" (*Feeling and Form*, 415).

5. See Landow, *Hypertext 2.0*, 7–10, on Bush and the Memex.

TWO | Virtual Reality as Dream and as Technology

Epigraph: Bricken, "Virtual Worlds: No Interface to Design."

1. Though *virtual reality* is the term that has captured the imagination of the general public, arguably because of the poetic appeal of its built-in oxymoron, the scientific community prefers terms such as *artificial reality* (the physico-spatial equivalent of artificial intelligence) or *virtual environments*. The official technical journal of the field, *Presence*, is subtitled *Teleoperators and Virtual Environments*.

2. See Bolter and Grusin, *Remediation*, chap. 16 ("The Virtual Self"), especially p. 252, for a survey of this controversy.

3. Lanier himself denies any significant inspiration from Gibson and his conception of cyberspace: "In these novels, like . . . *Neuromancer* and so forth, people don't do anything interesting with the artificial reality. . . . Cyberspace is the CB radio of Virtual Reality" (Zhai, *Get Real*, 194).

4. In *The Pearly Gates of Cyberspace,* her outstanding book on cultural conceptions of space, Margaret Wertheim notes (citing Michael Kubovy) that "when we look at a perspectival image from any position other than the center of projection, our minds automatically adjust and we mentally see the image as if we were looking from that point" (114). This phenomenon seems related to the "accommodation to virtual worlds" that helps VR users become immersed in the computer-generated world despite its sensory deficiency with respect to reality.

5. Walter Benjamin observed a similar disappearance of the technological equipment that generates images in the cinema: "Thus, for contemporary

man the representation of reality by the film is incomparably more significant than that of the painter, since it offers, precisely because of the thoroughgoing permeation of reality with mechanical equipment, an aspect of reality which is free of all equipment" ("Work of Art," 236). This explains why movies are the most immersive medium to date.

6. From an interview published in *Omni* 13 (January 1991).

7. One may, of course, imagine a system doing just that for the sake of aesthetic gratification: an interactive, multimedia implementation of surrealistic poetry deriving its effect from the incongruity of the metaphor. But in this case the user's action would aim toward magical transformation, not toward sinking golf balls into holes, and the response of the system would fulfill the user's intent.

8. See Lakoff and Johnson, *Philosophy in the Flesh*, 31–32, on the space-as-container-for-the-body schema.

9. This affinity of VR, or more generally digital representation, for dynamically unfolding space has been exploited in a number of installations and CD ROM artworks: for instance, in Marcos Novak's *Virtual Dervish* (1994–95), an installation that illustrates Novak's concept of "liquid architecture" (discussed in Heim, *Virtual Realism*, 55–58), and in a series of CD ROM and Internet works by the Australian electronic artist Paul Thomas on "emergent space" (see http://www.imago.com.au/spatial).

10. I owe these insights to an oral presentation by Annette Richards, Cornell University, January 1999.

11. An embryonic sense of this dynamic unfolding of space can be experienced on the Active Worlds site of the Internet (http://www.activeworlds .com), especially in ACDD, the world designed by the Art Center College of Design in Pasadena. See also Michael Heim's home page, http://www.mheim .com/worlds/worlds.html. [Both accessed 2/29/00.]

INTERLUDE | Virtual Realities of the Mind: Baudelaire, Huysmans, Coover

1. Jaron Lanier is explicit on his contempt of escapism: "People imagine Virtual Reality as being an escapist thing where people will be ever more removed from the real world and ever more insensitive. I think it's exactly the opposite; it will make us intensely aware of what it is to be human in the physical world, which we take for granted now because we're so immersed in it" (quoted in Zhai, *Get Real*, 188).

2. Last line of *L'Invitation au voyage*.

3. Last line of *Le Voyage*.

4. I use *reveal* rather than *create* in this case because Baudelaire believes in an underlying, absolutely objective order of things. It is the task of the poet to capture this order through a mathematically exact selection of words.

5. The text describes a few other scary drug experiences.

6. MOOs are typically designed as a city with many houses or a house with many rooms, one for each user, but they also contain some public meeting spaces.

THREE | The Text as World: Theories of Immersion

Epigraph: Wolfe, "The New Journalism."

1. Thus when Roman Ingarden, founding father of reader-response theory, uses the term *immersion,* he does not mean immersion in a world but immersion in the flow of language itself: "Once we are immersed in the flow of [sentence-thoughts], we are ready, after contemplating the thought of one sentence, to think out the 'continuation' " (quoted in Iser, "Reading Process," 54).

2. The standard argument invoked by theorists who oppose the assimilation of fictional worlds to possible worlds (e.g., Ronen, Walton, Lamarque and Olsen) is that possible worlds are ontologically complete while fictional worlds are not: the text that describes them cannot specify the truth value of every proposition. But since possible worlds are theoretical constructs, their alleged completeness is not a given but a property ascribed to them by the philosophical imagination. Now, if the imagination can construct possible worlds as ontologically complete, though it cannot run over the list of all propositions, there is no reason it cannot do the same with fictional worlds. The text may produce a necessarily incomplete image, but this does not prevent readers from imagining the textual referents as complete individuals. To take a classic example: We cannot decide how many children Lady Macbeth had, but we imagine her as a woman who had a specific number of children. The number itself is treated as unknown information, not as an ontological gap.

3. The term was first used by Leibniz, but its modern-day users deny any significant indebtedness to Leibniz's philosophy. "It may come as a surprise that this book on possible worlds . . . contains no discussion of the views of Leibniz," writes David Lewis. "Anything I might have to say about Leibniz would be amateurish, undeserving of another's attention, and better left unsaid" (*On the Plurality,* viii).

4. See my article "Possible Worlds in Recent Literary Theory" for an overview of these applications and interpretations.

5. This presentation summarizes ideas that I have presented in *Possible Worlds, Artificial Intelligence, and Narrative Theory.*

6. The same reasoning has been applied by Philip Zhai to the case of the possible worlds of VR: "We will realize the following symmetry: if we call the actual world real and the virtual world illusory when we are in the actual world, we can also call the actual world illusory and the virtual world real when we are immersed in the virtual world" (*Get Real,* 64).

7. This is not to say that fiction does not involve a moment of evaluation through which the reader distances herself from the image. But this evaluation concerns the art (or performance) of the author, not the truth of the representation. It may involve a comparison with other world-creating performances, but not one with the reader's representation of the actual world. This evaluation is therefore not performed from the point of view of a specific world.

INTERLUDE | The Discipline of Immersion: Ignatius of Loyola

1. I say "in part" because Walton quotes a fifteenth-century handbook for young girls that invites them to imagine the Passion in vivid detail: "The better to impress the story of the Passion on your mind, and to memorize each action of it more easily, it is helpful and necessary to fix the places and people in your mind." [This is followed by detailed instructions on how to imagine every character and object] (*Zardino de Oration*, Venice, 1494; quoted in Walton, *Mimesis as Make-Believe*, 27). Since this text predates Ignatius's *Exercises* by only half a century, one must assume that the idea of simulation as spiritual exercise was in the air in the early Renaissance. Note also that in this text, simulation is proposed not only as a way to relate spiritually to the biblical narrative but also as a mnemonic technique—another major preoccupation of the time. See Jonathan Spence, *The Memory Palace of Matteo Ricci.*

2. Here is an example:

THE THIRD APPARITION. St Matthew 28:8–10.

First Point. These Marys went away from the tomb with fear and great joy, eager to announce the Resurrection of the Lord to the disciples.

Second Point. Christ our Lord appeared to them on the way, and greeted them: "Hail to you." They approached, placed themselves at his feet, and adored him.

Third Point. Jesus says to them, "Do not be afraid. Go, tell my brothers to go into Galilee, and there they will see me." (*Spiritual Exercises*, 197)

I have found a few passages enclosed in quotes that involve reports of speech acts of nonspecified content: "The Shepherds returned, glorifying and praising God" (184).

3. The special treatment given to the spoken words of the Bible casts doubt upon Barthes's claim that for Ignatius, sight is the organ of knowledge par excellence: "Ignatius is a theorist of the image as 'view.' . . . These views can 'frame' tastes, odors, sounds, or feelings, but it is the 'visual' sight . . . which receives all of Ignatius' attention" (*Sade, Fourier, Loyola*, 54–55). Barthes argues that in the Middle Ages hearing was considered to be the "perceptive

sense par excellence," the one that "established the richest contact with the world" (followed by touch and sight), and that for Luther, the ear is the "organ of faith" (65). But at the time of Ignatius, and even more in the Baroque, sight began to emerge as the primary sense, the one that leads all perception. Barthes offers as evidence the fact that from the eighteenth century on, the *Exercises* were published in richly illustrated editions. While Barthes is right in pointing out the predominance of visual details in Ignatius, the fact that most exercises observe a similar progression leading from sight to hearing to smell to taste to touch suggests that the senses that involve the closest physical contact are the most reliable witnesses of the divinity. If sight has a privileged role in this progression, it is not because it is the "most complete" apprehension but because it is easier to add hearing, smell, and touch to a visual representation than the other way around. Sight may be a convenient initial step, but the purpose of the exercises is a contemplation that culminates in the participation of all the senses of the body and soul. In addition, the emphasis given to spoken utterances in the exercises of the fourth week suggests that Ignatius remains faithful to the medieval/Lutheran view that the ear is the "organ of faith."

FOUR | Presence of the Textual World: Spatial Immersion

Epigraphs: Welty, "Place in Fiction"; Birkerts, *The Gutenberg Elegies.*

1. In a satirical essay and mock stage play published in 1994 in the *New Yorker,* Ian Frazier quotes as epigrams half a dozen uses of the phrase: "Troy [New York] is the main character; it dominates the book"; "The novel is always about Bridgeport [Connecticut], which really is the central character"; and, expectedly of Joyce's *Ulysses,* "The city of Dublin is the novel's main character" (59).

2. The presence of a narrator in this novel is limited to the "I" that begins the book and the "me" that ends it.

3. For a description of these techniques, see Jonathan Spence, *The Memory Palace of Matteo Ricci.*

4. By contrast, my mental picture of the characters is very incomplete; my Eugénie wears her hair tied in a bun, but, as William Gass so wittily puts it, she crosses the novel without a nose, though of course I do not imagine her as a noseless person.

5. Balbec is an imaginary name for us but a real one for the narrator of Proust's novel.

6. I do not wish to suggest that this stance is rigidly maintained throughout the "natural" narrative performance. As the work of David Herman (*Ghost Stories*) and others has shown, the reduction of spatio-temporal distance at climactic moments through a variety of stylistic devices is a common feature of conversational narration.

7. This example was brought to my attention by Monika Fludernik's discussion in *Towards a "Natural" Narratology,* where it is quoted p. 193.

8. An opposite, equally important trend consists of multiplying stories in an attempt to stress the artificiality of narrative form; what is lost between these two extremes is the belief that narrative can be an authentic expression of life experience.

9. As the work of Uri Margolin has demonstrated.

10. See the special issue of *Style* edited by Monika Fludernik, *Second-Person Narrative,* for an overview of these variations.

11. This distinction between logical and imaginative recentering translates into my own model and terminology some concepts worked out by Zubin and Hewitt in their study of deixis in narrative ("Deictic Center").

FIVE | Immersive Paradoxes: Temporal and Emotional Immersion

Epigraph: Proust, *On Reading.*

1. For a sampling of this work, see the collection edited by Vorderer et al., *Suspense.*

2. The difference between suspense and surprise has been convincingly formulated by the acknowledged master of suspense, Alfred Hitchcock: "We are now having a very innocent little chat. Let us suppose that there is a bomb underneath this table between us. Nothing happens, and then all of a sudden, 'Boom!' There is an explosion. The public is surprised, but prior to this surprise, it has seen an absolutely ordinary scene [conversation between the people sitting at the table] of no special consequence. Now, let us take a suspense situation. The bomb is underneath the table and the public knows it. . . . In these conditions this same innocuous conversation becomes fascinating because the public is participating in the scene. The audience is longing to warn the characters on the screen: 'You shouldn't be talking about such trivial matters. There's a bomb beneath you and it's about to explode!' In the first case we have given the public fifteen seconds of surprise at the moment of the explosion. In the second case we have provided them with fifteen minutes of suspense" (Truffaut, *Hitchcock,* 52; quoted in Brewer, "Nature of Narrative Suspense," 114).

3. To my knowledge, the only "textualist" attempt to explain the phenomenon of emotional participation is Richard Walsh's essay "Why We Wept for Little Nell: Character and Emotional Involvement." Faithful to the structuralist view that characters are not human beings but textual constructs, Walsh advances the thesis that "the reading of fiction requires evaluation, but not belief or any simulacrum of belief" (312). In other words, the reader is invited to contemplate propositions, but not to become immersed in a virtual world. While the Victorian reader's tears for Little Nell can be partly explained by Victorian cultural habits, the modern reader's feelings of sadness for Dick-

ens's heroine are "a response to the idea of innocence . . . rather than to the innocent girl to which that idea contributes. . . . Instead of saying that readers' emotional responses to the fortunes of a character are the result of involvement with a represented person, [this account] assumes that their emotions attach to the particular complex of meanings constituting the character" (ibid.). In this account, the reader would be moved by the abstract idea that innocent little girls can die rather than by the death of one particular fictional individual. Along this line of thought, our evaluations of allegories (such as we find in *The Faerie Queene*) would be just as emotional as our responses to the characters of realistic fiction, or perhaps more, since allegories embody ideas more clearly than complex lifelike characters do. This sounds highly counterintuitive.

4. James Phelan *(Reading People)* distinguishes three components of character: "synthetic" (character as aggregate of features selected by writer), "mimetic" (character as person), and "thematic" (character as bearer of some sort of significance). The textual-world approach emphasizes the mimetic dimension, but it does not preclude the other two, since it maintains an ontological distinction between real and pseudo-persons. In this perspective, readers can either play the game of make-believe and regard characters as persons or adopt a distanced ("metafictional," "textualist") position and regard them as verbal creations fulfilling a thematic function. The textual and world approaches are complementary, but it is only in the world approach that emotional reactions can be justified.

5. Another example of fear that brings into play the thought that a fictional situation might be actualized and affect the spectator is arguably Aristotle's concept of terror. According to his translator Malcolm Heath (xxxix), this terror brings relief, and leads to pleasure, because it acquaints the spectator with the feeling and purges him of his tendency to surrender to excessive or inappropriate fear in real life. Through *katharsis,* tragedy takes the sting out of existential anguish.

6. In addition to the widely documented phenomenon of crying, reactions such as increased heartbeat and rise in body temperature have been clinically verified (Brewer, "Nature of Narrative Suspense," 117).

7. The question of the inherent realism of perspective drawings has been tested in a type of experiment mentioned by Bolter and Grusin in *Remediation:* "Although strict social constructionists and many other postmodern writers take it as dogma that linear-perspective representations are as artificial and arbitrary as any others, some psychologists and art historians still believe otherwise. An empirical test of the question has been to show perspective drawings, photographs, or films to subjects from cultures (often in Africa) that had never seen them. The results of the relatively few experiments have been mixed. When shown a photograph or perspective drawing for the first time, subjects sometimes had trouble interpreting the image, although

after a few minutes or a few tries they could handle the images more easily. In other experiments, subjects had little trouble understanding films that employ editing conventions" (72). It seems, however, that the subjects' initial trouble had at least as much to do with the novelty of the support—many of them had never seen paper before—as with perspective per se.

8. The contrast between the thoroughly moralizing reading of the defense lawyer and the "dangerous" reading of the prosecution lawyer is subtly analyzed by Dominick LaCapra in *Madame Bovary on Trial.*

INTERLUDE | Virtual Narration as Allegory of Immersion

1. Actually, since *If on a Winter's Night a Traveler* is an embedded novel within the book by the same name, we can say that a narrator of level 2, skipping one level, describes the contents of a mind of level 0.

2. If the referent of *you* is interpreted as a reader-character, the paradox occurs one level up: a narrator of level 2 records the mental operations of a reader-character of level 1.

SIX | From Immersion to Interactivity:
The Text as World versus the Text as Game

Epigraph: Carroll, *Through the Looking Glass.*

1. On these differences, see Wilson, *In Palamedes' Shadow,* chap. 3, especially 98–99.

2. This process of deciphering the rules has been invoked by Iser in support of the text/game analogy: "[The aesthetic pleasure of the text is heightened] when the text keeps its rules of play hidden, so that their discovery can become a game in itself.... [The] pleasure will increase when the rules found challenge the sensory and emotive faculties" (*The Fictive and the Imaginary,* 278).

3. The difficulty of fruitfully applying the notions of winning and losing to literature explains why the mathematical field of game theory and Jaakko Hintikka's game-theoretical semantics have had little impact on literary theory. Both are based on the idea of competition between players.

4. In treating *Afternoon* as a game I am disregarding this note on the package of the program: "Michael Joyce's *Afternoon* is a pioneering work of literature, a serious exploration of a new hypertextual medium. It is neither a game nor a puzzle." But why couldn't a "serious" use of hypertext, constituting a "pioneering" literary text, take the form of a game?

5. This concept is proposed in *Struktura khudozhestvennovo teksta (Structure of the Literary Text)* (Moscow: Iskutsstvo, 1970). See the entry for Lotman by Eva Le Grand in Makaryk, *Encyclopedia,* 407–10.

6. On the contrast between the metaphors of language as cube and as

mirror and on the importance of the latter for realism, see Furst, *All Is True*, 18–19.

7. Cf. Derrida's famous phrase "Il n'y a pas de hors-texte" (There is nothing outside the text).

8. The translation by Richard Howard has "functioning" here, but "playing" is closer to the French, "jouer" ("Ce lecteur est . . . plongé dans une sorte d'oisiveté, d'intransivité, et pour tout dire, de sérieux: au lieu de jouer lui-même, d'accéder pleinement à l'enchantement du signifiant, à la volupté de l'écriture, il ne lui reste plus en partage que la pauvre liberté de recevoir ou de rejeter le texte").

9. In the 2000 version, as well as in the NT versions that existed concurrently with DOS-based ones (95 and 98), Windows is no longer a graphical interface to DOS but the operating system itself. This means that it has become entirely similar to the Mac operating system.

SEVEN | Hypertext: The Functions and Effects of Selective Interactivity

Epigraphs: Borges, "An Examination of the Work of Herbert Quain," in *Ficciones;* Benjamin, "The Work of Art in the Age of Mechanical Reproduction."

1. In *Orality and Literacy* (136), Walter Ong argues that electronic media are bringing an age of "secondary orality" characterized by a participatory mystique, communal sense, concentration on the present moment, and use of formula. Writing in 1982, Ong could not have hypertext or the Internet in mind—he is thinking mostly of radio and TV—but each of the three points mentioned above can be illustrated by cyberculture phenomena: communal sense by electronic communities such as user interest groups and MOOs; concentration on the present moment by the real-time interaction of chat rooms; and use of formula by the increasing dependency of electronic writing on buzzwords, slogans, and fixed expressions so that documents posted on the 'Net can be easily found by search engines. For a more specific analysis of oral features in digital media, see Karin Wenz, "Formen der Mündlichkeit."

2. "The Materiality of Reading and Writing, 1450–1650," Cornell University lecture, March 1999.

3. In *Afternoon,* for instance, Michael Joyce uses guard fields to ensure that the reader cannot reach the screen that suggests Peter's possible responsibility for the accident that (perhaps) killed his son and ex-wife before Peter and the reader have gone through a therapy session with the psychologist Lolly. This reading is developed by J. Yellowlees Douglas in "How Do I Stop This Thing?"

4. Brian McHale's term (*Postmodernist Fiction,* chap. 7) for the postmodern practice of creating and destroying fictional worlds.

5. That it takes different abilities to handle language and to orchestrate the components of electronic texts is suggested by the division of labor in the

production of the interactive movie *I'm Your Man* (interlude to chapter 8). Different authors were responsible for the writing of the dialogue for the individual strands in the plot and for tying these strands together in a narratively (i.e., logically) coherent multiple-choice system. In the visual-arts domain of computer-aided creation, there is a dilemma as to whether art contests should be restricted to artists who do their own programming (cf. Bolter and Grusin, *Remediation*, 143).

6. For Cayley's work, see http://www.shadoof.net/ [accessed 2/29/00]. For Kac, http://www.ekac.org/ [accessed 2/29/00]. For Rosenberg, "Barrier Frames" and "Diffractions Through." On cyberpoetry, including works by Cayley and Kac, see issue 5 of *Electronic Book Review:* http://www.altx.com/ebr/ebr5/contents.htm [accessed 2/29/00].

7. See Mark Nunes, "Virtual Topographies: Smooth and Striated Cyberspace," for an enlightening application of these concepts to electronic culture.

8. Could there be more than one image? An example from print literature suggests that this could be the case. In Robert Coover's "The Babysitter" (in *Pricksongs and Descants*) the reader encounters a series of short paragraphs describing an evening spent by a teenager babysitting two kids. But it quickly turns out that the various fragments do not tell the same story: some form a sequence in which the teenager arranges a visit by her boyfriend at the children's home while the parents are away, one in which she is murdered by two intruders, one in which everything unfolds normally. As readers progress sequentially through the text, they sort out the narrative material and build several alternative possible worlds by assigning segments to the proper sequence. (A given segment may fit into more than one narrative script.) The text builds the various stories in a loose round-robin fashion, adding to one, then to another, but without disrupting chronological sequence.

INTERLUDE | Adventures in Hypertext: Michael Joyce's *Twelve Blue*

1. *Twelve Blue,* unlike *Afternoon,* does not offer a dictionary of all segments. This means that the reader is denied direct access to a given segment. Nor does it offer a map of the segments, as do other hypertexts, such as *Victory Garden* by Stuart Moulthrop.

2. Deleuze and Guattari associate the quilt, more particularly the crazy quilt, with the smooth, nomadic space that in their view forms the habitat of postmodern subjectivity: a decentered space that grows in all directions and whose structures consist of rhythmic repetitions of analogous elements rather than of rigid symmetries (*A Thousand Plateaus,* 476).

3. The underlining means that this segment of text is a hypertextual link.

4. In one strand of *Afternoon,* for instance, the narrator Peter witnesses an accident, but in another he causes it; in some screens he behaves as if this accident had killed his ex-wife and son, but in other screens he tries to get

information about the identity of the victims. At times the accident seems to be a mere fender-bender, and at others it is described as a fatal collision. Some readers may attempt to explain away these contradictions by interpreting them as epistemological uncertainties—the fictional world is logically coherent, Peter just doesn't know what the facts are, nor does the reader—or by regarding the screen where Peter is said to have caused the accident as a symbolic expression of his feeling of guilt. By these standards, however, there is hardly any contradictory narrative, short of nonsense verses, that cannot be naturalized.

5. It is nothing new to point out the dreamlike quality of hypertext; Coover has eloquently described it in his pioneering *New York Times Book Review* article, "Hyperfiction: Novels for the Computer": "As one moves through a hypertext, making one's choices, one has the sensation that just below the surface of the text is an almost inexhaustible reservoir of half-hidden story material waiting to be explored. This is not unlike the feeling one has in dreams that there are vast peripheral seas of imagery into which the dream sometimes slips, sometimes returning to the center, sometimes moving through parallel stories at the same time" (10).

6. This idea of collective consciousness has emerged as one of the dominant themes of electronic postmodernism, in part because it attempts to reconcile two conflicting political concerns of our times: a positive valuation of diversity and a sense of community. Pierre Lévy, for instance, regards information technologies and the phenomenon of computer networking as the breeding ground of a collective form of intelligence in which different minds (or processors) are linked together in a nonhierarchical structure, performing different tasks and passing information to each other. This collective intelligence reproduces the distributed processing of the brain on a higher level of organization.

7. For AI, see Douglas Hofstadter and the Fluid Analogies Research Group, *Fluid Concepts and Creative Analogies*, and for architecture, Marcos Novak, "Liquid Architecture in Cyberspace."

EIGHT | Can Coherence Be Saved? Selective Interactivity and Narrativity

Epigraph: Barnes, *Flaubert's Parrot*.

1. In a hypertextual database, a useful alternative to a complete graph is a network with a subset of nodes that are linked to every other one. The fully linked nodes will typically represent a table of contents. This configuration is implemented in those Web pages that present a constant list of options in a sidebar and a navigable hypertext in the main part of the screen. The constant list of links enables the user to return to the table in one trip from every point in the system.

2. Mark Bernstein, in "Patterns of Hypertext," has proposed a typology of

hypertext patterns that could be used to create further refinements within the network category. His catalog includes cycle; counterpoint; mirror worlds; tangle; sieve; montage; split/join; missing link; and feint. Split/join corresponds to what I describe below as the flow chart and sieve to the tree design. But not all of these patterns concern the basic shape of the network. The difference between counterpoint and mirror world is mainly thematic (different voices focused on different themes versus different voices presenting the same themes from different points of view); montage and feint are styles of visual presentation that seem compatible with several network configurations; and missing link is a matter of unfulfilled expectations. It is therefore impossible to subsume Bernstein's typology within the one I am presenting here, or vice versa.

3. An example of this strategy is a combinatory play written by the French Oulipo members Pierre Fournel and Jean-Pierre Enard (Motte, *Oulipo*, 156–58). By allowing crossover and closing the tree at the bottom, they were able to write a system that generates sixteen plays with only fifteen different "scenes," some of which allow no choice. They claim that their scheme saved them sixty-seven scenes, a substantial reduction of memorization for the actors, but a simple calculation shows that a binary tree of five levels with four decision points requires only thirty-one scenes ($16 + 8 + 4 + 2 + 1$). Fournel and Enard assume that sixteen plays of five scenes would take eighty scenes (16×5), but the sixteen plays would not form a combinatorial system, since they would be totally independent of each other.

4. Janet Murray (*Hamlet*, 158–59) describes a closely related idea, Alan Ayckbourn's play *The Norman Conquests*, which consists of three different plays taking place simultaneously in different rooms of the same house, but the production does not seem to allow the spectator to move from room to room. The plays are therefore parallel but not interactive.

5. *Hypercafé*, Landow tells us, is a video project created by Nitin Sawney, David Balcom, and Ian Smith at the Georgia Institute of Technology (*Hypertext 2.0*, 211).

6. William Gibson, the coiner of the term *cyberspace*, calls it a "nonspace of the mind" (*Neuromancer*, 57).

7. For instance, Nunes, "Virtual Topographies," and Moulthrop, "Rhizome and Resistance."

8. Far from being satisfied with her own diagnosis, Murray tries to make a case for the tragic and cathartic potential of electronic narrative by imagining three interactive ways of representing the journey of a young man toward suicide (*Hamlet*, 175–82); but while a gifted writer could conceivably manage to create emotional bonding with the character, this accomplishment would be more a matter of overcoming the limitations of the medium than of exploiting its distinctive properties.

9. As Espen Aarseth suggests in *Cybertext*, when he calls *Afternoon* a "game of narration" (94).

10. For instance, Stuart Moulthrop's *The Tomb Robbers* (http://raven.ubalt.edu/staff/moulthrop/hypertexts/tr/index.html) [accessed 2/29/00], which consisted of only a few screens at the time of this writing. For a list of hypertext sites using VRML on the World Wide Web, see Matthew Mirapaul, "Hypertext Fiction Adds a Third Dimension."

11. On the particularly strong immersive effect of the human voice, N. Katherine Hayles writes, "Whereas sight is always focused, sharp, and delineated, sound envelopes the body, as if it were an atmosphere to be experienced rather than an object to be dissected. Perhaps that is why researchers in virtual reality have found that sound is much more effective than sight in imparting emotional tonalities to their simulated worlds" (*How We Became Posthuman*, 219).

12. McLuhan's formula is widely interpreted as the expression of the self-referential character of postmodern art, but there is no mention of self-referentiality in the original context. For McLuhan, the formula expresses a variety of rather disparate ideas: (1) that a medium—for instance, electric light—has no intrinsic content: "The electric light is pure information. It is a medium without a message, as it were, unless it is used to spell out some verbal ad or name" (*Essential McLuhan*, 151); (2) that the message of a medium lies in its social impact: "For the 'message' of any medium or technology is the change of scale or pace or pattern that it introduces into human affairs" (152); (3) that the content of a medium is another medium: "The content of a movie is a novel or a play or an opera. . . . The 'content' of writing or print is speech" (159); and (4) that media shape perception: "For the medium determines the modes of perception and the matrix of assumptions within which objectives are set" (188).

INTERLUDE | *I'm Your Man:* Anatomy of an Interactive Movie

1. See Michael Spencer, *Michel Butor,* for a detailed description.

2. A satisfaction denied to the DVD user of *I'm Your Man,* since the map is shown on the package.

3. A proposal made in a 1998 MIT master's thesis by Philip R. Tiongson. See the Web page of the MIT Interactive Cinema Group at http://ic.www.media.mit.edu [accessed 2/29/00].

NINE | Participatory Interactivity from Life Situations to Drama

Epigraph: Langer, *Feeling and Form.*

1. I do not wish to convey the impression that VR developers and theorists regard the mutual compatibility of immersion and interactivity as totally unproblematic. Jonathan Steuer suggests that a very immersive VR display may "decrease the ability of subjects to mindfully interact with it in real time"

("Defining," 90). If a computer-generated environment is so rich in "fictional truths" that its exploration offers great rewards, why would the user bother to change this world? It is, however, through exploration that the user gains an appreciation for the immersive richness of the virtual world, and this exploration is itself a type of interactivity. Moreover, the problem suggested by Steuer would be a case of immersion interfering with interactivity—the familiar objection that immersion makes the user passive—while I am debating here the opposite question: Does interactivity favor immersion in three-dimensional VR environments?

2. Through this comparison I do not wish to deny that authors use materials from their own life experience. In the domain of art, literally speaking, there is no creation *ex nihilo.*

3. Painted in 1751 by Franz Josef Spiegler.

4. See Omar Calabrese, *Neo-Baroque: A Sign of the Times,* for a discussion of the affinities between Baroque art and postmodernism. The titles of the chapters tell much of the story: Rhythm and Repetition, Limit and Excess, Detail and Fragment, Instability and Metamorphosis, Disorder and Chaos, The Knot and the Labyrinth, Complexity and Dissipation, The Approximate and the Inexpressible, Distortion and Perversion.

5. This was Nietzsche's objection to the concept: "Now the serious events are supposed to prompt pity and fear to discharge themselves in a way that relieves us; now we are supposed to feel elated and inspired by the triumph of good and noble principles, at the sacrifice of the hero in the interest of a moral vision of the universe. I am sure that for countless men this, and only this is the effect of tragedy, but it plainly follows that all of these men, together with their interpreting aestheticians, have had no experience of tragedy as a supreme art" (*Birth of Tragedy,* sec. 22, p. 32).

6. As a counterexample to this principle, one should mention that privileged noblemen had a seat on stage in Parisian theaters of the time of Louis XIV, but this practice was a leftover from the Elizabethan age and would later disappear.

TEN | Participatory Interactivity in Electronic Media

Epigraph: Walser, "Spacemakers and the Art of the Cyberspace Playhouse."

1. The computer game Battle Chess, which offered vivid graphic sword fights every time a piece was taken by the opponent and a choice of perspectives on the board, added a mimetic dimension to the game that delighted certain users—children, fans of computer games—but annoyed serious chess players. The animation just wasn't in the spirit of the game.

2. See Aarseth's discussion of the 1982 game Deadline for an example of incoherence that leads to humorous effects (*Cybertext,* 115–24).

3. PacMan and Quake, the most popular games of the early 1980s and late

1990s, have, for instance, very similar narrative deep structures: in both games the user runs around a maze together with bad guys, trying to avoid being killed. The thematics of Quake are admittedly much more aggressive, since the Quake player gains points by killing his enemies while the PacMan player advances by gathering treasures, but these different means of survival do not impact the strategy. The main difference resides in the three-dimensionality and dynamic perspective of the Quake display, in the sophistication and variety of the resources available to the user, and in the possibility of playing Quake against, or with, other human players in a local network or over the Internet, a feature that leads to the formation of fanatical player communities.

4. I am talking here about games of problem solving, such as Myst and Zork, not about games of skills that pit human players against each other, such as Quake, or the player against himself, such as Tetris (the point of Tetris is to beat your previous best score). Games of this latter type are endlessly replayable, but the nature of the player's goal brings them closer to competitive sports than to art.

5. As reported by Julian Dibbell in "Net Prophet," interview with John Perry Barlow, *Details* 8 (1994). Quoted in Slouka, *War*, 174.

6. Quoted in Foner, "What's an Agent, Anyway?" sec. "A Sociological Look at Muds," 5.

7. My description is based on the vivid and moving account offered by Celia Pearce in *The Interactive Book* (413–20).

8. Archaeologists are still unsure of the precise social function of the kivas.

9. N. Katherine Hayles informs me that some users took as much as two hours to explore *Placeholder*. But since the future of this type of installation lies primarily in the theme-park industry, it is most likely that future applications will be designed for fifteen-minute visits.

10. On this interaction between users, Hayles writes, "Each can see the other in the simulation and hear the voice-filtered comments his or her companion makes. Participants can shapeshift by touching the appropriate totemic icon. Improvising on cues provided by the environment and each other, they create narrative" ("Embodied Virtuality," 19).

11. Only Snake's vision was implemented and, according to the authors, with only limited success: "The implementation was poor, in that it simply applied a red filter without increasing apparent luminance, thus effectively reducing rather than enhancing visibility" (Laurel et al., "Placeholder," 123).

12. On the achievements and strategies of AI in automated story generation, see chapter 11 of my book *Possible Worlds, Artificial Intelligence, and Narrative Theory* (233–57).

13. The project's focus on types of immersion other than spatial is demonstrated by the fact that it took place on a bare stage without props of any kind.

14. A precedent of the emergence of such a need can be found in the history of film. The first movie, by the brothers Lumière, was a documentary depicting a totally unremarkable scene of everyday life: the arrival of a train in La Ciotat. The scene was hardly worthy of attention in real life, but projected on a screen it became an object of utter fascination; some spectators were so stricken that they fled the room (Bolter and Grusin, *Remediation*, 155). The film industry depended for quite a while on this type of movie, but the novelty of cinematic immersion eventually wore off, and it became necessary to develop other themes and structures to maintain interest. It was then that cinema discovered the power of narrative, in both its epic and dramatic forms, and narrative has never let it down.

15. This is not to say that Aristotelian plot structures should be totally abandoned in VR installations. This description of the interactive theme park ride The Loch Ness Expedition by its creative director, Celia Pearce, shows that the ride was deliberately conceived as a three-act narrative: "Based on the legend of the Loch Ness Monster, this VR ride took the narrative element even farther [than other VR installations presented at the 1994 SIGGRAPH conference] by adding dramatic adventures, game logic, and multi-user interactivity to the mix. In *The Loch Ness Expedition*, players embark on an underwater mission to rescue the Loch Ness Monster's eggs from bounty hunters (other players, who look like good guys to themselves but bad guys to everybody else). *The Loch Ness Expedition* combined a three-act structure with basic tenets of game logic to create a hybrid movie/ride/game in which players became the heroes of the story" (*Interactive Book*, 515).

INTERLUDE | Dream of the Interactive Immersive Book: Neal Stephenson's *The Diamond Age*

1. Miranda's name, as Janet Murray reminds us (*Hamlet*, 121), is borrowed from the Shakespearean character who speaks of a "brave new world."

2. The inferior copies of the Chinese girls are connected to an automated voice.

3. The Primer experience presents interesting similarities indeed with the "Palace of Fates," a parable found at the end of Leibniz's *Theodicy*. In this parable the ancient Greek mathematician Theodorus is led by Athena into a palace built by Zeus, where he can contemplate all possible worlds. He follows the various destinies of one man, the Roman Sextus. Some of these destinies are happy and others are tragic; to his great surprise, Theodorus finds that in the best of all possible worlds Sextus is a vile criminal, for this was necessary to the harmony of the whole: "The crime of Sextus serves for great things: it renders Rome free; thence will arise a great empire" (Rescher, *G. W. Leibniz's Monadology*, 306). Moreover, as Theodorus examines the life of Sextus in each possible world, he does so by blowing up details from a

global picture, much in the way Nell fulfills her destiny by expanding the episodes of a predetermined script: "Theodorus saw the whole life of Sextus as at one glance, and as in a stage presentation. There was a great volume of writing in this hall: Theodorus could not refrain from asking what that meant. It is the history of the world that we are now visiting, the Goddess told him; it is the book of its fates. You have seen a number on the forehead of Sextus. Look in this book for the place which it indicates. Theodorus looked for it, and found there the history of Sextus in a form more ample than the outline that he had seen. Put your finger on any line you please, Pallas said to him, and you will see represented actually in all its detail that which the line broadly indicates. He obeyed, and he saw coming into view a portion of the life of Sextus" (ibid.). On the relevance of Leibniz to VR, see Eric Steinhart, "Leibniz's Palace of the Fates."

4. As N. Katherine Hayles has shown in *How We Became Posthuman* (272–79).

CONCLUSION | Literature in the Media Landscape

1. To extend Marshall McLuhan's definition of "hot medium" to VR, the restriction of "high definition" to data affecting a single sense must be loosened, so that multisensory media can also be covered. This interpretation seems to respect the spirit of the definition, since McLuhan categorizes film, a notoriously multisensory medium, in the hot category.

2. Project outline submitted to the Society for the Humanities, Cornell University, 1999.

WORKS CITED

PRINTED WORKS

Primary

Balzac, Honoré de. *Eugénie Grandet*. Trans. Marion Ayton Crawford. London: Penguin, 1955.

Barnes, Julian. *Flaubert's Parrot*. New York: Knopf, 1985.

Baudelaire, Charles. *Artificial Paradises*. Trans. Stacy Diamond. New York: Citadel Press, 1996.

Boccaccio, Giovanni. *The Decameron*. Trans. and ed. G. H. McWilliam. New York: Penguin, 1972.

Borges, Jorge Luis. *Ficciones*. Ed. and intro. Anthony Kerrigan. New York: Grove Weidenfeld, 1962.

Brontë, Charlotte. *Shirley*. Oxford: Clarendon, 1979.

Brontë, Emily. *Wuthering Heights*. New York: Bantam, 1983.

Calvino, Italo. *If on a Winter's Night a Traveler*. Trans. William Weaver. San Diego: Harcourt Brace, 1981.

Carroll, Lewis. *Alice's Adventures in Wonderland and Through the Looking Glass*. London: Puffin Books, 1948.

Conrad, Joseph. *The Nigger of the Narcissus*. London: Dent, 1974.

Coover, Robert. *Pricksongs and Descants*. New York: New American Library, 1969.

———. *The Universal Baseball Association, Inc., J. Henry Waugh, Prop.* New York: Plume Books, 1971.

Cortázar, Julio. "Continuity of Parks." *Blow-up and Other Stories*. Trans. Paul Blackburn. New York: Pantheon, 1967. 63–65.

de Cervantes Saavedra, Miguel. *Don Quixote of La Mancha*. Trans. Walter Starkie. New York: Signet, 1994.

Dickinson, Emily. *The Poems of Emily Dickinson*. Variorum edition. Ed. R. W. Franklin. Cambridge, Mass.: Belknap Press, 1998.

Duras, Marguerite. *L'Amant*. Paris: Minuit, 1984.

Flaubert, Gustave. *Madame Bovary*. Trans. Francis Steegmuller. New York: Random House, 1957.

———. *Oeuvres*. Ed. A. Thibaudet and R. Dumesnil. Paris: Bibliothèque de la Pléïade, 1958.

Fowles, John. *The French Lieutenant's Woman*. New York: Signet, 1970.

Genet, Jean. *The Balcony*. Trans. Bernard Frechtman. New York: Grove Press, 1966.

Gibson, William. *Neuromancer*. New York: Ace, 1994.

Huxley, Aldous. *Brave New World.* New York: HarperPerennial, 1989.

Huysmans, Joris-Karl. *Against Nature.* Trans. Margaret Mauldon. Oxford: Oxford UP, 1998.

Joyce, James. "Eveline." *Dubliners.* New York: Penguin Books, 1976. 36–41.

Loyola, Ignatius of. *The Spiritual Exercises and Selected Works.* Ed. George E. Ganss, S.J. New York: Paulist Press, 1991.

Morrison, Toni. *Beloved.* New York: Plume/Penguin, 1988.

Pavić, Milorad. *Dictionary of the Khazars: A Lexicon Novel in 100,000 Words.* Trans. Christina Pribicevic-Zoric. New York: Knopf, 1988.

Perec, Georges. *La Disparition.* Paris: Denoël, 1969.

———. *La Vie mode d'emploi.* Paris: Hachette, 1978.

Poe, Edgar Allan. "The Assignation." *The Portable Poe.* Ed. Philip Van Doren Stern. New York: Penguin-Viking, 1997. 192–207.

Proust, Marcel. *On Reading.* Trans. and ed. Jean Autret and William Burford. Intro. William Burford. New York: Macmillan, 1971.

———. *Remembrance of Things Past.* Vol. 1, *Swann's Way, Within a Budding Grove, The Guermantes Way.* Trans. C. K. Scott Moncrief. New York: Random House, 1934.

Queneau, Raymond. *Cent mille milliards de poèmes.* Paris: Gallimard, 1961.

Robbe-Grillet, Alain. *Two Novels by Robbe-Grillet: Jealousy and In the Labyrinth.* Trans. Richard Howard. New York: Grove Press, 1965.

Roubaud, Jacques. ∈. Paris: Gallimard, 1967.

Saporta, Marc. *Composition No 1.* Paris: Seuil, 1961.

Stendhal [Henri Beyle]. *Le Rouge et le noir: Chronique du XIXème siècle.* Ed. Henri Martineau. Paris: Garnier, 1958.

Stephenson, Neal. *The Diamond Age, or A Young Lady's Illustrated Primer.* New York: Bantam, 1996.

———. *Snow Crash.* New York: Bantam, 1992.

Woolf, Virginia. *The Waves.* London: Hogarth Press, 1963.

Secondary

Aarseth, Espen. "Aporia and Epiphany in Doom and The Speaking Clock: The Temporality of Ergodic Art." Ryan, *Cyberspace,* 31–41.

———. *Cybertext: Perspectives on Ergodic Literature.* Baltimore: Johns Hopkins UP, 1997.

Aristotle. *Poetics.* Trans. and intro. Malcolm Heath. New York: Penguin Books, 1996.

Artaud, Antonin. *Artaud on Theater.* Ed. Claude Schumacher. London: Methuen, 1989.

Auge, Marc. *Non-Places: Introduction to an Anthropology of Supermodernity.* New York: Verso, 1995.

Bachelard, Gaston. *The Poetics of Space.* Trans. Maria Jolas. Boston: Beacon Press, 1994.

Balsamo, Anne. *Technologies of the Gendered Body: Reading Cyborg Women.* Durham, N.C.: Duke UP, 1996.

Banfield, Ann. *Unspeakable Sentences: Narration and Representation in the Language of Fiction.* Boston: Routledge & Kegan Paul, 1982.

Barth, John. "The Literature of Exhaustion." *The Friday Book and Other Essays.* New York: Putnam, 1984. 62–76.

Barthes, Roland. "The Death of the Author." *Image, Music, Text.* Ed. and trans. Stephen Heath. New York: Hill & Wang, 1977. 142–48.

———. *Le Plaisir du texte.* Paris: Seuil, 1973.

———. "The Reality Effect." *French Literary Theory Today.* Ed. Tzvetan Todorov. Cambridge: Cambridge UP, 1982. 11–17.

———. *Sade, Fourier, Loyola.* Trans. Richard Miller. New York: Hill & Wang, 1976.

———. *S/Z.* Trans. Richard Howard. New York: Hill & Wang, 1974.

Bates, Joseph. "The Nature of Characters in Interactive Worlds and the Oz Project." Loeffler and Anderson, *Virtual Reality Casebook,* 99–102.

———. "The Role of Emotions in Believable Agents." *Communications of the ACM* 37, no. 7 (1994): 122–25.

Baudrillard, Jean. *Art and Artefact.* Ed. Nicholas Zurbrugg. London: Sage, 1997.

———. *The Perfect Crime.* Trans. Chris Turner. London: Verso, 1996.

———. "The Precession of Simulacra." *Simulacra and Simulation.* Trans. Sheila Faria Glaser. Ann Arbor: U of Michigan Press, 1994. 1–42.

Benedikt, Michael. "Introduction" and "Cyberspace: Some Proposals." Benedikt, *Cyberspace,* 1–16 and 119–224.

———, ed. *Cyberspace: First Steps.* Cambridge, Mass.: MIT Press, 1991.

Benjamin, Walter. "Epic Theater" and "The Work of Art in the Age of Mechanical Reproduction." *Illuminations.* Ed. Hannah Arendt, trans. Harry Zohn. New York: Schocken, 1969. 149–56 and 219–54.

Bernard, Michel. "Lire l'hypertexte." Vuillemin and Lenoble, *Littérature et informatique,* 313–25.

Bernstein, Mark. "Patterns of Hypertext." *Proceedings of Hypertext 98: The Ninth ACM Conference on Hypertext and Hypermedia.* Ed. Kaj Grønbæk, Elli Mylonas, and Frank Shipman III. N.p.: ACM, 1998. 21–29.

Biocca, Frank, and Ben Delaney. "Immersive Virtual Reality Technology." Biocca and Levy, *Communication,* 57–124.

Biocca, Frank, Taeyong Kim, and Mark R. Levy. "The Vision of Virtual Reality." Biocca and Levy, *Communication,* 3–14.

Biocca, Frank, and Mark R. Levy, eds. *Communication in the Age of Virtual Reality.* Hillsdale, N.J.: Lawrence Erlbaum, 1995.

Birkerts, Sven. *The Gutenberg Elegies: The Fate of Reading in an Electronic Age.* New York: Fawcett Columbine, 1994.

Bloch, R. Howard, and Carla Hesse. "Introduction." *Representations* 42 (1993): 1–12.

Bolter, Jay David. "Literature in the Electronic Writing Space." Tuman, *Literacy Online,* 19–42.

———. *Writing Space: The Computer, Hypertext, and the History of Writing.* Hillsdale, N.J.: Lawrence Erlbaum, 1991.

Bolter, Jay David, and Richard Grusin. *Remediation: Understanding New Media.* Cambridge, Mass.: MIT Press, 1999.

Booth, Wayne C. *The Rhetoric of Fiction.* Chicago: U of Chicago Press, 1961.

Bootz, Philippe. "Gestion du temps et du lecteur dans un poème dynamique." Vuillemin and Lenoble, *Littérature et informatique,* 233–47.

Brennan, Susan. "Conversation as Direct Manipulation: An Iconoclastic View." Laurel, *Art of Computer Interface Design,* 393–404.

Brewer, William F. "The Nature of Narrative Suspense and the Problem of Rereading." Vorderer, Wulff, and Friedrichsen, *Suspense,* 107–27.

Bricken, Meredith. "Virtual Worlds: No Interface to Design." Benedikt, *Cyberspace,* 363–82.

Caillois, Roger. *Men, Play, and Games.* Trans. Meyer Burasch. New York: Free Press, 1961.

Calabrese, Omar. *Neo-Baroque: A Sign of the Times.* Trans. Charles Lambert. Princeton, N.J.: Princeton UP, 1992.

Carroll, Noël. "The Paradox of Suspense." Vorderer, Wulff, and Friedrichsen, *Suspense,* 71–92.

Cayley, John. "Potentialities of Literary Cybertext." *Visible Language* 30, no. 3 (1996): 164–83.

Chatman, Seymour. *Coming to Terms: The Rhetoric of Narrative in Fiction and Film.* Ithaca: Cornell UP, 1990.

Coleridge, Samuel Taylor. *Biographia Literaria.* London: J. M. Dent, 1975.

Coover, Robert. "Hyperfiction: Novels for the Computer." *New York Times Book Review,* 29 August 1993, 1, 8–12.

Culler, Jonathan. *Structuralist Poetics.* Ithaca: Cornell UP, 1975.

Currie, Gregory. "Imagination and Simulation: Aesthetics Meets Cognitive Science." Davies and Stone, *Mental Simulation,* 151–69.

———. *The Nature of Fiction.* Cambridge: Cambridge UP, 1990.

Davenport, Glorianna. "The Care and Feeding of Users." *IEEE Multimedia* 4, no. 1 (1997): 8–13.

Davies, Martin, and Tony Stone. *Mental Simulation.* Oxford: Blackwell, 1995.

Deleuze, Gilles, and Félix Guattari. *A Thousand Plateaus: Capitalism and Schizophrenia.* Trans. Brian Massumi. Minneapolis: U of Minnesota Press, 1987.

Derrida, Jacques. "Structure, Sign, and Play in the Discourse of the Human

Sciences." *The Languages of Criticism and the Sciences of Man.* Ed. Richard Macksey and Eugenio Donato. Baltimore: Johns Hopkins Press, 1970. 247–65.

Dinkla, Söke. "From Participation to Interaction: Toward the Origins of Interactive Art." Leeson, *Clicking In,* 279–90.

Ditlea, Steve. "False Starts Aside, Virtual Reality Finds New Roles." *New York Times,* 23 March 1998, C3.

Doležel, Lubomír. *Heterocosmica: Fiction and Possible Worlds.* Baltimore: Johns Hopkins UP, 1998.

Douglas, J. Yellowlees. "How Do I Stop This Thing?" Landow, *Hyper/Text/Theory,* 159–88.

———. "Hypertext, Argument, and Relativism." Snyder, *Page to Screen,* 144–62.

Eagleton, Terry. *Literary Theory: An Introduction.* Oxford: Basil Blackwell, 1983.

Eco, Umberto. *The Open Work.* Cambridge, Mass.: Harvard UP, 1989.

———. "Possible Worlds and Text Pragmatics: 'Un Drame bien parisien.' " *The Role of the Reader: Explorations in the Semiotics of Texts.* Bloomington: Indiana UP, 1979. 200–260.

———. *Travels in Hyperreality: Essays.* Trans. William Weaver. San Diego: Harcourt Brace Jovanovich, 1986.

Eliade, Mircea. *Myths, Rites, Symbols: A Mircea Eliade Reader.* Ed. Wendell C. Beane and William G. Doty. Vol. 1. New York: Harper & Row, 1975.

Fischlin, Daniel, and Andrew Taylor. "Cybertheater, Postmodernism, and Virtual Reality: An Interview with Toni Dove and Michael Mackenzie." *Science-Fiction Studies* 21 (1994): 1–23.

Fleischman, Suzanne. *Tense and Narrativity: From Medieval Performance to Modern Fiction.* Austin: U of Texas Press, 1990.

Fludernik, Monika. *The Fictions of Language and the Languages of Fiction: The Linguistic Representation of Speech and Consciousness.* London: Routledge, 1993.

———. *Towards a "Natural" Narratology.* London: Routledge, 1996.

———, ed. *Second-Person Narrative.* Special issue of *Style* 28, no. 3 (1994).

Foucault, Michel. "What Is an Author?" *The Foucault Reader.* Ed. Paul Rabinow. New York: Pantheon, 1984. 101–20.

Fournel, Paul, and Jean-Pierre Enard. "The Theater Tree: A Combinatory Play." Motte, *Oulipo,* 159–62.

Frazier, Ian. "The Novel's Main Character." *New Yorker,* 5 September 1994, 59–60.

Friedman, Ted. "Making Sense of Software: Computer Games and Interactive Textuality." Jones, *Cybersociety,* 73–89.

Furst, Lilian R. *All Is True: The Claims and Strategies of Realist Fiction.* Durham, N.C.: Duke UP, 1995.

Gaggi, Silvio. *From Text to Hypertext: Decentering the Subject in Fiction, Film, the Visual Arts, and Electronic Media.* Philadelphia: U of Pennsylvania Press, 1997.

Garner, Stanton B., Jr. *Bodied Space: Phenomenology and Performance in Contemporary Drama.* Ithaca: Cornell UP, 1994.

Gass, William. "Representation and the War for Reality." *Habitations of the Word.* New York: Simon & Schuster, 1985. 73–112.

Genette, Gérard. *Narrative Discourse: An Essay in Method.* Trans. Jane E. Lewin. Ithaca: Cornell UP, 1980.

Gerrig, Richard J. *Experiencing Narrative Worlds: On the Psychological Activities of Reading.* New Haven: Yale UP, 1993.

———. "The Resiliency of Suspense." Vorderer, Wulff, and Friedrichsen, *Suspense,* 93–105.

Goodman, Nelson. "Reality Remade." *Philosophy Looks at the Arts.* Ed. Joseph Margolis. Philadelphia: Temple UP, 1987. 283–306.

———. *Ways of Worldmaking.* Indianapolis: Hackett, 1978.

Hayles, N. Katherine. "Artificial Life and Literary Culture." Ryan, *Cyberspace,* 205–23.

———. "The Condition of Virtuality." *The Digital Dialectic: New Essays on New Media.* Ed. Peter Lunenfeld. Cambridge, Mass.: MIT Press, 1999. 68–94.

———. "Embodied Virtuality, or How to Put Bodies Back into the Picture." Moser and McLeod, *Immersed,* 1–28.

———. *How We Became Posthuman: Virtual Bodies in Cybernetics, Literature, and Informatics.* Chicago: U of Chicago Press, 1999.

Heal, Jane. "How to Think about Thinking." Davies and Stone, *Mental Simulation,* 33–52.

Heim, Michael. "The Erotic Ontology of Cyberspace." Benedikt, *Cyberspace,* 59–80.

———. *The Metaphysics of Virtual Reality.* New York: Oxford UP, 1993.

———. *Virtual Realism.* New York: Oxford UP, 1998.

Herman, David. *North Carolina Ghost Stories: An Integrative Approach.* In preparation.

———. *Story Logic: Problems and Possibilities of Narrative.* Lincoln: U of Nebraska Press, 2001.

———. "Textual You and Double Deixis in Edna O'Brien's *A Pagan Place.*" *Style* 28, no. 3 (1994): 378–410.

———, ed. *Narratologies: New Perspectives in Narrative Analysis.* Columbus: Ohio State UP, 1999.

Hintikka, Jaakko. "Modality as Referential Multiplicity." *Ajatus* 20 (1957): 49–64.

Hofstadter, Douglas. *Gödel, Escher, Bach: An Eternal Golden Braid.* New York: Vintage, 1980.

------. *Metamagical Themas: Questing for the Essence of Mind and Pattern.* New York: Vintage, 1985.

Hofstadter, Douglas, and the Fluid Analogies Research Group. *Fluid Concepts and Creative Analogies: Computer Models of the Fundamental Mechanisms of Thought.* New York: Basic Books, 1995.

Hughes, Bob. *Dust or Magic: Secrets of Successful Multimedia Design.* Harlow, England: Addison-Wesley, 2000.

Huizinga, Johan. *Homo Ludens: A Study of the Play Element in Culture.* Boston: Beacon Press, 1955.

Ingarden, Roman. *The Literary Work of Art: An Investigation on the Borderlines of Ontology, Logic, and the Theory of Literature.* Trans. George G. Grabowicz. Evanston: Northwestern UP, 1973.

Iser, Wolfgang. *The Fictive and the Imaginary: Charting Literary Anthropology.* Baltimore: Johns Hopkins UP, 1993.

------. "The Play of the Text." *Prospecting: From Reader Response to Literary Anthropology.* Baltimore: Johns Hopkins UP, 1989. 249–60.

------. "The Reading Process." *Reader-Response Criticism: From Formalism to Poststructuralism.* Ed. Jane P. Tomkins. Baltimore: Johns Hopkins UP, 1980. 50–69.

Jameson, Fredric. *Postmodernism, or The Cultural Logic of Late Capitalism.* London: Verso, 1991.

Jones, Steven B., ed. *Cybersociety: Computer-Mediated Communication and Community.* Thousand Oaks, Calif.: Sage, 1995.

Joyce, Michael. *Of Two Minds: Hypertext, Pedagogy, and Poetics.* Ann Arbor: U of Michigan Press, 1995.

Keep, Christopher J. "The Disturbing Liveliness of Machines: Rethinking the Body in Hypertext Theory and Fiction." Ryan, *Cyberspace,* 164–81.

Kelso, Margaret Thomas, Peter Weyhrauch, and Joseph Bates. "Dramatic Presence." *Presence: Teleoperators and Virtual Environments* 2, no. 1 (1993): 1–15.

Kripke, Saul. "Semantic Considerations on Modal Logic." *Acta Philosophica Fennica* 16 (1963): 83–94.

Kroker, Arthur, and Michael A. Weinstein. *Data Trash: The Theory of the Virtual Class.* New York: St. Martin's Press, 1994.

LaCapra, Dominick. *Madame Bovary on Trial.* Ithaca: Cornell UP, 1982.

Lakoff, George, and Mark Johnson. *Philosophy in the Flesh: The Embodied Mind and Its Challenge to Western Thought.* New York: Basic Books, 1999.

Lamarque, Peter, and Stein Haugen Olsen. *Truth, Fiction, and Literature.* Oxford: Clarendon Press, 1994.

Landow, George P. *Hypertext 2.0: The Convergence of Contemporary Critical Theory and Technology.* Baltimore: Johns Hopkins UP, 1997.

------, ed. *Hyper/Text/Theory.* Baltimore: Johns Hopkins UP, 1994.

Langer, Susanne K. *Feeling and Form: A Theory of Art.* New York: Scribner's, 1953.

Lanham, Richard. *The Electronic Word: Democracy, Technology, and the Arts.* Chicago: U of Chicago Press, 1993.

Lanier, Jaron, and Frank Biocca. "An Insider's View of the Future of Virtual Reality." *Journal of Communications* 42, no. 4 (1992): 150–72.

Lasko-Harvill, Ann. "Identity and Mask in Virtual Reality." *Discourse* 14, no. 2 (1992): 222–34.

Laurel, Brenda. "Art and Activism in VR." *Wide Angle* 15, no. 4 (1993): 13–21.

——. *Computers as Theatre.* Menlo Park, Calif.: Addison Wesley, 1991.

——, ed. *The Art of Computer Interface Design.* Reading, Mass.: Addison-Wesley, 1990.

Laurel, Brenda, Rachel Strickland, and Rob Tow. "Placeholder: Landscape and Narrative in Virtual Environments." *Computer Graphics* 28, no. 2 (1994): 118–26.

Leeson, Lynn Hershman. *Clicking In: Hot Links to a Digital Culture.* Seattle: Bay Press, 1996.

Lévy, Pierre. *Becoming Virtual: Reality in the Digital Age.* Trans. Robert Bonono. New York: Plenum Trade, 1998.

——. *L'Idéographie dynamique.* Paris: La Découverte, 1991.

Lewis, David. *On the Plurality of Worlds.* Cambridge: Blackwell, 1986.

——. "Truth in Fiction." *American Philosophical Quarterly* 15 (1978): 37–46.

Liestøl, Gunnar. "Wittgenstein, Genette, and the Reader's Narrative in Hypertext." Landow, *Hyper/Text/Theory*, 87–120.

Loeffler, Carl Eugene, and Tim Anderson, eds. *The Virtual Reality Casebook.* New York: Van Nostrand Reinhold, 1994.

Lyotard, Jean-François. *The Postmodern Condition: A Report on Knowledge.* Trans. Geoff Bennington and Brian Massumi. Minneapolis: U of Minnesota Press, 1991.

Maître, Doreen. *Literature and Possible Worlds.* London: Middlesex Polytechnic Press, 1983.

Makaryk, Irena K., ed. *Encyclopedia of Contemporary Literary Theory: Approaches, Scholars, Terms.* Toronto: U of Toronto Press, 1993.

Maloney, Judith. "Fly Me to the Moon: A Survey of American Historical and Contemporary Simulation Entertainments." *Presence: Teleoperators and Virtual Environments* 6, no. 5 (1997): 565–80.

Margolin, Uri. "Of What Is Past, Is Passing or to Come: Temporality, Modality, Aspectuality, and the Nature of Literary Narrative." Herman, *Narratologies*, 142–66.

——. "Telling Our Story: On 'We' Literary Narratives." *Language and Literature* 5, no. 2 (1996): 115–33.

Martin, Wallace. *Recent Theories of Narrative.* Ithaca: Cornell UP, 1986.

Martínez-Bonati, Félix. *Fictive Discourse and the Structures of Literature: A*

Phenomenological Approach. Trans. Philip W. Silver. Ithaca: Cornell UP, 1981.

McHale, Brian. *Postmodernist Fiction.* London: Methuen, 1987.

McLuhan, Marshall. *Essential McLuhan.* Ed. Eric McLuhan and Frank Zingrone. New York: Basic Books, 1996.

Merleau-Ponty, Maurice. *The Phenomenology of Perception.* Trans. Colin Smith. London: Routledge & Kegan Paul, 1962.

———. *The Primacy of Perception.* Ed. James M. Edie. Chicago: Northwestern UP, 1964.

Miller, J. Hillis. *Topographies.* Stanford, Calif.: Stanford UP, 1995.

Moser, Mary Anne, and Douglas McLeod, eds. *Immersed in Technology: Art and Virtual Environments.* Cambridge, Mass.: MIT Press, 1996.

Mosher, Harold F. "Toward a Poetics of 'Descriptized' Narration." *Poetics Today* 12, no. 3 (1991): 425–46.

Motte, Warren F. *Playtexts: Ludics in Contemporary Literature.* Lincoln: U of Nebraska Press, 1995.

———, ed. and trans. *Oulipo: A Primer of Potential Literature.* Lincoln: U of Nebraska Press, 1986.

Moulthrop, Stuart. "No War Machine." Tabbi and Wutz, *Reading Matters,* 269–92.

———. "Rhizome and Resistance: Hypertext and the Dreams of a New Culture." Landow, *Hyper/Text/Theory,* 299–322.

Murray, Janet E. *Hamlet on the Holodeck: The Future of Narrative in Cyberspace.* New York: Free Press, 1997.

Nell, Victor. *Lost in a Book: The Psychology of Reading for Pleasure.* New Haven: Yale UP, 1988.

Nietzsche, Friedrich. *The Birth of Tragedy and the Case of Wagner.* Trans. Walter Kaufmann. New York: Random House, 1967.

Novak, Marcos. "Liquid Architecture in Cyberspace." Benedikt, *Cyberspace,* 225–54.

Nunes, Mark. "Virtual Topographies: Smooth and Striated Cyberspace." Ryan, *Cyberspace,* 61–77.

Ong, Walter J. *Orality and Literacy: The Technologizing of the Word.* London: Methuen, 1982.

Oren, Tim. "Designing a New Medium." Laurel, *Art of Computer Interface Design,* 467–79.

Paulos, John Allen. *Once Upon a Number: The Hidden Mathematical Logic of Stories.* New York: Basic Books, 1998.

Pavel, Thomas. *Fictional Worlds.* Cambridge, Mass.: Harvard UP, 1986.

Pearce, Celia. "The Ins and Outs of Nonlinear Storytelling." *Computer Graphics* 28, no. 2 (1994): 100–101.

———. *The Interactive Book.* Indianapolis: Macmillan Technical Publishing, 1997.

Penny, Simon. "Virtual Reality as the Completion of the Enlightenment." Loeffler and Anderson, *Virtual Reality Casebook*, 199–213.

Phelan, James. *Reading People, Reading Plots: Character, Progression, and the Interpretation of Narrative.* Chicago: U of Chicago Press, 1989.

Pimentel, Ken, and Kevin Teixeira. *Virtual Reality: Through the New Looking Glass.* New York: Intel/Windcrest McGraw Hill, 1993.

Popper, Frank. *Art—Action and Participation.* New York: New York UP, 1975.

Poster, Mark. "Theorizing Virtual Reality: Baudrillard and Derrida." Ryan, *Cyberspace*, 42–60.

Pratt, Mary Louise. *Toward a Speech Act Theory of Literary Discourse.* Bloomington: Indiana UP, 1977.

Prince, Gerald. *Dictionary of Narratology.* Lincoln: U of Nebraska Press, 1987.

Reid, Elizabeth. "Virtual Worlds: Culture and Imagination." Jones, *Cybersociety*, 164–83.

Renan, Sheldon. "The Net and the Future of Being Fictive." Leeson, *Clicking In*, 61–69.

Rescher, Nicholas. *G. W. Leibniz's Monadology: An Edition for Students.* Pittsburgh: U of Pittsburgh Press, 1991.

Rheingold, Howard. *Virtual Reality.* New York: Simon & Schuster, 1991.

Ronen, Ruth. *Possible Worlds in Literary Theory.* Cambridge: Cambridge UP, 1994.

Roose-Evans, James. *Experimental Theatre: From Stanislavsky to Peter Brooks.* London: Routledge, 1989.

Rotman, Brian. "Thinking Dia-Grams: Mathematics, Writing, and Virtual Reality." *South Atlantic Quarterly* 94, no. 2 (1995): 389–415.

Rousset, Jean. *L'Intérieur et l'extérieur: Essais sur la poésie et sur le théâtre au XVIIe siècle.* Paris: Corti, 1968.

Ryan, Marie-Laure. *Possible Worlds, Artificial Intelligence, and Narrative Theory.* Bloomington: Indiana UP, 1991.

———. "Possible Worlds in Recent Literary Theory." *Style* 26, no. 4 (1992): 528–53.

———, ed. *Cyberspace Textuality: Computer Technology and Literary Theory.* Bloomington: Indiana UP, 1999.

Savitch, Walter J. *Abstract Machines and Grammars.* Boston: Little, Brown & Co., 1982.

Searle, John. "The Logical Status of Fictional Discourse." *New Literary History* 6 (1974/75): 319–32.

———. *Speech Acts: An Essay in the Philosophy of Language.* Cambridge: Cambridge UP, 1969.

Semino, Elena. *Language and World Creation in Poems and Other Texts.* London: Longman, 1997.

Sheridan, Thomas B. "Musings on Telepresence and Virtual Presence." *Presence: Teleoperators and Virtual Environments* 1, no. 1 (1992): 120–25.

Slouka, Mark. *War of the Worlds.* New York: Basic Books, 1995.

Smith, Barbara Herrnstein. *On the Margins of Discourse.* Chicago: U of Chicago Press, 1978.

Snyder, Ilana. "Beyond the Hype: Reassessing Hypertext." Snyder, *Page to Screen,* 125–43.

———. *Hypertext: The Electronic Labyrinth.* Melbourne: Melbourne UP, 1996.

———, ed. *Page to Screen: Taking Literacy into the Electronic Era.* London: Routledge, 1998.

Spariosu, Mihai I. *Dionysus Reborn: Play and the Aesthetic Dimension in Modern Philosophical and Scientific Discourse.* Ithaca: Cornell UP, 1989.

Spence, Jonathan D. *The Memory Palace of Matteo Ricci.* New York: Viking, 1984.

Spencer, Michael. *Michel Butor.* New York: Twayne, 1974.

Steinhart, Eric. "Leibniz's Palace of the Fates: A Seventeenth-Century Virtual Reality System." *Presence: Teleoperators and Virtual Environments* 6, no. 1 (1997): 133–35.

Stenger, Nicole. "Mind Is a Leaky Rainbow." Benedikt, *Cyberspace,* 49–58.

Steuer, Jonathan. "Defining Virtual Reality: Dimensions Determining Telepresence." *Journal of Communications* 42, no. 4 (1992): 73–93.

Stich, Stephen, and Shaun Nichols. "Second Thoughts on Simulation." Davies and Stone, *Mental Simulation,* 87–108.

Stone, Allucquère Rosanne. "Will the Real Body Please Stand Up: Boundary Stories about Virtual Culture." Benedikt, *Cyberspace,* 81–118.

Suits, Bernard. "The Detective Story: A Case Study of Games in Literature." *Canadian Review of Comparative Literature* 12, no. 2 (1985): 200–219.

———. *The Grasshopper: Games, Life, and Utopia.* Toronto: U of Toronto Press, 1978.

Szilas, Nicolas. "Interactive Drama on Computer: Beyond Linear Narrative." *Narrative Intelligence: Papers from the 1999 AAAI Fall Symposium.* Technical Report FS-99-01. Menlo Park: AAAI (American Association for Artificial Intelligence) Press, 1999.

Tabbi, Joseph, and Michael Wutz. *Reading Matters: Narrative in the New Media Ecology.* Ithaca: Cornell UP, 1997.

Talin [David "Talin" Joiner]. "Real Interactivity in Interactive Environments." *Computer Graphics* 28, no. 2 (1994): 97–99.

Theall, Donald. *Beyond the Word: Reconstructing Sense in the Joyce Era of Technology, Culture, and Communication.* Toronto: U of Toronto Press, 1995.

Truffaut, François. *Hitchcock.* New York: Simon & Schuster, 1967.

Tuman, Myron C., ed. *Literacy Online: The Promise (and Perils) of Reading and Writing with Computers.* Pittsburgh: U of Pittsburgh Press, 1992.

Turkle, Sherry. *Life on the Screen: Identity in the Age of the Internet.* New York: Simon & Schuster, 1995.

Vorderer, Peter, Hans J. Wulff, and Mike Friedrichsen, eds. *Suspense: Conceptualizations, Theoretical Analyses, and Empirical Explorations.* Mahwah, N.J.: Lawrence Erlbaum, 1996.

Vuillemin, Alain, and Michel Lenoble, eds. *Littérature et informatique: La Littérature générée par ordinateur.* Arras: Artois, 1995.

Walser, Randall. "Spacemakers and the Art of the Cyberspace Playhouse." *Mondo 2000*, no. 2 (Summer 1990): 60–61.

Walsh, Richard. "Why We Wept for Little Nell: Character and Emotional Involvement." *Narrative* 5, no. 3 (1997): 306–21.

Walton, Kendall. *Mimesis as Make-Believe: On the Foundations of the Representational Arts.* Cambridge, Mass.: Harvard UP, 1990.

———. "Spelunking, Simulation, and Slime: On Being Moved by Fiction." *Emotion and the Arts.* Ed. Mette Hjort and Sue Laver. Oxford: Oxford UP, 1997. 37–49.

Welty, Eudora. "Place in Fiction." *South Atlantic Quarterly* 55 (1956): 57–72.

Wertheim, Margaret. *The Pearly Gates of Cyberspace: A History of Space from Dante to the Internet.* New York: Norton, 1999.

White, Hayden. "The Value of Narrativity in the Representation of Reality." *On Narrative.* Ed. W. J. T. Mitchell. Chicago: U of Chicago Press, 1981. 1–23.

Wilson, R. Rawdon. *In Palamedes' Shadow: Explorations in Play, Game, and Narrative Theory.* Boston: Northeastern UP, 1990.

Witmer, Bob G., and Michael J. Singer. "Measuring Presence in Virtual Environments: A Presence Questionnaire." *Presence: Teleoperators and Virtual Environments* 7, no. 3 (1998): 225–40.

Wittgenstein, Ludwig. *Philosophical Investigations.* Trans. G. E. M. Anscombe. New York: Macmillan, 1968.

Wolfe, Tom. "The New Journalism." *The New Journalism.* Ed. Tom Wolfe and E. W. Johnson. New York: Harper & Row, 1973. 1–52.

Zahorik, Pavel, and Rick L. Jenison. "Presence as Being-in-the-World." *Presence* 7, no. 1 (1998): 78–89.

Zhai, Philip. *Get Real: A Philosophical Adventure in Virtual Reality.* New York: Rowman & Littlefield, 1999.

Zizek, Slavoj. *Tarrying with the Negative: Kant, Hegel, and the Critique of Ideology.* Durham, N.C.: Duke UP, 1993.

Zubin, David A., and Lynne E. Hewitt. "The Deictic Center: A Theory of Deixis in Narrative." *Deixis in Narrative: A Cognitive Science Perspective.* Ed. Judith F. Duchan, Gail A. Bruder, and Lynne E. Hewitt. Hillsdale, N.J.: Lawrence Erlbaum, 1995. 129–55.

DIGITAL WORKS

Primary

Addison, Rita, Marcus Thiebaux, and David Zelter. *Detour: Brain Deconstruction Ahead.* VR installation, 1994. Description at: http://babelweb.org/virtualistes/galerie/detoura.htm [accessed 9/2/99].

Amerika, Mark. *Grammatron*. www.grammatron.com [accessed 9/9/99].

Bantock, Nick. *Ceremony of Innocence*. Producer Gerrie Villon. CD ROM. Real World MultiMedia Ltd., 1997.

Brown, Marc. *Arthur's Teacher Troubles*. CD ROM. Brøderbund, 1996.

Cayley, John. "Book Unbound." *Postmodern Culture* 7, no. 7 (1997). http://muse.jhu.edu/journals/postmodern_culture/voo7/7.3cayley.html [accessed 9/9/99].

Ceratini, Marc. *The Lurker Files*. http://www.randomhouse.com/lurkerfiles/ [accessed 6/12/98; no longer available].

Gibson, William, and Dennis Ashbaugh. *Agrippa (A Book of the Dead)*. New York: Kevin Begos, 1992.

Hegedüs, Agnes. *Things Spoken*. Interactive software. *Artintact 5*. Karlsruhe: ZKM/Zentrum für Kunst und Medientechnologie, 1999. Description at: http://www.zkm.de/surrogate/hegedues.html [accessed 9/2/99].

I'm Your Man. Dir. Bob Bejean. Perf. Mark Metcalf, Colleen Quinn, and Kevin M. Seal. A Choice Point Film. Presented by Planet Theory in association with DVD International. DVD edition produced by Bill Franzblau, 1998.

Joyce, Michael. *Afternoon, a Story*. Hypertext software. Cambridge, Mass.: Eastgate Systems, 1987.

———. *Twelve Blue: Story in Eight Bars*. World Wide Web hyperfiction. *Postmodern Culture* and Eastgate Systems, 1996 and 1997. http://www.eastgate.com/TwelveBlue/ [accessed 9/9/99].

Laurel, Brenda, Rachel Strickland, and Rob Tow. *Placeholder*. VR installation. Produced by Interval Research Corporation and the Banff Centre, 1993. Description at: http://www.interval.com/frameset.cgi?projects/placeholder/index.html [accessed 9/2/99].

Moulthrop, Stuart. *Hegirascope*. World Wide Web hyperfiction. http://raven.ubalt.edu/staff/moulthrop/hypertexts/hgs/ [accessed 9/20/99].

———. *Victory Garden*. Hypertext software. Cambridge, Mass.: Eastgate Systems, 1991.

Rosenberg, Jim. "The Barrier Frames" and "Diffractions Through." Software. *Eastgate Quarterly Review of Hypertext*. Cambridge, Mass.: Eastgate Systems, 1996.

Shaw, Jeffrey, and Dick Groeneveld. *De Leesbare Stad*. VR installation. Karlsruhe Museum of Electronic Art, 1991.

Secondary

Foner, Leonard. "What's an Agent, Anyway? A Sociological Case Study." http://foner.www.media.mit.edu/people/foner/Julia/Julia.html [accessed 9/8/99].

Holland, Norman N. "Eliza Meets the Postmodern." *EJournal* 4, no. 1. http://

www.hanover.edu/philos/ejournal/archive/ej-4--1.txt [accessed 9/8/99].

Mirapaul, Matthew. "Hypertext Fiction Adds a Third Dimension." *New York Times on the Web,* 10 December 1998. http://www.nytimes.com/library/tech/98/12/cyber/artsatlarge/10artsatlarge.html [accessed 12/12/98].

Rees, Gareth. "Tree Fiction on the World Wide Web." http://lucilia.ebc.ee/~enok/tree-fiction.html [accessed 9/9/99].

Risden, Kirsten. "Can Theories of Traditional Narrative Understanding Inform the Design of Hypermedia Stories." http://www-iet.open.ac.uk/iet/MENO/ht97/kirsten/html [accessed 1/21/98].

Ulmer, Gregory. "A Response to Twelve Blue by Michael Joyce." *Postmodern Culture* 8, no. 1 (1998). http://muse.jhu.edu/journals/postmodern_culture/v008/8.1ulmer.html [accessed 9/8/99].

Wenz, Karin. "Formen der Mündlichkeit und Schriftlichkeit in digitalen Medien." http://viadrina.euv-frankfurt-o.de/wjournal/wenz.htm [accessed 9/8/99].

INDEX

Aarseth, Espen, 18, 183, 184, 206, 209, 217, 233, 251, 308, 309, 369n. 9, 371n.2
abstract art, 3
accessibility relations, 100
Active Worlds (Web site), 359n. 11
actuality: conceptions of, 100; indexical definition, 101; vs. possibility, 99–100
actual world. *See* actuality
Addison, Rita, 317–18
Adventure (computer game), 310
affordance, 71
Afternoon (Michael Joyce), 183, 222, 265, 285, 357n. 1, 365n. 4, 366n. 3, 367n. 1, 367n. 4, 369n. 9
Against Nature (Huysmans), 81–85; and artificiality, 82; and poetry, 83; and sexuality, 83
Agrippa (A Book of the Dead) (Gibson), 268
AI agents. *See* intelligent agents
algorithms, underlying electronic texts, 215
Alice in Wonderland (L. Carroll), 50, 121; and AI agents, 315–16; and interactive drama, 330
Allen, Woody, 50
Allende, Isabel, 256
Alternate World Syndrome (AWS), 10
Amerika, Mark, 8, 9, 216, 219, 269
amour (Bootz), 269
Anderson, Antje Schaum, 151
animation in electronic texts, 216
"Any where out of the world" (Baudelaire), 75
Apollinaire, Guillaume, 216
Apollo XIII (movie), 146
aporia, 251
arbitrariness: in computer codes, 58–60; of conventions, 160; of linguistic signs, 187
architectural space, 290–92
A Rebours. See Against Nature
Aristotle: and *catharsis*, 148, 296; and dramatic plot structure, 64, 239, 243, 245, 246, 272–75, 326, 330; on imitation, 40; and mental simulation, 113, 115; on probability, 44, 157
Artaud, Antonin, 55, 215, 302–4

Arthur's Teacher Troubles (electronic text), 249, 258
artificial intelligence (AI), 312–17, 368n. 7
artificial life, 347
Artificial Paradises (Baudelaire), 75–85
artificial realities, non-digital, 75–85
as if, philosophy of, 176
"Assignation, The" (Poe), 145
Austen, Jane, 208
author: death of, 8, 344; perspective on textual world, 283
authorship: in electronic texts, 215; subversion of, 203
Ayckbourn, Alan, 369n. 4

"Babysitter, The" (Coover), 367n. 8
Bachelard, Gaston, 122, 123, 127
Bakhtin, Mikhail, 6, 186
Balcony, The (Genet), 288
Balsamo, Anne, 52
Balzac, Honoré de, 125–26, 129, 144, 161, 176
Banfield, Ann, 158
Bantock, Nick, 310
Barlow, John Perry, 312
Barnes, Julian, 242
Baroque art, 3; and immersion, 299–300; as immersive/interactive experience, 290–93
Barrier Frames, The (Rosenberg), 269
Barth, John, 349
Barthes, Roland, 6, 8, 144, 191; on Loyola, 115, 118, 361n. 3; on readerly and writerly, 195–96; on reality effect, 130
Bates, Joseph, 313, 314, 317. *See also* Kelso, Margaret
Battle Chess (computer game), 371n. 1
Baudelaire, Charles, 75–81, 84, 85, 359n. 4; description of drug experience, 76–80; and esoteric doctrines, 79; and nature, 75–76; and randomness, 76
Baudrillard, Jean, 13, 36, 37, 40, 288, 351–52; on evolution of images, 29; and interactivity, 31; on "radical thought," 33–34; on simulacra and simulation, 29, 62–63; on transparency, 30; and virtuality, 27–35

Library of Congress Cataloging-in-Publication Data

Ryan, Marie-Laure, 1946–
   Narrative as virtual reality : immersion and interactivity in
literature and electronic media / Marie-Laure Ryan.
     p.   cm. — (Parallax)
   Includes bibliographical references (p.   ) and index.
   ISBN 0-8018-6487-9 (alk. paper)
    1. Books and reading. 2. Interactive multimedia. 3. Virtual reality.
I. Title. II. Parallax (Baltimore, Md.)
   Z1003.R97 2001
   028'.9—dc21   00-008955

Printed in the United Kingdom
by Lightning Source UK Ltd.
106157UKS00001B/100